REACTIVE INTERMEDIATES IN THE GAS PHASE

Generation and Monitoring

Contributors

LARRY I. BONE

MICHAEL A. A. CLYNE

N. DJEU

WILLIAM FELDER

ARTHUR FONTIJN

J. H. KOLTS

M. C. LIN

J. R. McDONALD

WING S. NIP

D. W. SETSER

REACTIVE INTERMEDIATES IN THE GAS PHASE

Generation and Monitoring

Edited by
D.W. SETSER

Department of Chemistry
Kansas State University
Manhattan, Kansas

 1979

ACADEMIC PRESS

A Subsidiary of Harcourt Brace Jovanovich, Publishers

New York   London   Toronto   Sydney   San Francisco

ACADEMIC PRESS, INC.
111 Fifth Avenue, New York, New York 10003

United Kingdom Edition published by
ACADEMIC PRESS, INC. (LONDON) LTD.
24/28 Oval Road, London NW1 7DX

Library of Congress Cataloging in Publication Data

Main entry under title:

Reactive intermediates in the gas phase.

   Includes bibliographies and index.
   1. Chemical reaction, Conditions and laws of--
Addresses, essays, lectures. 2. Chemistry, Physical
organic--Addresses, essays, lectures. I. Setser, D. W.
QD501.R346   541'.39   79–51698
ISBN 0–12–637450–3

PRINTED IN THE UNITED STATES OF AMERICA

79 80 81 82   9 8 7 6 5 4 3 2 1

Contents

4. Production and Detection of Reactive Species with Lasers in Static Systems

M. C. Lin and J. R. McDonald

5. Production of Small Positive Ions in a Mass Spectrometer

Larry I. Bone

6. Discharge-Excited Rare Gas Halide Lasers

N. Djeu

List of Contributors

Numbers in parenthesis indicate the pages on which the authors' contributions begin.

LARRY I. BONE* (305), Department of Chemistry, Appalachian State University, Boone, North Carolina 28608

MICHAEL A. A. CLYNE (1), Department of Chemistry, Queen Mary College, University of London, London E1 4NS, United Kingdom

N. DJEU (323), Laser Physics Branch, Naval Research Laboratory, Washington, D.C. 20375

WILLIAM FELDER (59), AeroChem Research Laboratories, Inc., Princeton, New Jersey 08540

ARTHUR FONTIJN (59), AeroChem Research Laboratories, Inc., Princeton, New Jersey 08540

J. H. KOLTS† (151), Department of Chemistry, Kansas State University, Manhattan, Kansas 66506

M. C. LIN (233), Chemical Diagnostics Branch, Chemistry Division, Naval Research Laboratory, Washington, D.C. 20375

J. R. McDONALD (233), Chemical Diagnostics Branch, Chemistry Division, Naval Research Laboratory, Washington, D.C. 20375

WING S. NIP (1), Division of Chemistry, National Research Council of Canada, Ottawa, Ontario K1A OR6, Canada

D. W. SETSER (151), Department of Chemistry, Kansas State University, Manhattan, Kansas 66506

*Present address: Dow Chemical Co., Freeport, Texas 77541.
†Present address: Research and Development Department, Phillips Petroleum Company, Bartlesville, Oklahoma 74004.

Preface

An understanding of the time evolution of complex chemical systems can be achieved only from knowledge of the individual elementary reaction steps comprising the system. Frequently some of those elementary steps involve reactive intermediate chemical species. To study the reactive, transient intermediates, methods must be developed for isolating the species and observing their thermodynamic, kinetic, and structural properties. The need for knowledge of the macroscopic chemical properties of reactive intermediates frequently coincides with a desire for observation of state-to-state chemical dynamics of their elementary reactions. Fortunately, sensitive analytical tools for measuring the properties of a given chemical species are available. The limitation to the study of reactive intermediates is often the development of methods to generate the species in an environment favorable for quantitative measurements. A description of such methods for reactive intermediates in the gas phase is the central theme of this book.

The book was planned so that the chapters would enable readers to learn about techniques for generating reactive intermediates, as well as to review the literature on various reactive species. Insofar as possible, pertinent experimental detail also will be emphasized in the chapters. Writing on these topics is not as glamorous as applying modern dynamical theory to state-to-state data or interpreting a mechanism of reaction. However, this book was intended to provide insight into production and isolation of reactive intermediates, rather than to discuss reaction dynamics. The need for a book emphasizing experimental techniques is accentuated by the lack of experimental detail given in most journal articles.

A balance regarding presentation of techniques and review of interesting species was sought. The first three chapters deal with flow techniques, and the discussion proceeds from ground state atoms and radicals, normally at room temperature, to high temperature conditions and species, and finally to long-lived electronically excited states. The fourth chapter is devoted to discussion of pulsed excitation systems with emphasis on use of lasers for both generating and monitoring the chemical species. The authors of this chapter had an especially challenging task because of the revolution occurring with the introduction of lasers into the field. The fifth chapter discusses positive ions with an emphasis on generation by photo-ionization methods. The last chapter discusses rare gas halide discharge lasers and their applications.

I wish to thank the authors for their efforts in presenting pertinent experimental detail and yet remaining sufficiently general in scope to cover a variety of chemical species.

1

Generation and Measurement
of Atom and Radical Concentrations
in Flow Systems

MICHAEL A. A. CLYNE

Department of Chemistry
Queen Mary College
University of London
London
United Kingdom

WING S. NIP

Division of Chemistry
National Research Council of Canada
Ottawa, Ontario
Canada

1

I. INTRODUCTION

The discharge-flow method is established as one of the main quantitative techniques for kinetic studies of gaseous elementary reactions. Since the very innovative work of Kaufman (from 1958 onward), which essentially established the method, a series of major developments of the discharge-flow method have taken place. These developments included the extension of the scope of the method from the three common atoms O, H, and N to many other atoms, and to molecular free radicals including OH, ClO, BrO, FO, and SO. In parallel, the first radical detection methods employed—chemiluminescence, thermal gauges, etc., have been gradually superseded by novel detectors, or by ones adapted from other branches of experimental chemical physics. Foremost among the direct and highly sensitive detection techniques now available are atomic resonance in the vacuum ultraviolet, mass spectrometry, EPR and laser magnetic resonance, and laser-induced fluorescence.

Possibly the most well-known recent application of the discharge-flow technique has been to the kinetics of atmospherically significant elementary reactions. Virtually all the direct measurements of rate constants for such reactions have been and are being carried out using either the discharge-flow or flash-photolysis methods. For instance, the first direct determinations at 298 K of the rate constants for the following and many other stratospheric reactions were made using the discharge-flow technique: $Cl + O_3 \rightarrow ClO + O_2$, $O + ClO \rightarrow Cl + O_2$, $ClO + NO_2 + M \rightarrow ClNO_3 + M$, $HO_2 + NO \rightarrow OH + NO_2$, $H + O_3 \rightarrow OH + O_2$. There are numerous other major applications areas to which the discharge-flow method is currently contributing fundamental kinetic data. These include: chemical lasers, excimer lasers, combustion chemistry, and free-radical spectroscopy. In the present article, we survey the scope of the discharge-flow method. Because of its particular usefulness as a highly sensitive detection method, we include a more detailed account of atomic resonance in the vacuum ultraviolet. Other major detection techniques are considered in less detail.

II. DESCRIPTION OF DISCHARGE FLOW SYSTEMS

The basic components of a discharge-flow system are shown in Fig. 1. Atoms and radicals are formed either directly from a microwave (or radiofrequency) discharge in tube D, or by rapid reaction of an atom with a

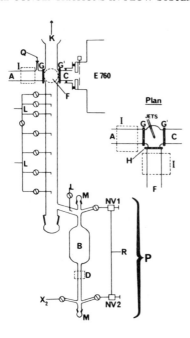

Fig. 1. Resonance absorption and fluorescence detection of atoms in a discharge-flow system. Note discharge-bypass system P, comprising needle valves NV1, NV2 for carrier gas, molecular gas inlet X_2, microwave discharge D, recombination volume B, joints M for poisoning discharge tube with H_3PO_4. Flow tube includes reagent inlets L. Absorption (A) fluorescence (F) cell includes LiF windows G, G′, H, microwave discharge I, connecting tube C to 1-m vacuum monochromator (Hilger E760), K to pump. [After Clyne and Cruse (1972a).]

molecule. The active species are passed along the flow tube (of cylindrical or rectangular cross section) where they may be mixed with another reagent from fixed or moveable inlet jets L. Measurements of atom and radical concentrations are made downstream of the mixing zone.

In this section, we indicate various parameters which govern the operation of a discharge-flow system. (The methods of generation and detection of atoms and radicals will be described later.) Stirred flow reactors have been used to a limited extent, and their operation has been described elsewhere (Mulcahy and Williams, 1961; Mulcahy *et al.*, 1967, 1969).

Under plug flow conditions, the elapsed time in a flow system is simply equal to $x/\bar{v}$, where x is the displacement along the flow tube and $\bar{v}$ is the flow velocity. Flow tubes (of pyrex or silica glass) are typically 20–50 mm in internal diameter and 10–200 cm in length. A total pressure P of 0.5–10 Torr in the flow tube is maintained, usually by fast conventional rotary pumps (typically 300–1000 liter min^{-1}) using high-conductance cold traps. The partial pressure of each component p_i in the flow tube is then related to the individual flow rates (F_i) into the system by the expression $p_i = PF_i/\Sigma F_i$, where ΣF_i is the total mass flow rate (mainly carrier gas).

Although the flow system is versatile in the production of atoms and radicals, several aspects of its operation must be carefully considered in order to obtain accurate kinetic information.

A. Measurements of Pressure and Flow Rates

For a reaction which is first order in reagent A, the rate constant k' is given by $k' = -\bar{v}\, d\ln[A]/dx = -RT(\Sigma F_i/aP)\, d\ln[A]/dx$ where $\bar{v}$ is the plug flow velocity, a is the cross section area of the flow tube, and x is the displacement along the flow tube. Thus, a realistic error of $\pm 2\%$ in P and $\pm 3\%$ in ΣF_i will lead to a $\pm 5\%$ uncertainty in the measurement of k'. The uncertainty in higher order reactions will be progressively greater (Clyne, 1973). Thus, the uncertainty in the bimolecular rate constant of an atom (A) + molecule (B) reaction determined under pseudo first-order conditions will be $\pm 13\%$.

Calibrations of flow rates are usually done by following the rate of fall in pressure, $-dP/dt$, of gas in a calibrated volume V, $F_i = -(V/RT)\, dP/dt$. As pointed out by Clyne (1973), P is a strongly nonlinear function of t; and it is desirable that $-dP/dt$ be measured from the almost linear plot of $1/P$ versus t: $-dP/dt = P^2\, d(1/P)/dt$. The mean flow rate of a condensable reagent can also be determined by trapping material downstream of the needle valve over a known period of time and measuring the total mass collected. "Bubble meters" have also been used to calibrate flows from > 1 atm backing pressure. It is realistic to expect an accuracy no better than $\pm 3\%$ using existing flow calibration methods.

Errors in total pressure (P) measurements in flow systems arises from two factors. (a) Errors in P itself. At a typical pressure of 1 Torr, a 2% accuracy implies measuring the pressure to 0.02 Torr, which is marginal using oil manometers. This problem has been solved by the availability of sensitive and accurate pressure transducers such as capacitance manometers. (b) The pressure gradient along the flow tube. According to the Poiseuille flow equation, $(P_1 - P_2) = 8\eta RT/\Sigma F_i/P_1\pi r^4$, where η is the viscosity of the mixture, the relative pressure drop $(P_1 - P_2)/P_1$ along a 2-cm-i.d. flow tube would be 0.3 per 100 cm length with an argon flow of 2×10^{-4} mole sec^{-1} near 300 K and $P_1 \approx 1$ Torr. The effect may be minimized by linear interpolation between the pressures measured at both ends of the flow tube. According to the Poiseuille equation, the relative pressure drop increases with increase in total flow rate and with decrease in pressure and, particularly, with tube radius r.

B. Mixing of Reagents

It is desirable in a kinetic experiment to achieve complete mixing between the reagent flow from the inlet jets with the main flow in as short a time as possible. Specifically, the mixing time must be short compared with the half-life of the reaction. To increase the radial velocity component of the incoming flow, reagents may be admitted to the flow tube through multi-perforated fine jets located on the tube axis, preferably in a direction opposite to the main flow. The degree of mixing can be checked by the attainment of homogeneous optical density when a light-absorbing reagent is added through the inlet jet (Clyne and Cruse, 1970a). Alternatively, a chemilumine-scent reaction such as the addition of NO_2 to a large excess of O atoms may be used.

C. Diffusion

The problem of back diffusion and radical diffusion in flow systems is a difficult one in fluid dynamics and it is beyond the scope of this chapter to present an analytical solution to the flow equations. However, approximate solutions have been given by Kaufman (1961), Poirier and Carr (1971, and references therein), Walker (1961), Mulcahy and Pethard (1963), among others. We note here that radial concentration gradients are not significant under typical flow tube conditions of P about 1 Torr and r about 1 cm, unless the first-order wall recombination rate is very fast as for ions (Ferguson et al., 1969) and metals (see Chapter 2). The conditions for negligible back diffusion are more stringent and are given approximately by $Dk'/\bar{v}^2 \gg 1$ where D is the diffusion coefficient of radicals and k' is the first-order rate constant for radical decay.

Clearly, all the sources of error cannot be minimized simultaneously. Thus, increasing the pressure in the flow tube decreases the pressure gradient along the flow tube, but it also increases the radial concentration gradients. Such effects severely limit the range of parameters over which a flow tube should be used for kinetic studies: pressures should be in the 0.5–5 Torr range and flow velocities < 3000 cm sec^{-1}.

D. Analysis of Kinetic Data

Most kinetic studies in discharge-flow systems were performed under pseudo-first-order conditions, i.e., the decay of the radical concentration [A] is monitored when various excesses of the stable reagent B are introduced or vice versa. It has been shown by Clyne and Thrush (1963) and later by Westenberg and de Haas (1967) that, for the geometry of fixed observation point and fixed point of radical formation, the slope of a $\ln[A]$ versus time plot gives the apparent first-order rate constant $k_I[B]$ for the reaction. A necessary condition is that the destruction of radicals on the flow tube wall is also first order and independent of the presence of B. Thus,

$$d \ln\{[A]_{B=0}/[A]\}/dt = k_I[B].$$

This expression is very useful because it does not require a knowledge of the wall destruction rate or a knowledge of the rate of second-order atom (radical) recombination provided this is small. The analysis can also be extended to experiments where a section of the flow tube is kept at a different temperature. In this case,

$$\ln\{[A]_{B=0}/[A]\} = k_I[B]t + K,$$

where K is a constant.

The above analysis fails when the wall destruction is irreproducible or depends on [B]. It may be used in a slightly different form when the rate of wall destruction varies along the tube but otherwise satisfies the stated criteria (Clyne and Coxon, 1966; Clyne et al., 1976a). In these cases,

$$\ln\{[A]_{B=0}/[A]\} = k_I[B]t + K',$$

where K' describes the variation of k_w along the flow tube. A slightly different analysis is then used to determine the bimolecular rate constant (Clyne and Coxon, 1966). $\ln[A]$ is first plotted as a function of [B], at constant t. Since K' is constant at constant t (i.e., constant x), the plots are linear. The slopes of these graphs $(-\partial \ln[A]/\partial[B])_t$, when plotted against t, give values of k_I. Clyne et al. (1976a) have also shown that the wall effect in the $O + Cl_2$ reaction can be decreased by increasing the pressure in the flow tube.

E. Effects of Wall Coatings

In cases where the wall reactions of atoms and radicals interfere with the experiment, it is common practice to "poison" the flow tube against heterogeneous recombination. Kaufman (1961) has compiled an extensive list of surface recombination coefficients γ for O atoms, and data are also available for other atoms (e.g., Ogryzlo, 1961; Clyne and Stedman, 1968b). Although the absolute magnitudes of γ are not certain, it is evident that acids, especially oxyacids such as H_3PO_4, H_2SO_4, and $HClO_4$ are effective in reducing surface recombinations. Surface destruction of O, H, Cl, Br, N, and OH can easily be reduced to acceptable values by rinsing the tube with H_3PO_4. Various detailed procedures have been described, for example, Anderson et al.

(1974) found that k_w for OH can be reduced to values as low as 15 sec^{-1} by baking the surface under vacuum at 450 K for several days (cf., k_w = 50 sec^{-1} by only rinsing the flow tube with H_3PO_4).

The use of fluorocarbon coatings such as Teflon (e.g., Debely, 1970) and "Fluoro-Kote" (Badachhape et al., 1976) has become popular in recent years. For some cases, these coatings are superior to oxyacid coatings because of their inert nature. This is particularly significant in cases where halogen atoms are involved (Takacs and Glass, 1973a,b,c; Badachhape et al., 1976; Clyne and Nip, 1977b). Fluorocarbon coatings are particularly desirable for kinetic work with F atoms. Although the efficiency of fluorocarbon coatings toward most atom recombination is high, there is no agreement that they reduce the wall recombination of OH radicals. Whereas Glass and co-workers (references above) have reported successful use of a "fluoro-koted" surface, Clyne and Holt (1979) found Teflon to be an inferior coating to H_3PO_4.

F. The Discharge Bypass System

As described in the next section, atoms in a flow tube are usually produced by passing a dilute mixture of the parent molecule in an inert gas (Ar or He) through a microwave discharge. Quantities of H, O, and N are usually formed from a discharge in traces of water and air impurities present in the carrier gas. However, the impurity level can be reduced if only a very small fraction of the carrier flow is discharged (Fig. 1), and the bulk of the carrier flow bypasses the discharge. The formation of impurity atoms is decreased because the bulk of the gas does not pass through the discharge. The low flow velocity through the discharge gives the impurities enough time to react with suitable undissociated molecules (e.g., $O + Cl_2$ in the production of Cl from Cl_2), or to partially recombine. The flow time of the discharged gas, before rejoining the main stream, can be increased further by the use of a wide-bore glass tube as a recombination volume. This system has been successful in reducing H, O, and N concentrations to low values, i.e., $<5 \times 10^{10}$ cm^{-3} (Clyne and Cruse, 1972a). The bypass system necessitates the use of a relatively high flow of parent molecules through the discharge, because a large fraction of the atoms under study will also be destroyed in the recombination volume. This effect is, however, beneficial in producing low concentration of atoms, because it tends to stabilize the atom flow. By manipulating the flow of the parent molecule, the high and low flows of the carrier gas, one can admit controlled amounts of more than one reactive species into the flow tube. This approach proved to be very useful in our study of the $O + ClO$ reaction (Clyne and Nip, 1976b). The discharge bypass also has proved useful as a clean source of SO radicals, using a discharge in dilute mixture of SO_2 with He (Clyne and MacRobert, 1979).

III. PRODUCTION OF ATOMS

The most convenient way of producing atoms in a flow system is to pass a stream of parent molecules, usually the parent homonuclear diatomic, through an electrodeless discharge. Radiofrequency (typically ~ 20 MHz, ≤ 100 W) or microwave (typically 2450 MHz, 25–100 W) power can be coupled to a discharge tube via external electrodes, inductance coils (in the case of rf discharge), or cavities (in the case of microwave discharge) (e.g., Clyne and Thrush, 1961a; Fehsenfeld et al., 1965). Both types of discharges are efficient in the production of atoms (Clyne, 1973) but the microwave discharge is slightly more favored because of its ease of operation (Fehsenfeld et al., 1965). The degree of dissociation when a diatomic molecule is passed through an electrodeless discharge depends on factors such as the power of the discharge and the gas flow rate. The condition of the discharge tube and the impurity level of the gas also play important roles in determining the atom yield (Campbell and Thrush, 1965; Clyne, 1973). The effect of impurities is marked and it is possible that ultrapure gases do not dissociate appreciably (Kaufman, 1969; Brocklehurst and Jennings, 1967). The scope of this review excludes the chemistry within the discharge plasma (Kaufman, 1969). We note that the degree of dissociation increases with: (i) increase in power, (ii) decrease in concentration of the parent molecule, (iii) increase in flow rate, and (iv) increase in impurity concentration.

We will now consider the different ways of producing the common ground state atoms, and will describe some well established titration reactions for the determination of absolute concentration of atoms.

A. H(^{2}S)

Hydrogen atoms can be produced by passing H_2 or $H_2 + Ar(He)$ mixtures through a microwave discharge, provided that the gases are not ultrapure (Mitchell and Le Roy, 1977). Low concentrations ($\sim 1 \times 10^{12}$ cm^{-3}) can be produced by passing commercial grade Ar or He through a discharge, thus utilizing the dissociation of trace impurities such as H_2O and hydrocarbons.

If the discharge tube is coated with phosphoric acid, impurity [H] as high as 2×10^{12} cm^{-3} has been observed (Clyne and Nip, 1977a). Traces of O and N are usually present with H atoms produced in this manner, but they can easily be eliminated by a bypass system after the discharge (see above). Metastable states of H_2 and H have not been observed in discharge flow systems, although the presence of vibrationally excited H_2 has been inferred (Stedman et al., 1970; Heidner and Kasper, 1972).

Thermal decomposition of H_2 on hot tungsten filaments is another clean source of $H(^2S)$ (Trainor et al., 1973). The degree of dissociation is limited by thermodynamics and impurity levels are usually low. The amount of vibrationally excited H_2 formed this way will depend on the temperature of the filament (Birely et al., 1975).

Recently, Bemand and Clyne (1977) have produced known concentrations of H from the reaction of $O + H_2$ at 780 K. It has also been suggested that H atoms free from H_2 can be produced by the rapid reaction $F + H_2$ (Warnatz et al., 1971a). They showed that the reaction products contain only HF and unreacted H at the end point where $[F]_0 = [H_2]_0$,

The hydrogen atom concentration in a flow has been determined by isothermal calorimetry (e.g., Trainor et al., 1973; Mitchell and Le Roy, 1977; Larkin and Thrush, 1964) or by titration reactions with NO_2 or NOCl. Both reactions are very fast:

$$H + NO_2 \longrightarrow NO + OH, \quad k_{298} = 1.3 \times 10^{-10} \text{ cm}^3 \text{ molecule}^{-1} \text{ s}^{-1}$$

(Bemand and Clyne, 1977; Clyne and Monkhouse, 1977; Wagner et al., 1976b);

$$H + NOCl \longrightarrow NO + NCl, \quad k_{298} = 1.6 \times 10^{-11} \text{ cm}^3 \text{ molecule}^{-1} \text{ s}^{-1}$$

(Wagner et al., 1976b). The $H + NO_2$ titration has been extensively used but secondary reactions may be important in this system. The stoichiometry $\Delta[NO_2]/\Delta[H]$ changes from unity at low [H] ($< 1 \times 10^{12} \text{ cm}^{-3}$) and short reaction time to 1.5 at high [H] or long reaction time, due to the secondary reactions,

$$OH + OH \longrightarrow H_2O + O \quad \text{and} \quad O + OH \longrightarrow O_2 + H$$

(Phillips and Schiff, 1962); although $\Delta[H]/\Delta[NO]$ remains unity with [H] as high as 10^{15} cm^{-3} at reaction time $\simeq 3$ msec (McKenzie et al., 1974; Slemř and Warneck, 1977). The $H + NO_2$ titration reaction may also be inapplicable if the gas flow contains a high concentration of H_2, because the step,

$$OH + H_2 \longrightarrow H_2O + H, \quad k_{298} = 1.6 \times 10^{-15} \text{ cm}^3 \text{ molecule}^{-1} \text{ sec}^{-1},$$

(Baulch et al., 1968), can regenerate H at an appreciable rate.

The $H + NOCl$ titration appears to be free from secondary reactions, provided the flow tube wall is coated with phosphoric acid to minimize the recombination of Cl atoms, and hence to suppress the fast $H + Cl_2 \rightarrow HCl + Cl$ reaction, which has a rate constant of $2.2 \times 10^{-11} \text{ cm}^3 \text{ molecule}^{-1} \text{ sec}^{-1}$ at 298 K (Bemand and Clyne, 1977; Wagner et al., 1976b). The unity stoichiometry of $\Delta[H]/\Delta[NOCl]$ has been verified by Clyne and Stedman (1966) and also by Dunn et al. (1971).

B. $O(^3P_J)$

A discharge in pure O_2 or O_2 + Ar(He) mixtures provides a convenient source of $O(^3P_J)$ atoms. As in the case of $H(^2S)$, a discharge in cylinder Ar or He produces $\sim 1 \times 10^{12}$ cm^{-3} of $O(^3P_J)$ from H_2O and O_2 impurities. This has been used as a source of oxygen atoms in atomic resonance fluorescence studies (Clyne et al., 1976a). The degree of dissociation of O_2 by microwave discharge approaches 25–50%.

Although experimentally simple, production of $O(^3P_J)$ by a direct discharge through a stream of O_2 or O_2 + Ar(He) is often objectionable for kinetic studies because of the high concentrations of metastable O_2 (O_2^*) that are also produced. Indeed this is a good method for producing $O_2(^1\Delta_g)$ (Wayne, 1969, and references therein); smaller concentrations of $O_2(^1\Sigma_g^+)$ have also been identified (Wayne, 1969). Mathias and Schiff (1964), among others, have pointed out the importance of O_2^*, through the reactions,

$$O + O_2 + O_2 \longrightarrow O_3 + O_2,$$

$$O + O_3 \longrightarrow 2O_2,$$

and

$$O_2^* + O_3 \longrightarrow O_2 + O_2 + O,$$

in a stream of electrically discharged O_2. However, there has been some evidence that metastable O_2 is not important when very dilute mixtures (10^{-3}–10^{-4}%) of highly purified O_2 in Ar are discharged (Clyne et al., 1965).

A preferred method of producing $O(^3P_J)$ in a flow system is by the reaction

$$N + NO \longrightarrow N_2 + O, \qquad k_{298} \sim 2 \times 10^{-11} \text{ cm}^3 \text{ molecule}^{-1} \text{ sec}^{-1}$$

(Clyne and McDermid, 1975; Lee et al., 1978). Depending on the purpose of the experiment, one can add excess NO to trace amounts of N atoms to produce O atoms for kinetic studies, or add trace amounts of NO to excess N to produce a known concentration of $O(^3P_J)$ for calibration purposes. It has been shown (Felder and Young, 1972) that the production of $O(^3P_J)$ from N + NO is at least ten times faster than that of any $O(^1D)$ formed. The main impurities are trace amounts of $H(^2S)$ (from H_2O and hydrocarbons), $N(^2P)$, $N(^2D)$ (Morse and Kaufman, 1965; Lin and Kaufman, 1971), $N_2(A^3\Sigma_u^+)$ (e.g., Campbell and Thrush, 1968), and vibrationally excited N_2 (Kaufman and Kelso, 1958), but none of these species are expected to be significant in most kinetic studies carried out in fast flow systems (Clyne, 1973).

In cases where impurity atoms and molecules are unacceptable, e.g., in kinetic studies of the O + O_3 reaction, thermal decomposition has been used as a source of $O(^3P_J)$. Up to 2% decomposition of N_2O and O_2 has been achieved by passing these gases through a Nernst glower (Lundell *et al.*, 1969). High temperature decomposition of ozone is another clean source of $O(^3P_J)$ atoms (Bemand, 1974; Kaufman and Kelso, 1964). Using the rate constant expression of Jones and Davidson (1962), >99% decomposition of O_3 can be expected in 5 msec at 1100 K and 1 Torr total pressure.

A common titration method for $O(^3P_J)$ atoms is the reaction $O + NO_2 \rightarrow NO + O_2$. Although the stoichiometry is simple, its rate constant at 298 K is only about 9.5×10^{-12} cm^3 molecule^{-1} sec^{-1} (Bemand *et al.*, 1974, and references therein). If $O(^3P_J)$ is monitored, the $O + NO_2$ titration is not suitable for concentrations much below 5×10^{12} cm^{-3} under most flow conditions ($t < 100$ msec). However, if NO_2 is monitored, e.g., by mass spectrometry, the $[O(^3P_J)]$ can be deduced from the decrease in NO_2 (in excess) when the discharge is activated; and the limit of detectability is only limited by the accuracy with which the NO_2 decrease can be monitored (Bemand *et al.*, 1973, 1974).

C. N(^{4}S)

All the reported kinetic studies on N(^{4}S) have used electric discharges in N_2 or $N_2 + Ar(He)$ as sources of atoms. There is only one known reaction for the production of N atoms in discharge flow systems:

$$H + NF \longrightarrow HF + N, \quad k_{298} = 2.5 \times 10^{-13} \text{ cm}^3 \text{ molecules}^{-1} \text{ s}^{-1}$$

(Clyne and White, 1970; Cheah *et al.*, 1979). The reaction products are mainly ground-state N(^{4}S) atoms, but excited metastable N(^{2}D) (and to a lesser extent N(^{2}P)) also have been detected by resonance fluorescence (Cheah *et al.*, 1979). The N(^{2}D) atoms are believed to be formed through a state-selective reaction of H atoms with excited metastable $NF(a^1\Delta)$ radicals.

$$O + CN \longrightarrow CO + N(^2D) \quad (0.85)$$
$$\longrightarrow CO + N(^4S) \quad (0.15)$$

in flash photolysis.

The degree of dissociation of N_2 by microwave discharge is low (typically $\ll 5\%$), and it is extremely sensitive to impurities. Nitrogen atom concentrations can be increased up to a factor of twenty when various impurities such as O_2, NO, H_2, and SF_6 are added to the purified N_2 (Herron, 1965; Young *et al.*, 1964).

The main active species in the product of a microwave discharge on N_2 or $N_2 + Ar(He)$ is N(^{4}S). Although appreciable concentrations of ^{2}D ($\sim 10\%$) and ^{2}P ($\sim 4\%$) metastable excited atoms have been found near the discharge plasma (Foner and Hudson, 1962; Lin and Kaufman; 1971), these excited atoms, as well as $N_2(A^3\Sigma_u^+)$ (e.g., Setser and Stedman, 1970; Young and St. John, 1968), are destroyed within a few milliseconds. Vibrationally excited N_2 is an active species present in the flow system which probably does not affect kinetic studies of $N(^4S)$ (Evans and Winkler, 1956; Kaufman and Kelso, 1958).

The most common titration reaction for $N(^4S)$ is $N + NO$. As mentioned above, this reaction is fast and the stoichiometry is simple. It is one of the very few gas titration reactions with a visual end point (Kaufman, 1961; Brocklehurst and Jennings, 1967). The titration reaction $N + C_2H_4$ proposed by Verbeke and Winkler (1960) is now considered unreliable, although the precise sources of error are unknown (Fersht and Back, 1965; Brocklehurst and Jennings, 1967). Recently, Michael and Lee (1977) reported that $N(^4S)$ atoms do not directly react with C_2H_4 ($k_{298} < 5 \times 10^{-16}$ cm^3 molecule^{-1} sec^{-1}); Safrany (1971) proposed that the rate of the reaction of N with C_2H_4 is controlled by trace impurities such as H or OH.

D. $S(^3P_J)$

Clyne and Townsend (1975) passed a $SO_2 + Ar$ mixture through microwave discharge and successfully generated sulphur atoms, but the main products were SO and O. A discharge bypass system (in which most of the O and SO recombine to form SO_2) has been used by Clyne and Whitefield (1979) to minimize the ratio of O atoms and SO radicals to S atoms. In this way, these workers measured the rate constants for the $S + NO_2$ and $S + O_2$ reactions over a range of temperatures. The rate constants of $S(^3P_J)$ reactions are one or more orders of magnitude higher than those of the corresponding reactions of SO and O (Clyne and Townsend, 1975), for example, in the case of the reactions $S + NO_2$, $S + O_3$ and $S + Cl_2$. $S(^3P_J)$ has also been identified in the product of a microwave discharge in SF_6 (Kley and Broida, 1976), but no kinetic studies have been made using this source. The possibilities of other precursors such as elemental sulphur and CS_2 should be explored.

Fair and Thrush (1969b,c) have used the reaction sequence $H + H_2S \rightarrow H_2 + HS$, followed by $H + HS \rightarrow H_2 + S$, to generate $S(^3P_J)$ atoms. The rate constants for these reactions have been subsequently determined to be $\sim 1 \times 10^{-12}$ cm^3 molecule^{-1} sec^{-1} (Perner and Franken, 1969; Kurylo et al. 1971; Cupitt and Glass, 1975) and 2.5×10^{-11} cm^3 molecule^{-1} sec^{-1} (Cupitt and Glass, 1975) respectively. In order to produce $S(^3P_J)$ from these reactions which have low rate constants, high concentrations of $H(> 5 \times 10^{14}$ cm^{-3}) were used in order to give complete conversion within a short period of time (say <5 msec). This high ratio of $[H(^2S)]/[S(^3P_J)]$ would severely limit the usefulness of such a source of $S(^3P_J)$ for kinetic studies.

The concentration of $S(^3P_J)$ atoms from a discharge in SO_2 + Ar has been determined (Clyne and Townsend, 1975; Clyne and Whitefield, 1979) by converting the sulphur atoms to $O(^3P_J)$ through the reaction

$$S + O_2 \longrightarrow SO + O, \quad k_{298} \simeq 2 \times 10^{-12} \text{ cm}^3 \text{ molecule}^{-1} \text{ sec}^{-1}$$

(Clyne and Townsend, 1975, and references therein) and then monitoring $[O(^3P_J)]$ formed. Since the reaction

$$SO + O_2 \longrightarrow SO_2 + O, \quad k_{298} = 3.3 \times 10^{-16} \text{ cm}^3 \text{ molecule}^{-1} \text{ sec}^{-1}$$

is slow (Homann et al., 1968), the increase in $[O(^3P_J)]$ can be determined by atomic resonance fluorescence (see below). Considering that the product of the SO_2 + Ar discharge contains a high concentration of $O(^3P_J)$, such a method for the determination of $S(^3P_J)$ is quite difficult.

E. $C(^3P_J)$

Ground state carbon atoms are formed when carbonaceous compounds are passed through a microwave discharge. Kley et al. (1972a,b) have demonstrated that stoichiometric amounts of $C(^3P_J)$ are produced when small quantities of C_2N_2 are added to an excess of active nitrogen. Although the mechanism of the formation is not clear, this has so far been the only method used to generate $C(^3P_J)$ for kinetic studies in flow systems (Ogryzlo et al., 1973). Johnson and Fontijn (1973) and Fontijn and Johnson (1973) have also observe $C(^3P_J)$ (with $\simeq 10\%$ $C(^1D)$) in the reactions of N + C_2F_4 and O + C_2H_2 by resonance fluorescence, but no kinetic investigations were made.

Kley et al. (1972a,b) have demonstrated that the reaction

$$C(^3P_J) + O_2 \longrightarrow CO + O, \quad k_{298} = 3.3 \times 10^{-11} \text{ cm}^3 \text{ molecule}^{-1} \text{ sec}^{-1}$$

(Braun et al., 1969; Husain and Kirsch, 1971) is suitable as a titration reaction for $C(^3P_J)$ atoms. As in the case of $S(^3P_J)$, a clean source of ground state carbon atoms is required before systematic kinetic investigations can be made.

F. $F(^2P_J)$

The degree of dissociation of F_2 is high ($\geq 70-80\%$) when a mixture of F_2 + He(Ar) is passed through a microwave discharge (Appelman and Clyne, 1978; Warnatz et al., 1971a; Goldberg and Schneider, 1976; Kolb and Kaufman, 1972; Rosner and Allendorf, 1971). One possible problem with this method for producing $F(^2P_J)$ is the impurities, mainly $O(^3P_J)$, O_2, and SiF_4, resulting from attack of molecular fluorine on the silica discharge tube.

The use of passivated alumina discharge tubes eliminates the silicon impurities (Kolb and Kaufman, 1972; Rosner and Allendorf, 1971), with some reduction in the degree of dissociation; however, some oxygen impurity persists. Nevertheless, these are convenient ways of producing atomic fluorine. Coating of the discharge tube by oxyacids (a practice common for the production of other atoms, particularly Cl and Br) is not recommended. This practice only serves to reduce the atom yield drastically and to introduce additional impurities into the flow from the reactions of atomic or molecular fluorine on the coating (Clyne et al., 1973; Watson, 1974a; Polanyi and Woodall, 1972). On the other hand, the coating of the flow tube by teflon (Debely, 1970) is very efficient in protecting the flow tube from attack by F and F_2 (Foon and Kaufman, 1975; Clyne and Nip, 1977b) and in reducing the surface loss of fluorine atoms. Coating of the discharge tube with teflon is not practicable. Discharge in fluorine precursors such as SF_6 and CF_4 have been used to produce ground state $F(^2P_J)$ atoms (e.g., Johnson and Lovas, 1972; Persky, 1972; Kolb and Kaufman, 1972; Setser and Sang, 1977) from microwave discharges. Although easier to handle than F_2, these compounds unavoidably introduce a large number of impurity species (e.g., see Kley and Broida, 1976).

$F(^2P_J)$ atoms have also been produced chemically in flow systems. Several laboratories have used the reaction of $N + NF_2$ for this purpose (Homann et al., 1970; Warnatz et al., 1971b; Clyne and White, 1970):

$$N + NF_2 \longrightarrow N_2 + 2F.$$

The rate constant of this reaction is not known, although it is expected to be high by analogy with other atom–radical reactions. NF_2, from thermal decomposition of N_2F_4 at ~ 500 K, is mixed with a stream of $N(^4S)$ atoms from a microwave discharge in N_2. Under typical flow conditions ($P < 5$ Torr and low reagent concentrations), mass spectrometric studies have confirmed the stoichiometries $\Delta[N]/\Delta[NF_2] = 1$ and $\Delta[F]/\Delta[NF_2] = \frac{1}{2}$ (Homann et al., 1970; Clyne and Watson, 1974a; Clyne and Connor, 1972); thus the occurrence of significant side reactions such as

$$F + NF_2 + M \longrightarrow NF_3 + M, \quad k_{298} = 1.4 \times 10^{-31} \text{ cm}^6 \text{ molecule}^{-2} \text{ sec}^{-2}$$

(Warnatz et al., 1972; Clyne and Watson 1974a) and

$$NF_2 + NF_2 + M \longrightarrow N_2F_4 + M, \quad k_{298} = 1.3 \times 10^{-32} \text{ cm}^6 \text{ molecule}^{-2} \text{ sec}^{-2}$$

(Clyne and Connor, 1972) are not likely. The good agreement between the kinetic data derived from this and other sources of $F(^2P_J)$ confirms the validity of the $N + NF_2$ method.

Another reaction that has been used to produce $F(^2P_J)$ is $NO + F_2 \rightarrow$ FNO + F, particularly in chemical laser studies (Cool and Stephens, 1969, 1970). Although the rate constant for this reaction is quite small [between

10^{-14} and 10^{-15} cm^3 molecule^{-1} sec^{-1} at 298 K (Rapp and Johnston, 1960; Kim et al., 1972)], the method has the advantage that both reactants are stable species and the $F(^2P_J)$ formed is self-monitoring by means of the yellow chemiluminescent reaction of F with NO

$$F + NO + M \longrightarrow FNO^* + M,$$
$$FNO^* \longrightarrow FNO + h\nu$$

(Johnston and Bertin, 1959). With a few exceptions, the kinetic data from this source of $F(^2P_J)$ are in fair agreement with those derived from other sources (Clyne and Nip, 1978; Foon and Kaufman, 1975).

Thermal decomposition of F_2 has not been used to generate $F(^2P_J)$ atoms in flow systems. However, a method, which involves passing pure F_2 through a nickel oven at 1000 K (Parson and Lee, 1972), has been developed for molecular beam studies and could be useful in flow system kinetics.

There exist several titration methods for the determination of $F(^2P_J)$ concentrations which, at 298 K, consist of 7% $F(^2P_{1/3})$ and 93% $F(^2P_{3/2})$ atoms. The earlier proposals, using $F + H_2 \to HF + H$ (Bozzelli and Kaufman, 1973; Kolb and Kaufman, 1972) and $F + NOCl \to FCl + NO$ (Homann et al., 1970; Warnatz et al., 1971b), were applicable only in the absence of substantial concentrations of F_2. In the case of $F + H_2 \to HF + H$, the secondary reaction

$$H + F_2 \longrightarrow HF + F, \qquad k_{298} \simeq 4 \times 10^{-12} \text{ cm}^3 \text{ molecule}^{-1} \text{ sec}^{-1}$$

(Goldberg and Schnieder, 1976, and references therein) may complicate the titration reaction. The F + NOCl reaction also suffers from possible complications from $NO + F_2 \to FNO + F$, which regenerates $F(^2P_J)$ atoms. Even in the absence of F_2, this titration may be complicated by the contribution of the reaction sheme

$$F + NOCl \longrightarrow FNO + Cl,$$
$$Cl + NOCl \longrightarrow NO + Cl_2,$$
$$F + Cl_2 \longrightarrow FCl + Cl.$$

This scheme may influence the stoichiometry under some conditions.

More recently, other titration reactions were developed which include the reaction $F + Cl_2 \to FCl + Cl$ (Nordine, 1974; Nordine and Rosner, 1976; Schatz and Kaufman, 1972; Ganguli and Kaufman, 1974; Appelman and Clyne, 1975; Bemand and Clyne, 1976; Clyne and Nip, 1977b). Another reaction is $F + Br_2 \to FBr + Br$ (Bemand and Clyne, 1976; Strattan and Kaufman, 1977) as well as $F + HCl \to HF + Cl$ (Ultee and Bonczyk, 1977) and possibly the reaction $F + HBr \to HF + Br$ (Strattan and Kaufman, 1977). All these titration reactions are fast (Foon and Kaufman, 1975; Sung and Setser, 1977); and they are virtually free from secondary reactions, provided that third-order recombinations are unimportant (Nordine and Rosner, 1976).

However, a cautionary comment is made regarding those titrations which utilize the Cl_2^* chemiluminescence from $Cl + Cl + M \rightarrow Cl_2^* + M$ to indicate the end point (Kaufman and co-workers; Nordine and co-workers) because of the complexity of the halogen recombination reactions (Clyne and Stedman, 1968a). The spectrum of the chemiluminescence exhibits a red shift with decrease in $[Cl(^2P_J)]$, hence resulting in a change with wavelength in the order of reaction (i.e., in the exponent n in the equation $I_{\text{chemiluminescence}} = k[Cl]^n$). This problem can be partly overcome using log–log plots of $I_{\text{chemiluminescence}}$ versus $[Cl_2]_{\text{added}}$ (taking the $F + Cl_2$ titration reaction as an example) at a particular wavelength. The surprising observation that the slopes of these "titration curves" decrease below 600 nm instead of increase as predicted by Clyne and Stedman's (1968a) investigation, while the chemiluminescence extends to below 530 nm, suggests the possible presence of another source of chemiluminescence in this wavelength region. This would also explain why a sharp end point could not be obtained at wavelengths below 530 nm and the unusual shapes of the titration curves (Nordine, 1974). The (possible) chemiluminescences of $F + F + M$ and $F + Cl + M$ have not been observed, but it is conceivable that $F + Cl + M$ could have emission maxima below 680 nm (Clyne et al., 1971, 1972a; Clyne and Coxon, 1967, 1968b; Clyne and Stedman, 1968a). Judging from the evidence, this titration method, with detection of Cl_2^* chemiluminescence, is probably only valid if wavelengths longer than 600 nm are used to observe emission.

G. $Cl(^2P_J)$

Ground state chlorine atoms (0.7% $Cl(^2P_{1/2}) + 99.3\%$ $Cl(^2P_{3/2})$ at 298 K) can be produced by microwave discharge in Cl_2 or a $Cl_2 + Ar(He)$ mixture. The high recombination rate of Cl on silica necessitates coating with oxyacids of the discharge tube. Some workers also coat the flow tube (e.g., Zahniser and Kaufman, 1977) with an oxyacid, typically phosphoric acid. This unavoidably introduces traces of $O(^3P_J)$ impurities into the products of the discharge, although the $H(^2S)$ impurity is normally scavenged by undissociated Cl_2.

Chemically, $Cl(^2P_J)$ atoms can be produced by a wide variety of reactions, such as by the addition of chlorine dioxide OClO into an excess of $O(^3P_J)$ atoms (Clyne et al., 1972b):

$$O + OClO \longrightarrow O_2 + ClO, \quad k_{298} = 5 \times 10^{-13} \text{ cm}^3 \text{ molecule}^{-1} \text{ sec}^{-1};$$

(Bemand et al., 1972); followed by

$$O + ClO \longrightarrow Cl + O_2, \quad k_{298} \simeq 5 \times 10^{-11} \text{ cm}^3 \text{ molecule}^{-1} \text{ sec}^{-1};$$

(Bemand et al., 1972; Clyne and Nip, 1976a; Zahniser and Kaufman, 1977). This gives, overall, the stoichiometry

$$2O + OClO \longrightarrow Cl + 2O_2.$$

Similarly, stoichiometric conversions to Cl have been reported in the reaction of N with OClO (Clyne and Cruse, 1971). An approach which does not involve any atom source is the reaction of nitric oxide with OClO:

$$NO + OClO \longrightarrow NO_2 + ClO, \qquad k_{298} = 3.4 \times 10^{-13} \text{ cm}^3 \text{ molecule}^{-1} \text{ sec}^{-1}$$

(Bemand et al., 1972); followed by

$$NO + ClO \longrightarrow Cl + NO_2, \qquad k_{298} \simeq 2 \times 10^{-11} \text{ cm}^3 \text{ molecule}^{-1} \text{ sec}^{-1}$$

(Clyne and Watson, 1974b; Zahniser and Kaufman, 1977); giving, overall,

$$2NO + OClO \longrightarrow Cl + 2NO_2.$$

In this way, known concentrations of $Cl(^2P_J)$ could be produced without a microwave discharge. One disadvantage of the above chemical methods is the instability of OClO, which tends to decompose when stored as dilute mixtures ($<1\%$ OClO) (Bemand et al., 1972), although concentrated mixtures have been reported to last for several days (Coon et al., 1962; Clyne and White, 1971). The OClO concentration can be monitored readily by absorption spectroscopy at 366 nm (Coon et al., 1962). The third-order recombinations of NO and NO_2 with Cl, i.e.,

$$NO_2 + Cl + M \longrightarrow NO_2Cl + M, \qquad k_{298, Ar} = 7.2 \times 10^{-31} \text{ cm}^6 \text{ molecule}^{-2} \text{ sec}^{-1}$$

(M. A. A. Clyne and I. F. White, unpublished data); and

$$NO + Cl + M \longrightarrow ClNO + M, \qquad k_{298, Ar} = 9.3 \times 10^{-32} \text{ cm}^6 \text{ molecule}^{-2} \text{ sec}^{-1}$$

(Clyne and Stedman, 1968b); may tend to reduce the stoichiometry $\Delta[Cl]/[OClO]$ below unity (Clyne and Cruse, 1971).

If the use of fluorine is acceptable, the reactions

$$F + Cl_2 \longrightarrow FCl + Cl, \qquad k_{298} \sim 1.5 \times 10^{-10} \text{ cm}^3 \text{ molecule}^{-1} \text{ sec}^{-1}$$

(Appelman and Clyne, 1975, and references therein) and

$$F + HCl \longrightarrow HF + Cl, \qquad k_{298} \simeq 1.2 \times 10^{-11} \text{ cm}^3 \text{ molecule}^{-1} \text{ sec}^{-1}$$

(Clyne and Nip, 1977b and references therein) are also good sources of $Cl(^2P_J)$ (Clyne and Nip, 1977b). As mentioned above, for the titration of $F(^2P_J)$ atoms, these reactions have simple 1:1 stoichiometries. Whereas the reaction $F + Cl_2$ produces a Boltzmann distribution of $[Cl(^2P_{1/2})]$ and $[Cl(^2P_{3/2})]$, the reaction $F + HCl$ gives a higher fraction of $Cl(^2P_{1/2})$, namely $([Cl(^2P_{1/2})]/([Cl(^2P_{1/2})] + [Cl(^2P_{3/2})]) \simeq 0.1)$ (Clyne and Nip, 1977b); $F + HCl$ should thus be a source of J-excited $(^2P_{1/2})$ chlorine atoms. Further comment on this reaction will be made in Chapter 3 (Kolts and Setser).

Known concentrations of $Cl(^2P_J)$ have also been produced by reacting Cl_2 with excess H. Although the reaction $H + Cl_2$ is fast $[k_{298} \simeq 2.1 \times 10^{-11} \, cm^3 \, molecule^{-1} \, sec^{-1}$ (Bemand and Clyne, 1977; Wagner et al., 1976a)], the secondary reaction

$$H + HCl \longrightarrow H_2 + Cl, \quad k_{298} = 5.9 \times 10^{-14} \, cm^3 \, molecule^{-1} \, sec^{-1}$$

(Watson, 1974b) is important at high [H] and a knowledge of the exact concentration of $H(^2S)$ is required to correct the stoichiometry (Clyne and Nip, 1977a).

It is thus straightforward to produce known concentrations of $Cl(^2P_J)$. In cases where an unknown $[Cl(^2P_J)]$ has to be determined, the most well established titration reaction is

$$Cl + ClNO \longrightarrow Cl_2 + NO, \quad k_{298} = 3 \times 10^{-11} \, cm^3 \, molecule^{-1} \, sec^{-1}$$

(Clyne and Cruse, 1972a; Ogryzlo, 1961; Poulet et al., 1974; Bader and Ogryzlo, 1964; Hutton and Wright, 1965; Clyne and Stedman, 1968a; Leu and De More, 1976). The $1:1$ stoichiometry has been confirmed by Clyne et al. (1972b) up to $[ClNO] = [Cl(^2P_J)] = 2.4 \times 10^{13} \, cm^{-3}$. Recently, the reaction sequence

$$Cl + Br_2 \longrightarrow BrCl + Br, \quad k_{298} = 1.9 \times 10^{-10} \, cm^3 \, molecule^{-1} \, sec^{-1};$$

(Bemand and Clyne, 1975)

$$Cl + BrCl \longrightarrow Cl_2 + Br, \quad k_{298} = 1.4 \times 10^{-11} \, cm^3 \, molecule^{-1} \, sec^{-1};$$

(Clyne and Cruse, 1972b) overall

$$2Cl + Br_2 \longrightarrow Cl_2 + 2Br,$$

has been used to titrate $[Cl(^2P_J)]$ (Clyne and Smith, 1978). In this case, $[Cl]_0 = \frac{1}{2}[Br_2]_{endpoint}$. Since it involves a secondary reaction, the $Cl + Br_2$ system is expected to show a change in stoichiometry at low $[Cl(^2P_J)]$.

If $Cl(^2P_J)$ atoms are produced by microwave discharge of Cl_2 or Cl_2 + $Ar(He)$ mixtures, a simple method for the estimation of the atom concentration is by monitoring the decrease in Cl_2 concentration when the discharge is activated. Molecular chlorine has an easily accessible absorption continuum centered at around 340 nm $[\sigma_{340 \, nm} = 1.0 \times 10^{-19} \, cm^2$ (Seery and Britton, 1964)]. Alternatively, the change in Cl_2 concentration can be monitored by a mass spectrometer (Leu and De More, 1976). This method assumes negligible atom recombination and is probably only valid when a phosphoric acid-coated flow tube is used.

H. $Br(^2P_J)$

Most of the methods of producing $Br(^2P_J)$ are the analogs of those for the production of $Cl(^2P_J)$. Thus, the passage of a Br_2 or $Br_2 + Ar(He)$ flow through a microwave discharge in an oxyacid-poisoned tube produces up to 50% dissociation. The $Br(^2P_J)$ atoms produced in this way are usually free of the impurities N, O, and H, since these species are scavenged rapidly by undissociated Br_2 (Clyne and Cruse, 1971).

$Br(^2P_J)$ can also be produced by displacement reactions with $F(^2P_J)$, e.g., by

$$F + Br_2 \longrightarrow FBr + Br, \quad k_{298} = 2 \times 10^{-10} \text{ cm}^3 \text{ molecule}^{-1} \text{ sec}^{-1}$$

(Bemand and Clyne, 1976; Appelman and Clyne, 1975) and

$$F + HBr \longrightarrow HF + Br, \quad k_{298} \simeq 6 \times 10^{-11} \text{ cm}^3 \text{ molecule}^{-1} \text{ sec}^{-1}$$

(Sung and Setser, 1977, and references therein). As in the case of $Cl(^2P_J)$ atoms from $F + HCl$, a non-Boltzmann distribution of $[Br(^2P_{1/2})]/([Br(^2P_{1/2})] + [Br(^2P_{3/2})]) = 0.1$ has been reported for the reaction $F + HBr$ (Sung and Setser, 1977). The displacement reactions: $Cl + Br_2 \rightarrow BrCl + Br$ and $Cl + BrCl \rightarrow Cl_2 + Br$ (see above), as well as

$$Cl + HBr \longrightarrow HCl + Br, \quad k_{298} = 8 \times 10^{-12} \text{ cm}^3 \text{ molecule}^{-1} \text{ sec}^{-1}$$

(Bergmann and Moore, 1975; Mei and Moore, 1977), can also be used for the production of $Br(^2P_J)$ atoms.

However, unlike the $O + Cl_2$ reaction, the $O + Br_2$ reaction is fast $[k_{298} = 1.4 \times 10^{-11} \text{ cm}^3 \text{ molecule}^{-1} \text{ sec}^{-1}$ (Clyne et al., 1976a)] and can be used as a source of $Br(^2P_J)$ when trace amounts of Br_2 are added to an excess of O atoms (Clyne and Townsend, 1974)

$$O + Br_2 \longrightarrow BrO + Br,$$
$$O + BrO \longrightarrow Br + O_2, \quad k_{298} = 2.5 \times 10^{-11} \text{ cm}^3 \text{ molecule}^{-1} \text{ sec}^{-1}$$

(Clyne et al., 1976a); with the overall stoichiometry

$$2O + Br_2 \longrightarrow 2Br + O_2.$$

As for $Cl(^2P_J)$ atoms, $Br(^2P_J)$ atom concentrations can be determined from the decrease in $[Br_2]$ when the microwave discharge is activated. The monitor of Br_2 is either by light absorption near 415 nm (Passchier et al.,

1967) or mass spectrometry (Leu and De More, 1977a). A titration can be made with the reaction

$$Br + ClNO \longrightarrow BrCl + NO, \quad k_{298} = 1 \times 10^{-11} \text{ cm}^3 \text{ molecule}^{-1} \text{ sec}^{-1}$$

(Clyne and Cruse, 1972a), which has unit stoichiometry (Clyne et al., 1972b).

I. $I(^2P_J)$

Production of iodine atoms by microwave discharge is inefficient (typically $< 1\%$ decomposition) (Browne and Ogryzlo, 1970; Ogryzlo, 1961; Clyne and Cruse, 1972a). Thus, $I(^2P_{3/2})$ atoms in flow systems have been produced by chemical reactions. Typical sources of $I(^2P_{3/2})$ atoms are

$$Cl + ICl \longrightarrow I + Cl_2, \quad k_{298} = 8 \times 10^{-12} \text{ cm}^3 \text{ molecule}^{-1} \text{ sec}^{-1}$$

(Clyne and Cruse, 1972b), and

$$Br + IBr \longrightarrow I + Br_2, \quad k_{298} = 3.5 \times 10^{-11} \text{ cm}^3 \text{ molecule}^{-1} \text{ sec}^{-1}$$

(Clyne and Cruse, 1972b). The simple stoichiometries of both these reactions have been confirmed (Clyne and Cruse, 1972b). The reactions

$$F + HI \longrightarrow HF + I, \quad k_{298} \simeq 7.5 \times 10^{-11} \text{ cm}^3 \text{ molecule}^{-1} \text{ sec}^{-1}$$

(Sung and Setser, 1977, and references therein) and

$$Cl + HI \longrightarrow HCl + I, \quad k_{298} \simeq 1.5 \times 10^{-10} \text{ cm}^3 \text{ molecule}^{-1} \text{ sec}^{-1}$$

(Bergmann and Moore, 1975; Mei and Moore, 1977) are other possible sources of $I(^2P_J)$. There have been conflicting reports in the literature as to whether the reaction $F + HI$ produces a non-Boltzmann distribution in $I(^2P_J)$. Whereas Sung and Setser (1977) could not observe a non-Boltzmann distribution in $I(^2P_J)$ from $F + HI$, Burak and Eyal (1977) reported that 50% of the I atoms formed in the same reaction are in the $^2P_{1/2}$ state. Unlike the reactions of F with bromine and chlorine, the reaction $F + I_2$ is not a good source of $I(^2P_J)$, in spite of its high rate constant [$k_{298} = 4.3 \times 10^{-10}$ cm^3 molecule^{-1} sec^{-1} (Appelman and Clyne, 1975)]. This is because of the formation of solid $(IF)_x$ deposits on the flow tube wall following the initial $F + I_2$ reaction (Appelman and Clyne, 1975). Because of the low bond energy in HI, reactions of $I(^2P_{3/2})$ with RH species are slow at the 298 K, and we are not aware of any titration reactions for iodine atoms.

IV. PRODUCTION OF MOLECULAR RADICALS

The discharge-flow method has been used to generate a wide variety of molecular (diatomic and triatomic) free radicals. Carrington and co-workers have observed the EPR spectra of many free radicals in this way. However,

Table I

Production of Selected Molecular Free Radicals

Radical	State[a]	Source of radical[b]	Detection[c]	Reference
OH	$X^2\Pi$	$H + NO_2 \longrightarrow OH + NO$	RF, LF	Anderson and Kaufman (1972)
OD	$X^2\Pi$	$D + NO_2 \longrightarrow OD + NO$	RF	Clyne and Williamson (1979)
ClO	$X^2\Pi_{3/2,1/2}$	$Cl + OClO \longrightarrow 2ClO$ or	MS, UV	Clyne and Watson (1974b)
		$Cl + O_3 \longrightarrow ClO + O_2$		Leu and De More (1977a)
BrO	$X^2\Pi$	$Br + O_3 \longrightarrow BrO + O_2$ or	MS, UV	Clyne and Watson (1975)
		$O + Br_2 \longrightarrow BrO + Br$		Clyne and Cruse (1970a,b)
IO	$X^2\Pi$	$I + O_3 \longrightarrow IO + O_2$	MS	Clyne and Cruse (1970a,b)
				Wagner et al. (1972)
FO	$X^2\Pi$	$F + O_3 \longrightarrow FO + O_2$	MS	Clyne and Watson (1974b)
SO	$X^3\Sigma^-$	Discharge in $SO_2 + He$	CL	Halstead and Thrush (1966a,b)
			MS	Clyne and MacRobert (1979)
CN	$X^2\Sigma^+$	Various		Boden and Thrush (1968)
NF_2	X	$N_2F_4 + M \longrightarrow 2NF_2 + M$	UV, MS	Clyne and Connor (1972)
NF	$X^3\Sigma^-$, $a^1\Delta$, $b^1\Sigma^+$	$H + NF_2 \longrightarrow$	CL	Clyne and Watson (1974a)
			EPR	Clyne and White (1970)
			MS	Clough et al. (1971)
N_3	$X^2\Pi$	$Cl + N_3Cl \longrightarrow Cl_2 + N_3$	CL, UV	Combourieu et al. (1977)
NCl_2	X	$Cl + NCl_3 \longrightarrow NCl_2 + Cl_2$	UV, MS	Clark and Clyne (1970b)
NCl	$X^3\Sigma^-$, $b^1\Sigma^+$	$Cl + N_3 \longrightarrow NCl + N_2$	MS, CL	Clark and Clyne (1970a)
HO_2	X	Various	LMR	Burrows et al. (1977)
				Howard and Evenson (1977)

[a] X is the ground state.

[b] Reactant in ground state unless stated.

[c] RF, resonance fluorescence; MS, mass spectrometry; EPR, electron paramagnetic resonance; CL, chemiluminescence; UV, ultraviolet absorption spectroscopy; LF, laser fluorescence; LMR, laser magnetic resonance.

21

the number of radicals whose reactions are suitable for quantitative kinetic study is much more limited. We summarize some of these radicals, for which at least limited quantitative kinetic data have been obtained, in table I.

By far, the most extensive kinetic studies have been carried out on the reactions of OH, ClO, and BrO. Particularly for the former two species, their kinetic behavior can be now considered well established, at least at 298 K. Recently, systematic studies of elementary reactions of HO_2 radicals have been commenced, using laser magnetic resonance (see below) (Howard and Evenson, 1977; Burrows et al., 1977). Kinetic studies of SO radical reactions using mass spectrometry also are in progress (Clyne and MacRobert, 1979).

V. DETECTION OF ATOMS AND RADICALS

In this section, we describe some principal detection techniques for atoms and radicals: (A) optical spectrophotometry, (B) mass spectrometry, (C) electron paramagnetic resonance and laser magnetic resonance, and (D) chemiluminescence. Calibration of the monitoring systems, either by production of known atom (radical) concentrations or by titration, is usually required to convert the measured signals into absolute concentrations. The titration reactions described in the previous sections may also be used to chemically convert atoms and radicals into more readily detectable forms. For example, ClO radicals, which do not themselves fluoresce (Clyne et al., 1976b), have been detected by Cl atom resonance fluorescence by converting ClO into an equivalent concentration of Cl by reaction with added NO (Zahniser and Kaufman, 1977; Anderson et al., 1977; Clyne and Watson, 1977).

A. Optical Spectrophotometry, Including Resonance Fluorescence and Absorption

Since the review in 1973 by Clyne, optical detection of atom and radical concentrations in flow systems, particularly atomic resonance, has become a most important technique for measuring rate constants. New lamp designs and the availability of simple photon counting systems have contributed to making optical resonance techniques the most sensitive, if not the most versatile, detection technique at the present time. The use of laser induced fluorescence, which is described in the chapter by Lin and McDonald, has enabled OH concentrations as low as 5×10^6 cm^{-3} to be detected (Wang and Davis, 1974; Wu et al., 1976).

Even without the use of a laser, similar sensitivity has been claimed by

Anderson (1976) in his measurements of stratospheric OH concentrations using resonance fluorescence. However, the monitoring of OH concentrations under laboratory conditions appears to be considerably less sensitive (Chang and Kaufman, 1977).

Figure 1 (after Clyne and Cruse, 1972a) shows a typical experimental arrangement for resonance absorption and fluorescence detection of atoms in the vacuum ultraviolet.

1. Optical Absorption—Continuum Sources

Optical absorption with continuum sources has been extensively used in flash photolysis experiments, and this technique has been used to some extent also in discharge-flow systems. However, an unfavorable combination often exist: the short path lengths across flow systems (typically 2–3 cm), and the low absorption cross sections of many molecular radicals. The problem has been solved for kinetic studies of ClO, BrO, NCl_2, and N_3 radicals (see Fig. 2) by using multiple traversals (8 or more) across a flow tube of rectangular cross section (Clark and Clyne, 1969, 1970a,b; Clyne and Coxon, 1968a; Clyne and Cruse, 1970a,b; Clyne et al., 1971); but the number of traversals in the ultraviolet seldom exceeds 12 before the alignment becomes too difficult and the spatial resolution (hence time resolution in kinetic studies)

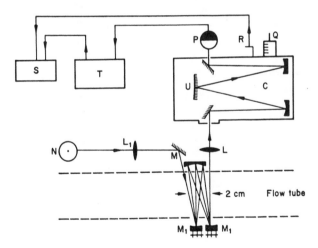

Fig. 2. Schematic diagram of optical system for detection of molecular radicals by absorption spectrophotometry, using a continuous source. C, 0.3-m Czerny–Turner monochromator; L, L_1, biconvex silica lenses; M, M_1, multiple reflection cell (8 traversals shown); N, D_2 arc lamp; P, photomultiplier cell; Q, spectrometer wavelength readout; S, recorder; T, dc amplifier; U, grating. [After Clark and Clyne (1969).]

Table II

Lowest Energy Allowed Transitions of H, O, N, C, S, F, Cl, Br, I

Atom	Lowest energy resonance transition	λ(nm)
H	$^2P-^2S$	121.6
O	$^3S-^3P_2$	130.2
N	$^4P_{1/2}-^4S$	120.1
C	$^3P_1-^3P_0$	165.7
S	$^3S-^3P_2$	180.7
F	$^4P_{5/2}-^2P_{3/2}$	97.65
Cl	$^4P_{5/2}-^2P_{3/2}$	139.0
Br	$^4P_{5/2}-^2P_{3/2}$	157.7
I	$^4P_{5/2}-^2P_{3/2}$	183.0

along the flow tube becomes unacceptable. The method is not suitable for most nonmetallic atoms which absorb in the vacuum ultraviolet (see Table II).

2. Optical Absorption—Resonance Line Source

The development of resonance line sources has greatly increased the sensitivity of optical absorption, due to the close matching between the source line profile and absorption line profile. The theory of resonance absorption has been described (Mitchell and Zemansky, 1934) and applied to particular kinetic problems (e.g., Kaufman and Parkes, 1970; Lin *et al.*, 1970; Bemand and Clyne, 1973a; Clyne and Townsend, 1974; Clyne and Piper, 1976).

The principles of atomic resonance spectroscopy will be summarized for the case of absorption of a single resonance line, in order to illustrate factors which affect the sensitivity of the technique. The extension of the theory to molecular absorption has been indicated elsewhere (Del Greco and Kaufman, 1962; Boden and Thrush, 1968).

In flow systems where pressures are no more than a few Torr, pressure broadening (Holtsmark and Lorentz broadening), as well as Stark broadening and natural broadening, are normally small compared with Doppler broadening Δv_D of the absorption line. The extinction coefficient at frequency v, k_v, can then be expressed in terms of the absorption coefficient at the line center k_o by

$$k_v = k_o \exp(-\omega^2),$$

where ω is a reduced frequency function given by

$$\omega = 2\sqrt{\ln 2}(v - v_0)/\Delta v_D \quad \text{and} \quad \Delta v_D = v_0 2\sqrt{2R \ln 2}(T/M)^{1/2}/c,$$

and T, M, c are temperature, atomic mass, and the velocity of light. The optical depth $k_o l$ of the absorber is given by

$$k_o l = \frac{2fN}{\Delta v_D} \left(\frac{\ln 2}{\pi}\right)^{1/2} \frac{\pi e^2 l}{mc},$$

where l is the length of the absorption path, N the concentration of absorber, e the electronic charge, m the electronic mass, and f the oscillator strength. Here f is defined as

$$f = \frac{1}{\tau}\left(\frac{mc^2}{8\pi^2 e^2 v_0^2}\right)\left(\frac{g_2}{g_1}\right) = 1.51 \frac{g_2}{g_1} \lambda_0^2,$$

where g_2 and g_1 are the statistical weights of the excited and ground states, respectively; and τ is the radiative lifetime, which is the reciprocal of the Einstein coefficient for spontaneous emission.

The fraction of absorbed energy of an emitting line, $A = I_{abs}/I_0$, is given by

$$A = \frac{\int_{+\infty}^{-\infty} I(\omega)\{1 - \exp[-k_o\, l\, \exp(-\omega^2)]\}\, d\omega}{\int_{-\infty}^{+\infty} I(\omega)\, d\omega},$$

where $I(\omega)$ is the emitter line profile. If the source line is assumed to have a Doppler profile (see below), then $I(\omega) = \text{const} \times \exp[-(\omega/\alpha)^2]$, where $\alpha = (T_{source}/T_{abs})^{1/2}$. Thus, A is given by

$$A = \frac{\int_{-\infty}^{+\infty} \exp[-(\omega/\alpha)^2]\{1 - \exp[-k_o\, l\, \exp(-\omega^2)]\}\, d\omega}{\int_{-\infty}^{+\infty} \exp[-(\omega/\alpha)^2]\, d\omega}.$$

Any appreciable self-absorption (self-reversal) within the source will modify the source profile $I(\omega)$. In the case of a two-layer model, where the primary emitting layer is followed by a secondary, self-reversing layer whose optical density is $k_o m$, then

$$I(\omega) = \text{const}\, \exp[-(\omega/\alpha)^2]\, \exp[-k_o m \exp(-\omega/\beta)^2].$$

In this expression, β lies between α and unity since $\beta = (T_{source}/T_{self\text{-}reversing})^{1/2}$. Hence, the fraction of absorbed energy in a resonance line can be calculated in terms of absorber concentration, provided the source line profile is known. Figure 3 illustrates the dependence of A on $k_o l\, (\propto N)$ under different lamp conditions.

Although the Doppler model has certain limitations, it provides an adequate quantitative description for many experimental studies at low pressures (<5 Torr). In cases where oscillator strength values were derived from resonance absorption measurements (Lin et al., 1970; Kikuchi, 1971; Clyne and Piper, 1976; Clyne and Nip, 1977a,b), the results are consistent

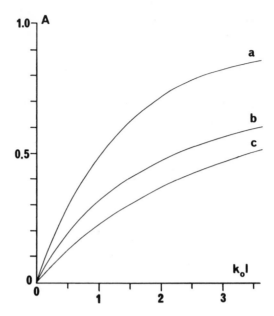

Fig. 3. Variation of fractional absorption A with optical depth $k_0 l$. Typical curves are shown. (i) no self-reversal, optical depth in source $k_0 m = 0.0$: (a) $\alpha = 1.0$, (b) $\alpha = 2.0$. (ii) with self-reversal $k_0 m = 5.0$, (c) $\alpha = 1.0$.

with those derived from other techniques such as phase shift and lifetime measurements.

For a Doppler line, α gives the width of the emitter line, but, for other cases, it can be used as an empirical parameter to describe a complex line profile by a simple model. Thus, for instance, for the $ns(^3S)-2p^4(^3P_2)$ multiplets of O, the use of a constant value of $\alpha = 1.8$ gave good linearity of $k_0 l$ versus N plots (Clyne and Piper, 1976), thereby verifying the use of this model.

On the other hand, thermalization is slow for H atom systems, and recent direct measurements by Clyne and Monkhouse (1976) have shown that the Lyman-α (121.6 nm) line profile from an H_2 + Ar discharge is not Doppler, but rectangular. However, in all cases, the fractional absorption of a resonance line by a certain absorber concentration decreases with increase in self-absorption (i.e., with $k_0 m$), and with increase in emitter temperature (i.e., source line width). Hence, desirable parameters in the design of resonance absorption lamps are minimal self-reversal and temperature near ambient.

3. Resonance Absorption Lamps

A simple method for the production of atomic resonance radiation is by a microwave discharge in a trace of the parent molecule mixed with inert gas.

Lamps powered by electrodeless discharges are less intense than those pro-duced by arc-glow discharge, but give few excited ions (Samson, 1967; Harteck et al., 1964; Thompson et al., 1965). Their ease of operation and low levels of continuum emission are advantages in kinetic studies with resonance absorption.

Lamps using flowing gas (Davis and Braun, 1968; Morse and Kaufman, 1965) are more convenient than sealed-off lamps with getters (Okabe, 1964, 1969; Washida et al., 1970; McNesby and Okabe, 1964; Anderson, 1975; Zahniser et al., 1976; Dieke and Cunningham, 1952; Dickenson et al., 1974). There have not been any comparisons between flowing and static lamps under identical conditions and it is noted that transmission efficiencies of vacuum ultraviolet radiations are sensitive to the condition of the window material. However, it is probable that both versions give comparable intensities of inert gas resonance lines (Kr, Xe, and Ar) and H Lyman-α (121.6 nm). For other atomic lines (N, O, Cl, Br, S, C), the flowing gas lamps can be one or more orders of magnitude more intense than the sealed-off lamps (Davis and Braun, 1968). There have also been some reports on the dependence of emitter line intensities on the carrier gas in flowing gas lamps. It was noted (Bemand, 1974; Bemand and Clyne, 1973b) that the intensities of N(I) triplets near 120.0 and 113.5 nm emitted from He lamps (~ 1 Torr pressure) are an order of magnitude greater than those from Ar lamps (also near 1 Torr pressure). Similar increases were also observed for Cl(I) lines with He lamps around 135 nm and F(I) lines around 95 nm. Clyne and Nip (1977b) also noted a difference in Cl(I) line intensity distributions from He and Ar lamps. These different intensities are probably due to the roles of metastable species He($^3S_1$) and Ar($^3P_{2,0}$) in excitation within microwave excited plasmas (Kaufman, 1969). On the other hand, H Lyman-α (121.6 nm) radiation from H_2 + Ar lamps has been reported to be more intense than that from H_2 + He lamps (Bemand, 1974). It is unclear whether this is due to the higher impurity levels in Ar or to less self-absorption in H_2 + Ar lamps because of its broad emission line (see below).

As mentioned above, it is desirable to use nonreversed lamps, giving thermal emission, in resonance absorption studies, although these may not always be Doppler-type line sources (Mitchell and Zemansky, 1971; Lin et al., 1970; Morse and Kaufman, 1965; Lynch et al., 1976; Braun et al., 1970). The absence of self-reversal in the source can be achieved by working with very low atom concentrations in the lamp. For lamps of atoms with more than one J level in the ground electronic state, the ratio of emission intensities from the same excited state to various J levels in the ground state provides an estimate of self-reversal. For example, self-absorption of the O(I) 130.2 nm line can be assumed to be negligible if the intensity ratio $I_{130.2 \text{ nm}} : I_{130.5 \text{ nm}} : I_{130.6 \text{ nm}}$ is very close to the ratio of the quantities $(2J + 1)A_{kl}$ for each line in the triplet, which is $5:3:1$ in this case (Clyne and Piper, 1976; Morse and Kaufman, 1965). In cases, such as halogens, where the appropriate oscillator strengths are not known, the concentrations of parent molecules in the lamps are reduced until the appropriate intensity ratio becomes constant (Clyne and Nip, 1977b).

In recent years, some nonreversed sources, discussed below, have been developed in the Queen Mary College laboratory (Clyne and Townsend, 1974; Clyne and Nip, 1976a).

a. Optically Thin Resonance Fluorescence. Figure 4 shows the arrangement for the use of resonance fluorescence from Br ("source 2") as the source for absorption measurements. The primary source for excitation of resonance fluorescence ("source 1") is a powerful, but self-reversed microwave Br lamp. Provided that the concentration of atoms in source 2 is low (i.e., source 2 is optically thin, $k_o l \leq 0.3$), the resonance fluorescence can be used as a non-reversed source for the detection of atoms in a flow tube.

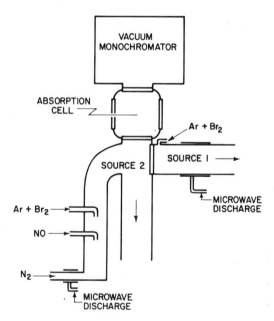

Fig. 4. Resonance absorption with a nonreversed source. Source 1 is a usual self-reversed Br atom microwave-excited lamp. Source 2 is a flow containing low concentrations of Br $^2P_{3/2}$ atoms, which emits optically thin resonance fluorescence. This is the source for absorption measurements in the kinetic flow system. Br $^2P_{3/2}$ atoms in source 2 are produced by the reaction sequence $N + NO \rightarrow N_2 + O, O + Br_2 \rightarrow BrO + Br, O + BrO \rightarrow O_2 + Br.$ [After Clyne and Townsend (1974).] Setup for photolytic double source is similar, except that source 1 is microwave-excited H Lyman-α lamp, and source 2 is a flow of Br_2.

b. Photolytic Double Source. Figure 4 also shows this source, which is based on photolysis of Br_2 (or ICl) by Lyman-α radiation (121.6 nm) as the source of excited Br (or I), which emit atomic resonance radiation (Clyne and Cruse, 1971; Bemand and Clyne, 1972). Clyne and Townsend (1974) have used this method to generate I(I) resonance lines from the photolysis of ICl by Lyman-α radiation.

The concentration of atomic species (the sum of the ground and excited states) was estimated to be of the order of 10^{10} cm^{-3}, which is low enough to eliminate self-reversal in the resonance lamp. However, it is possible that the excited atom can be translationally hot, unless the energy defect is zero. Thus, the emission line width could be appreciably broader than the thermal value.

c. Rare Gas Metastable Lamp. This lamp is based on the interaction of metastable, excited $He^*(^3S_1)$ or $Ar^*(^3P_{2,0})$ atoms with ground state molecules. The lamp design is based on the work of Stedman and Setser (1971) and King and Setser (1976) (Fig. 5). A flow of metastable argon or helium atoms can be produced by passing the purified gas through a dc hollow-cathode discharge operating near 240 V and a few milliamps. Since impurities such as H_2O and air are efficient quenchers for Ar^* and He^* (Piper *et al.*, 1972), they must be removed by cooled molecular sieve traps before the discharge. The inert gas is flowed for 1–2 hr in order to achieve a steady concentration of Ar^* and He^*. The Ar^* or He^* metastables are then mixed with a flow of a substrate molecules in front of the window of the absorption cell, e.g., Ar^* + $Cl_2 \rightarrow Ar + Cl^* + Cl$. Since the concentration of Ar^* produced by this

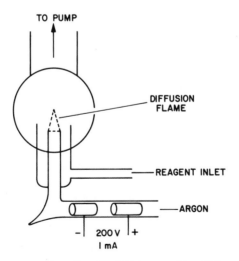

Fig. 5. Cell for spectroscopic studies of Ar^*, He^* metastable collisions. [After Stedman and Setser (1971) and King and Setser (1976).]

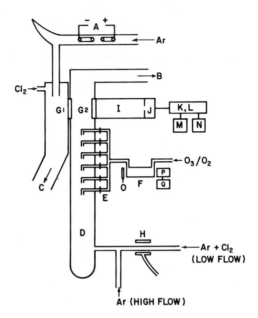

Fig. 6. Resonance absorption studies in a flow system using an Ar* metastable lamp. System shown is for studies of reaction Cl + O_3, using Ar* + Cl_2. A, 260 V dc discharge across cylindrical tantalum electrodes; B, C, connections to vacuum pumps; D, 26-mm-i.d. Flow tube; E, inlet jet manifold; F, uv absorption cell for characterization of reagents (O_3); G1, G2, cleaved LiF windows; H, microwave discharge cavity; I, Hilger 1-m normal incidence monochromator; J, E.M.I. 9789QA photomultiplier cell; K, L, M, N, photon counting system; and O, P, UV source and photomultiplier for ozone analysis. [After Clyne and Nip (1976a).]

method is only $\sim 10^{10}$ cm^{-3} (Piper *et al.*, 1973), the Cl or other ground state atom concentration is less than this value, ensuring the absence of self-reversal. The Ar* + Cl_2 and Ar* + Br_2 reactions have been used to produce nonreversed Cl(4s–3p^5) and Br(5s–4p^5) lines (Clyne and Nip, 1976a; Clyne and Smith, 1978) and extension of the technique to other resonance lines should be straightforward. Figure 6 shows a schematic of the arrangement for using an Ar* metastable lamp for resonance absorption studies.

Except for Cl* atom emission from Ar* + Cl_2, the profiles of many atomic lines excited in metastable lamps may be expected to be non-Doppler on account of incomplete thermalization of kinetic energy. Thus, Clyne and Monkhouse (1976) have directly measured the emission line profile from Ar* + H collisions, and have found the profile to be almost rectangular rather than Gaussian. Subsequent measurement of the Lyman-α line width from a H_2 + Ar microwave discharge has revealed a similar profile. The similarity lends some support to the speculation (Lin *et al.*, 1970; Kaufman, 1969) that the reactions of Ar* are important in discharges. The almost rectangular line shape from Ar* + H fits closely to the Biondi model (Rogers and Biondi, 1964; Connor and Biondi, 1965). As an example of the strong dependence of the source line profile on the nature of the excitation process, Clyne and Monkhouse (1976) found that a microwave discharge in a H_2 + He mixture emits a narrow Lyman-α line. This difference between Ar and He discharges had been inferred also for O atom emission $(3s(^3S)-2p^4(^3P_J))$ at 130.2 nm) by Parkes and Kaufman (1970). Some Br(I) lines (those with short lifetimes and large energy discrepancies) from Ar* + Br_2 have also been reported to have Biondi profiles (Clyne and Smith, 1978).

4. Applications of Resonance Absorption Spectrometry

The Doppler line model often is used as a useful approximation for the computation of fractional absorption of a resonance line as a function of absorber concentration (Myerson and Watt, 1968; Lynch et al., 1976; Barker and Michael, 1968; Michael and Weston, 1966; Kaufman and Parkes, 1970; Parkes et al., 1967; Braun et al., 1970; Clyne and Townsend, 1974; Clyne and Piper, 1976; Bemand and Clyne, 1973a; Clyne and Nip, 1977a,b; Tellinghuisen and Clyne, 1976; Braun and Carrington, 1969). Provided that the oscillator strength and the source line profile are known, together with the path length l, the absolute absorber concentration may thus be inferred from the measured fractional absorption, independent of experimental variables.

The ability to dispense with a calibration procedure for measuring absolute atom concentrations is an advantage of resonance absorption over other detection techniques. In principle, the approach may be extended to self-reversed sources, but the treatment is liable to severe errors and is not straightforward (Kaufman and Parkes, 1970; Braun et al., 1970). Therefore, it is advisable to calibrate the atom concentration independently in these cases (Clyne and Cruse, 1971: Clyne et al., 1972b).

The use in resonance absorption calculations of tables of oscillator strengths (e.g., Wiese et al., 1966, 1969) for atoms with nuclear hyperfine splittings such as F, Cl, Br, and I requires special care. The nuclear hyperfine structure often serves as a major line broadening effect, and corrections must be made (Tellinghuisen and Clyne, 1976; Bemand and Clyne, 1976; Clyne and Nip, 1977b) to determine the curve of growth.

A modified Beer–Lambert formulation, $\log_{10}(1 - A) = -\zeta(k_o l)^\gamma$, with $1 \geq \gamma \geq 0$ and where ζ is a constant, has been used to describe empirically the relationship between fractional absorption of a resonance line and absorber concentration (Donovan et al., 1970; Davis et al., 1970; Deakin and Husain, 1972; Husain and Kirsch, 1971; Husain and Norris, 1977). Although both the Doppler model and the Biondi model will reduce to a Beer–Lambert relationship with $\gamma = 1$ when $A \to 0$ (Bemand and Clyne, 1973a; Clyne and Smith,

1978), there is no theoretical justification for the modified form, and the exponent γ is an empirical parameter which depends on both the lamp characteristics and on the range of absorbance under consideration. The use of a certain parameter γ outside its calibrated range is likely to lead to erroneous results (Phillips, 1976a; Bemand and Clyne, 1973a).

5. Molecular Resonance Absorption

Most of the kinetic investigations using resonance absorption have involved reactions of atoms. Both resonance absorption and fluorescence are, in principle, applicable to monitoring molecular radicals (e.g., Golden et al., 1963; Del Greco and Kaufman, 1962; Kaufman, 1964; Boden and Thrush, 1968). However, the low line strengths for resonance absorption by most molecules, and the difficulty in finding convenient light sources, except for a few cases such as OH, CN, NH, have been a problem. Tunable dye lasers in the vacuum ultraviolet are likely to help somewhat in providing sources of resonance radiation. Of course, an inherent limitation in resonance fluorescence detection of molecules is that the excited state must be stable.

6. Limits of Detection

The lower limit of detection by resonance absorption is set fundamentally by the path length and the oscillator strength of the relevant transition. It also depends on the source line intensity and profile, the stability of the lamp, and the background intensity from the lamp. In practice, it is reasonable to expect that a fractional absorption of 0.01 could be measured unless the line is very weak. At high concentrations, there are different considerations applicable. As can be seen from Fig. 3, the fractional absorption A becomes very insensitive to $k_0 l$ (and hence to absorber concentration) for $k_0 l$ in excess of two. Depending on the discrimination of the detection system against resonantly scattered light, the simple resonance absorption model may break down before the curve of growth flattens out. In their study of fractional absorption versus [Br] (using a 0.3-m McPherson vacuum spectrometer), Clyne and Townsend (1974) found deviations from the model at $k_0 l > 1.2$ ($A > 0.55$). One technique which almost eliminates scattered light is the use of collimated hole structures (Clyne and Piper, 1976) (Fig. 7); however, the intensity is reduced considerably.

The range of concentrations that can be detected with resonance absorption, using a given line, is about a factor of 100. However, many-electron atoms and radicals have manifolds of energy levels, transitions from which to the ground state possess a variety of oscillator strengths. Thus, different concentration ranges can be covered by switching from one resonance line to another. For example, various resonance lines from the $Cl(4s-3p^5)$ manifold

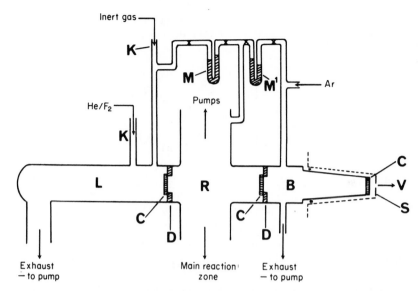

Fig. 7. Windowless system for resonance absorption studies in the far vacuum uv ($\lambda <$ 110 nm). System shown is used for studies of $F(^2P_J)$ reaction kinetics, and of $O(^3P_J)$ and $N(^4S)$ oscillator strengths. B, buffer chamber; C, collimated hole structures, two of which were mounted on glass disks, D; K, inlet to lamp L; M, M′, differential manometers; R, section of flow tube; S, spectrometer slit, sealed via an O-ring seal to the silica apparatus; and V, vacuum spectrometer. [After Bemand and Clyne (1976).]

enabled Cl concentrations from 1×10^{10} to $>2 \times 10^{14}$ cm^{-3} to be monitored in a 2.5-cm-diam flow tube (Clyne and Nip, 1977a). The use of transitions involving J excited states is another approach for extending the useful range of resonance absorption for the measurement of the concentrations of thermal distributions of atoms (Bemand and Clyne, 1973a; Morse and Kaufman, 1965).

7. Resonance Fluorescence

The theory of resonance fluroescence follows closely that of resonance absorption. The processes forming and removing excited atoms or molecules A* are as follows:

$$A \xrightarrow{I_{abs}} A^*, \qquad \text{absorption,}$$

$$A^* \xrightarrow{k_R} A + h\nu \qquad \text{fluorescence,}$$

$$A^* \xrightarrow{k_D} B + C \qquad \text{predissociation or preionization,}$$

$$A^* + M \xrightarrow{k_q} A + M \qquad \text{quenching.}$$

In the absence of significant radiation trapping (Phillips, 1973, 1975, 1976a,b; Van Volkenburgh and Carrington, 1971), the fluorescence intensity is given by

$$I_F = K\left(\frac{k_R}{k_R + k_D + k_q[M]}\right)I_{abs},$$

$$= K\left(\frac{k_R}{k_R + k_D + k_q[M]}\right)AI_o.$$

In the above equations, K is a constant ($\ll 1$) dependent on the geometry of the system, and k_R has the value of τ^{-1} where τ is the natural radiative lifetime of the excited state A*. The quantum yield Φ of fluorescence is $k_R/(k_R + k_D + k_q[M])$. In atomic resonance fluorescence at pressures ~ 1 Torr, k_R often dominates in the denominator, but the reverse is usually true for molecular fluorescence where the lifetime is often longer. In cases where the upper electronic state is unstable, for instance because of predissociation, e.g., for ClO (Clyne et al., 1976b), fluorescence will be totally or partially absent.

Unlike resonance absorption, where the fractional absorption is obtained from a small difference between two large signals, it is easy to detect low fluorescence intensities provided the background scattered light is reasonably low. The use of long integrating times associated with a flow system is particularly advantageous. Bemand and Clyne (1973a) have demonstrated that fluorescence intensities corresponding to $A \geq 0.0001$ could be measured for the fully allowed transition $O(^3S_1-^3P_2)$ at 130.2 nm. This corresponds to a lower detection limit for $O(^3P_J)$ atoms of $\sim 1 \times 10^{10}$ cm^{-3}. Comparable sensitivities in the 10^9-10^{11} cm^{-3} range have been achieved in resonance fluorescence studies of H, S, Cl, and Br (Clyne and Monkhouse, 1977; Bemand and Clyne, 1975, 1977; Clyne and Townsend, 1975; Clyne and Cruse, 1972a,b; Clyne et al., 1976a; Zahniser and Kaufman, 1977). In general, the sensitivity of the resonance fluorescence technique is so high that the fractional absorption of the incident light can be much less than 0.02. Under these circumstances, $A \propto N$ where N is the atom (radical) concentration in the fluorescence cell (see above), and thus $I_F \propto N$.

The high sensitivity coupled with the linear dependence of fluorescence intensity on atom (radical) concentrations make the technique highly suitable for kinetic investigations of atoms and radicals. Normally, the pseudo-first-order decay of the atom or radical in a known excess of stable reagent is followed. In these studies, the absolute concentration of the transient species need not be known exactly, so long as it is low enough to ensure pseudo-first-order kinetics and to preclude secondary reactions.

8. Design of Resonance Fluorescence Apparatus

The source of radiation in resonance fluorescence frequently is a microwave-powered lamp similar to that used for resonance absorption experiments. However, unlike resonance absorption, there is no clear advantage in working with a nonreversed source because the fluorescence intensity is proportional to the product AI_0 (see above) and I_0 is usually higher for self-reversed sources. The general practice is to trade off some fractional absorption by working with a reversed source against an increase in incident light intensity. Similarly, there is no decisive advantage in using the resonance line with the highest oscillator strength. An intense resonance line with low oscillator strength may prove to be as favorable as a weak line with high oscillator strength. For example, in their resonance fluorescence study of Cl atoms, Bemand et al. (1975) found the 137.9 nm line, which is the most intense line in the $4s-3p^5$ manifold, to be preferable to the 134.7 nm line, although the oscillator strength of the 134.7 nm line is 40 times higher than that of the 137.9 nm line (Clyne and Nip, 1977a). On the other hand, Zahniser et al. (1976) have found the fluorescence from the 134.7 nm transition to be up to five times stronger than that at 137.9 nm. The difference has been attributed to the different degrees of self-reversal from the two lamps.

The reduction of light scattered off surfaces in the detection system normally improves the sensitivity of a resonance fluorescence system. There are several approaches. Collimation of the source and of the detected fluorescence is particularly effective but reduces the light flux greatly. It is also advantageous to remove windows as far as possible away from the zone from which fluorescence is collected. Also, a Wood's horn may be placed opposite the lamp as illustrated in Fig. 8. The scattered light from the microwave discharge in the flow tube should also be carefully trapped by Wood's horns.

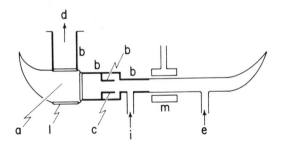

Fig. 8. Collimated resonance fluorescence lamp for studies of H-atom kinetics. a, axis of flow tube; b, areas internally blackened; c, collimating tube; d, to monochromator and detection system; e, lamp exhaust; i, lamp inlet; m, microwave cavity; l, LiF window. [After Clyne and Monkhouse (1977).]

Although unnecessary in some cases for resonance absorption, a calibration is always required to convert fluorescence signals into absolute concentrations. The reproducibility of the resonance lamp is then very important because the fluorescence intensity from a certain atom (radical) concentration in the cell depends both on the degree of self-reversal in the lamp and the intensity of the emitter line source. The self-reversal in the lamp can be reproduced satisfactorily by maintaining a constant composition of gas mixture in the lamp and the same microwave discharge power. The inevitable (but small) day-to-day variation in the source intensity should be allowed for, preferably by calibrating the ratio I_F/I_S against concentration N, instead of I_F against N, because I_S, the (nonresonant) scattered light intensity, is also proportional to the source intensity. It is a major advantage of using pseudo-first-order kinetics, that absolute concentrations of the atom (or radical) need not be known accurately in most kinetic studies (see above).

Although it is desirable to work with wavelength-resolved resonance fluorescence such a practice is not essential in some cases. In the conventional ultraviolet and the visible regions, the spectrometer can be replaced by an interference filter, e.g., for the detection of OH (Zellner et al., 1977; Clyne and Holt, 1979). Similarly, the Lyman-α (121.6 nm) transition of H can be isolated easily using an O_2 filter (Samson, 1967; Wagner et al., 1976a; Monkhouse, 1976). Another approach is to use a set of four dichroic mirrors in order to isolate the 180 nm lines of $S(^3P_J)$ in resonance fluorescence with high side band rejection (Clyne and Whitefield, 1979). By putting a CaF_2 window in front of a specially modified electron multiplier with Be–Cu dynodes, which is effectively blind at $\lambda > 150$ nm, Clyne et al. (1976a) have monitored the fluorescence signal at 130 nm from the unresolved triplet of $O(3s–2p^4)$. The unresolved $O(3s–2p^4)$ triplet has also been monitored with far ultraviolet interference filters (Anderson, 1975).

There is no doubt that, wherever possible, wavelength selection via a monochromator is a preferred technique in order to carry out source diagnostic checks, and to avoid ambiguity in the nature of the fluorescing state.

Figure 9 shows a typical energy level diagram, in this case for the transitions to the $2p^5(^2P_{3/2, 1/2})$ states of F. Note the J splitting of the ground states of the halogen atoms. Figure 10 shows the spectrum emitted from a rather self-reversed F-atom lamp using microwave excitation with He carrier gas. Note the comparable intensities of lines involving the $^4P–^2P$ and the $^2P–^2P$ transitions. A nonreversed F-atom lamp shows much more intense $^2P–^2P$ lines, compared with the $^4P–^2P$ lines (Clyne and Nip, 1977b).

9. Functional Dependence of I_F on N

One major problem associated with the use of resonance fluorescence as a detection technique, particularly for atoms, is reabsorption of the fluorescent radiation in the fluorescence cell. The gas in the cell is absorbing from a kinetically hot and a sometimes self-reversed source and is emitting thermal fluorescence. The gas to be analyzed can thus be optically thin with respect to the source but optically thick with respect to reabsorption. The discussions provided by Mitchell and Zemansky (1971) and Bemand and Clyne (1973a) are used in the following notes on the functional dependence of I_F upon N, including the upper limit of proportional dependence of I_F upon N.

The energy E_v emitted at frequency v from a resonance fluorescence cell, having finite optical depth $k_o l'$, is given by

$$E_v = C[1 - \exp(-k_v l')],$$

where C is a frequency-independent constant, equal to the total energy absorbed from the exciting beam. Thus, the total fluorescence intensity is given by

$$I_F = \int_{-\infty}^{+\infty} E_v \, dv = C \int_{-\infty}^{+\infty} [1 - \exp(-k_v l')] \, dv.$$

Using $k_v = k_o \exp(-\omega^2)$, where $\omega = 2\sqrt{\ln 2}(v - v_0)/\Delta v_D$ (see above), and expanding the expression into the Taylor series, one obtains

$$I_F = C'\left[k_o l' \int_{-\infty}^{+\infty} e^{-\omega^2} d\omega - \tfrac{1}{2}(k_o l')^2 \int_{-\infty}^{+\infty} e^{-2\omega^2} d\omega \right.$$

$$\left. + \tfrac{1}{6}(k_o l')^3 \int_{-\infty}^{+\infty} e^{-3\omega^2} d\omega + \cdots \right],$$

where C' is now also a function of the geometry of the optical system. Upon substituting the numerical values for the definite integrals;

$$I_F = 1.772C'[(k_o l') - 0.345(k_o l')^2 + 0.0962(k_o l')^3 - 0.0208(k_o l')^4 + \cdots].$$

Thus, a plot of I_F against $k_o l'$ (which is proportional to N) will initially be a proportional (linear) dependence at low N, then show a negative curvature with increase in N. For a given atom, the resonance line with the highest oscillator strength will have the highest sensitivity $\Delta I_F/\Delta N$ only if a non-reversed thermal source is used. However, such a resonance transition will show the most limited range of concentration before deviation from the linear proportional dependence of I_F on N occurs. Therefore, one should not use resonance fluorescence at absorber optical depths above (and normally much smaller than) $k_o l' = 0.3$, unless a nonlinear calibration graph of I_F against N is explicitly available.

In the cases of $O(^3P)$, $Cl(^2P)$, and $F(^2P)$ atoms, for instance, the useful range of the technique can be extended upward by monitoring resonance fluorescence transitions to J excited sublevels of the ground state. The population of these J excited states are low at 298 K, and self-reversal of resonance fluorescence is less severe than for transitions to the J ground state.

For transitions with long radiative life times, e.g., the OH(A − X) or OD(A − X) transitions ($\tau_R \sim 1$ μs) (Bennett and Dalby, 1964; German and Zare, 1969; Becker and Haaks, 1973), electronic quenching of the excited state may significantly reduce the sensitivity of resonance fluorescence.

In a recent kinetic study of reactions of OH with chlorofluorocarbons, Clyne and Holt (1979) found that the quenching of excited OH by virtually all these compounds was collisional, and because of the slow rates of their reactions with ground state OH radicals, allowance for quenching was essential in determining rate constants from these experiments. Quenching would also be significant in cases where the radiative lifetimes are lengthened by trapping of radiation when the absorbing gas is optically thick (Bemand and Clyne, 1973a). In extreme cases, this can show up as a maximum in the I_F versus N plot (Bemand and Clyne, 1973a, 1977; Phillips, 1973, 1975, 1976a,b,c).

10. Resonance Fluorescence Using Laser Excitation

Lin and McDonald's review elsewhere in this volume provides a detailed account of the detection of transient species using infrared, visible, and ultraviolet lasers. The use of tunable dye lasers and excimer lasers in the visible and ultraviolet, in particular, is an extremely promising method of monitoring labile species in flow systems. Particularly suitable labile molecules for detection with high sensitivity include: OH, SO, CN, and NH radicals in the ultraviolet and NH_2, PH_2, BO_2, IF, and BrF in the visible regions. NO_2, I_2 Br_2, Cl_2, BrCl, ICl are examples of stable molecules whose sensitive detection by laser induced fluorescence can be very useful in discharge-flow kinetics. Work up to about 1975 has been reviewed by Walther (1976). Typical more recent literature citations include: McAfee and Hozack (1976) and Clyne and McDermid (1976, 1977, 1978a–d).

B. Mass Spectrometry

The application of mass spectrometry to kinetic studies implies that every species in the reaction can, in principle, be monitored simultaneously as a function of time. However, mass spectrometric detection involves the problem that a representative sample of gas must be taken from the reaction zone for analysis. Clearly, only small mass flows ($\sim 10^{-8}$ mole sec^{-1}) can be admitted into the mass spectrometer, in order to operate below 10^{-5} Torr pressure. On the other hand, the sample analyzed by the mass spectrometer should be representative of the reaction mixture.

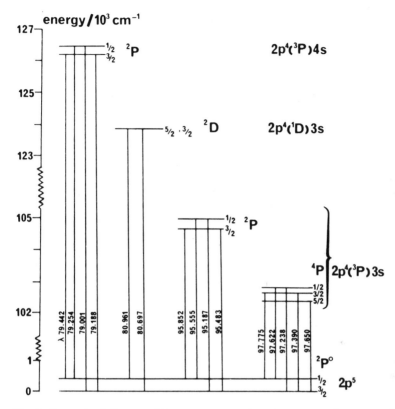

Fig. 9. Energy level diagram of F. Note close spacing of levels in $3s(^2P_J)$ and $2p(^2P_J)$ manifold; λ is in nm.

Successful sampling has been a major obstacle in the application of mass spectrometry for detecting atoms and radicals in discharge-flow reactors. The problem of satisfactory sampling explains, in part, some discrepancies between quantitative results from early experiments (e.g., Phillips and Schiff, 1962a,b; Kistiakowsky and Volpi, 1957; Jackson and Schiff, 1953; Klein and Herron, 1964) and those from more recent results. However, earlier workers were successful in identifying many reaction intermediates such as CH_3 (Eltenton, 1947; Lossing and Tickner, 1952), HO_2 (Ingold and Bryce, 1956; Foner and Hudson, 1953b), and the monitoring of atomic species such as H (Phillips and Schiff, 1962b), N (Phillips and Schiff, 1962a), and O (Klein and Herron, 1964; Jackson and Schiff, 1953).

Sampling systems have been described by Foner (1966) and more recently by Anderson and Bauer (1977). It is now generally accepted that direct sampling of the reaction mixture into the mass spectrometer by effusive flow

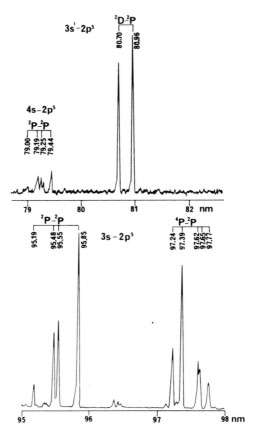

Fig. 10. Emission spectrum of F atom resonance lamp. Flowing gas, He + F_2 mixture, 0.4 Torr. 25-W incident microwave power. Note self-reversal between $^2P_{3/2}$ and $^2P_{1/2}$ lines. 1-m normal-incidence spectrometer, first order of 600 line/mm grating blazed at 90 nm, 5 μm slits. [After Bemand and Clyne (1976).]

(generally referred to as single-stage sampling system) based on the design by Eltenton (1947) is not suitable for quantitative measurements of atom and radical concentrations. This is because of the poor signal-to-noise ratio resulting from scattering of gas molecules in the ion source, and the severe loss of the reactive species through collisions with the stainless steel walls of the mass spectrometer. However, a single-stage system has been used successfully for kinetic studies of methyl radicals when coupled to a photoionization mass spectrometer (Washida and Bayes, 1976; Jones and Bayes, 1973).

The sensitivity of detection can greatly be improved by the use of a multistage sampling system, such as the one illustrated in Fig. 11 (after Homann *et al.*, 1970). The reaction mixture expands out of the first sampling

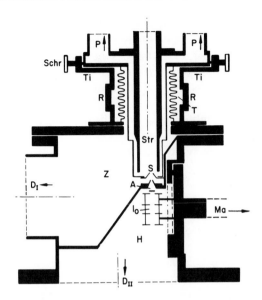

Fig. 11. Collision free mass spectrometry for kinetic studies in a discharge flow system. A, skimmer; D_I, 2×250 liter sec^{-1} diffusion pumps; D_{II}, trapped 700 liter sec^{-1} diffusion pump. H, high vacuum chamber maintained at 10^{-6} Torr by D_{II}. I_o, ion source; P, flow tube pump; S, 0.2-mm hole in quartz thimble; Str, flow system; Schr, adjusting screws; Z, intermediate vacuum chamber maintained at 5×10^{-4} Torr by D_I. [After Homann *et al.* (1970).]

cone at supersonic velocities in the forward direction (Anderson *et al.*, 1965). The skimmer A then selects the central portion of this beam for onward transmission to the ion source. Because of the high degree of collimation and the supersonic velocity of the beam, collisions between radicals themselves and between radicals and the stainless steel walls of the mass spectrometer are insignificant. This type of sampling system is often referred to as "collision-free." Wagner and co-workers have used it extensively in their studies of atom–molecule reactions (e.g., Gering *et al.*, 1969; Homann *et al.*, 1971; Wagner *et al.*, 1972). Recently, De More and co-workers (Leu and De More, 1976; Leu *et al.*, 1977) have used a similar design to detect HO_2 from $Cl +$ H_2O_2 and ClO from $Cl + O_3$.

An alternative approach is to use effusive molecular beam sampling. In these cases, the sample cone *S* and the skimmer A are replaced by simple orifices whose diameters are not large compared with the mean free path of the sample gas. Although the designs are much simpler than the supersonic beam sampling, the gas densities in effusive beams are not as high as super-sonic beams for similar conditions. However, the signal-to-noise ratios in these effusive beams are often adequate for the kinetic study of fairly stable radicals such as FO, ClO, and BrO (e.g., Clyne and Watson, 1974a,b, 1975, 1977; Birks et al., 1977), SO radicals (Clyne and MacRobert, 1979), and N atoms (Clyne and McDermid, 1975). Besides the much higher signal-to-noise ratios compared with a single-stage system, the larger apertures used in multistage pumping systems are not so readily blocked by foreign solid particles.

A realistic lower concentration limit for the detection of most molecular radicals and stable species is $\sim 10^{10}$ cm^{-3}, with that for atoms about an order of magnitude higher. An increase in discrimination against background ion peaks can be achieved by modulating the molecular beam (Leu and De More, 1976; Foner and Hudson, 1953a, 1958; Homann et al., 1970). Recently, Blumenberg et al. (1977) have successfully modulated the atom concentration in the flow tube by switching the microwave discharge.

A second problem associated with the detection of atoms and radicals by mass spectrometry is interference by fragmentation of the parent ions using 70 eV electrons. Thus, a peak at $m/e = 14$ can be derived from either atomic or molecular nitrogen. To minimize this interference, it is desirable to work at low ionization energies (preferably below the appearance potential of the fragment ion). Unfortunately, this mode of operation gives very low ionization cross sections, and hence small ion count rates.

An alternative approach for discriminating against interfering fragment ions is to monitor the doubly discharged species. Thus, Cl and Br were monitored from Cl^{2+} and Br^{2+} peaks at 45 and 70 eV, respectively (Clyne and Watson, 1974a). Oxygen atoms can also be monitored by observing O^{2+} at $m/e = 8$ at electron energy of 80–85 eV (Wong and Potter, 1963).

With one or the other of the above approaches, the following atoms have been identified by mass spectrometry: N (e.g., Clyne and Watson, 1974a; Clyne and McDermid, 1975; Jackson and Schiff, 1953; Phillips and Schiff, 1962a); O (e.g., Clyne and Watson, 1974; Klein and Herron, 1964); H (e.g., Herron and Penzhorn, 1969; Phillips and Schiff, 1962b; Wagner and Zellner, 1972); F (Wagner et al., 1976a; Clyne and Watson, 1974a); Cl, Br, and I (Clyne and Watson, 1974a).

Detection of molecular radicals by mass spectrometry is usually easier than detection of atoms because of the higher ionization cross sections associated with radicals (Otvos and Stevenson, 1956). In many cases, it is still necessary to work at reduced electron energies to reduce interference from fragmentation ions. Following this approach, Combourieu et al. (1977) have studied the reactions of N$_3$, NCl, and NCl$_2$ radicals in the Cl + N$_3$Cl system with a quadrupole mass spectrometer operating at 20 eV electron energy. Also, by ensuring that the reactants are all consumed, or by using reactants lacking the bond present in the radical product, 70 eV electrons can be used quite satisfactorily without interference from fragmentation. Such a case is the detection of ClO formed from Cl + O$_3$ → ClO + O$_2$ (Clyne and Watson,

1974a); or $Cl + OClO \rightarrow 2ClO$ (Clyne and Watson, 1974a), the latter reaction being fast and thus complete (Bemand et al., 1973).

In kinetic studies of atom–molecule reactions, an alternative to monitoring atom concentrations is to monitor the pseudo-first-order decay of the reactant molecule in a known excess concentration of atoms. The initial atom concentration can be determined either (i) from the decrease in parent molecular ion or (ii) by carrying out a titration reaction. For example, the concentration of Cl in a flow can be determined either by the decrease in the Cl_2^+ peak height when the microwave discharge is activated ($[Cl] = 2([Cl_2]_{off} - [Cl_2]_{on})$); or by a titration with NOCl (Clyne et al., 1972b; Leu and De More, 1976). A similar approach applies to the determination of radical concentrations. For example, the concentration of BrO can be determined by the formation of NO_2 when excess NO is added: $NO + BrO \rightarrow NO_2 + Br$ (Clyne and Watson, 1975).

Mass spectrometric detection of SO radicals at reduced electron energies has been used to study several elementary reactions, including $SO + NO_2 \rightarrow SO_2 + NO$, $SO + OClO \rightarrow SO_2 + ClO$, $SO + ClO \rightarrow SO_2 + Cl$ and $SO + BrO \rightarrow SO_2 + Br$ (Clyne and MacRobert 1979). Calibration of SO ion currents was carried out using titration with NO_2, with measurement of the concentration of SO_2 formed in this reaction.

A recent application of mass spectrometry is the identification of primary reaction channels in chemical reactions. These reactions were carried out, either in a conventional flow tube coupled to a high resolution mass spectrometer through a Laval nozzle (Blumenberg et al., 1977), or by crossing a beam of atoms with a beam of molecules in a high vacuum chamber; the products were identified by photoionization mass spectrometry (Kanofsky et al., 1973; Pruss et al., 1974; Slagle et al., 1975; Graham and Gutman, 1977; Gilbert et al., 1976; Foner and Hudson, 1970). (Electron impact ionization was used in this last study for the detection of radicals formed in the reaction of $H + N_2H_4$).)

Related studies were made by Jones and Bayes (1973) who used a single-stage sampling system and photoionization. Ionization by photons from monochromatic high intensity vacuum ultraviolet sources provides excellent selectivity, which is very valuable for the analysis of free radicals with similar structures. Table III gives the range of lamps used by Gutman and co-workers (after Kanofsky et al., 1973). Unfortunately, photon energies > 11.83 eV (as in the Ar lamp) involve technical difficulties because the wavelength of radiation is below the LiF cutoff ($\lambda \sim 105$ nm), and the necessity to use collimated hole structures severely reduces the light intensity.

The high resolution mass spectrometer–conventional flow tube system used by Blumenberg et al. (1977) compares favorably with the photoionization mass spectrometry–crossed-jet system, although the selectivity of species is poorer because of the finite (typically ~ 0.3 eV) spread in electron energy.

Table III

Photon Energies Associated with Different Resonance Lamps.

Emitting element	Resonance energy (eV)	Lamp gas composition and pressure (Torr)	Lamp window	Intensity (photons sec^{-1})
Xe	8.44	1 % Xe in He (1)	Sapphire	6×10^{12}
O	9.5	1 % O_2 in He (700)	CaF$_2$	2×10^{13}
H	10.2	8.5 % H_2 in Ne (1)	MgF$_2$	4×10^{13}
Ar	11.62, 11.83	Pure Ar (0.3)	LiF	2×10^{12}
Ne	16.67, 16.85	Pure Ne (0.1)	Collimated hole structure	8×10^{11}

C. Electron Paramagnetic Resonance (EPR) and Laser Magnetic Resonance (LMR)

The use of EPR for the detection of atoms and radicals in the gas phase has been reviewed recently by Westenberg (1973). Briefly, the technique is based on Zeeman shifts of the energy levels of atoms and radicals when placed in strong magnetic fields. The resonance is usually detected by absorption of electromagnetic radiation. The interaction is either between the magnetic vector of the radiation and the magnetic dipole moment of the test species, i.e., a magnetic dipole transition; or between the electric vector of the radiation and the electric dipole moment of a polar molecule, i.e., an electric dipole transition. Since electric dipole moments are usually of greater magnitude than magnetic dipole moments, the electric dipole transition usually dominates the magnetic dipole transition in cases where both occur.

Usually EPR spectrometers operate in the X-band microwave range with a fixed frequency between 9 and 10 GHz, and with a scanning magnetic field of 3–4 kG (sometimes as high as 10 kG). In the absence of significant saturation, i.e., nonequilibrium distribution between energy states, the signal is proportional to the square root of the microwave power but continuous increase in power causes the signal eventually to go through a maximum, and it then decreases in height and broadens (Westenberg, 1973, and references therein).

For the development of EPR as a method of monitoring concentrations of atoms and radicals, calibration gases are usually used, particularly when absolute concentrations are required. Thus, by measuring the relative integrated magnetic dipole transition intensities of species A, and of a known concentration of $O_2(X^3\Sigma_g^-)$ under identical cavity conditions, the absolute concentration of A can be obtained using the transition probabilities of A and

O_2 (Krongelb and Strandberg, 1959). Similar calibration for polar radicals (e.g., OH), which undergo electric dipole transitions, have been developed, based on $NO(X^2\Pi)$ as the calibrant gas (Westenberg, 1965; Westenberg and Wilson, 1966; Westenberg and de Haas, 1965; Dixon-Lewis et al., 1966).

Thus, the absolute concentrations of several transient species can be followed in the same experiment. This has been useful in kinetic studies of reactions with complex stoichiometries (e.g., Brown and Thrush, 1967; Brown et al., 1966; Westenberg and de Haas, 1967; Goldberg and Schneider, 1976; Spencer and Glass, 1977a,b; Cupitt and Glass, 1970, 1975; see also Clyne, 1973). For example, Cupitt and Glass (1975) were able to follow the absolute concentrations of S, H, O, SH, and SO simultaneously in their study of the $H + H_2S + O_2$ reaction system; and Brown and Thrush (1967) determined the stoichiometries $\Delta[O]/\Delta[C_2H_2] = 1.8$ and $\Delta[H]/\Delta[O] = 0.95$ in their study of the $O + C_2H_2$ reaction. The advantage of being able to follow the absolute concentrations of several species offsets somewhat the relatively poor sensitivity of the EPR technique, especially toward atoms. The lower limits of detection are at least one to two orders of magnitude higher than those for resonance fluorescence. For example, recent publications gave detection limits of $\sim 3 \times 10^{12}$ cm^{-3} for [F] (Goldberg and Schneider, 1976); 1×10^{12} cm^{-3} for [O], [SO], and [H] (Cupitt and Glass, 1975; Ambidge et al., 1976); and 2×10^{10} cm^{-3} for [OH] (Spencer and Glass, 1977a,b). Poor spatial resolution, leading to possibly inadequate time resolution in flow systems (see Fig. 12), is a drawback in EPR detection in fast kinetics studies.

One useful feature of EPR is the ability to detect vibrationally excited radicals. The EPR spectrum of vibrationally excited OH($v = 1$–4) has been observed from the reaction $H + O_3 \to OH + O_2$ (Clough et al., 1970,

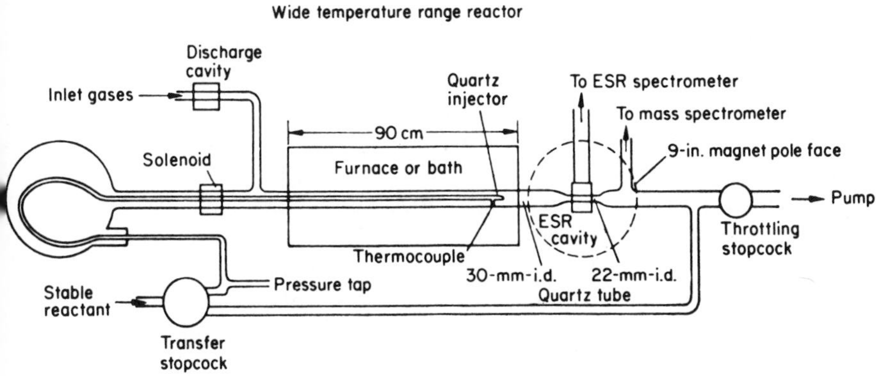

Fig. 12. EPR spectrometry in a discharge-flow system. [After Westenberg and De Haas (1969).]

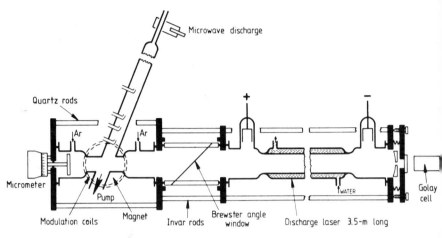

Fig. 13. LMR detection of radicals. [After Burrows *et al.* (1977).]

1971) and recently Glass and co-workers (Spencer and Glass 1977a,b; Spencer *et al.*, 1977) have used EPR to follow the reactions of OH ($v = 1$ and 2) generated from the reactions O + HBr and H + NO$_2$.

The availability of cw infrared lasers has promoted the development of laser magnetic resonance, LMR (or laser Zeeman spectroscopy), which can be regarded as a development of EPR. In fact the term "laser EPR" was used in early publications on the method (Evenson *et al.*, 1968; Boettcher *et al.*, 1968). In LMR, an infrared laser of fixed frequency is used as the electromagnetic radiation instead of the microwave power used in EPR, and as in EPR, a strong magnetic field is used to shift the energy gap into resonance with the laser frequency. The absorption of radiation at resonance within the cavity is measured as a decrease in laser output. Figure 13 gives a schematic of the apparatus. Unlike EPR, which involves transitions of Zeeman-split levels of one rotational state, the energy levels involved in LMR normally involve different rotational states or spin–orbit states in the case of an atom.

One main advantage of LMR over EPR is its greater sensitivity. Two factors in favor of LMR are the frequency of radiation and the equilibrium population of the two energy levels involved. Other factors being similar, the spectrometer sensitivity increases with the first or second power of the frequency depending on saturation (Miller, 1976). Since most infrared lasers have frequencies > 30 cm^{-1} compared with microwave frequencies of ~ 0.3 cm^{-1} as used in EPR, an increase of two to four orders of magnitude in sensitivity can be expected. Saturation effects, which lead to nonequilibrium distribution between energy states, are usually less severe in LMR than in EPR. This is because of the larger energy transition in LMR, which gives a significant difference in the populations of the two states involved. The lower limit of 2×10^8 cm^{-3} reported for OH concentrations using LMR (Radford et al., 1974) can be compared with that of 2×10^{10} cm^{-3} reported for OH using EPR (Spencer and Glass, 1977a).

One limitation of LMR as a general detection technique is its reliance on near coincidence between laser lines and molecular (or atomic) spectral lines. The "mismatch" cannot be more than ~ 1 cm^{-1} under most experimental conditions (Davies, 1977). Thus, the pioneering experiments of Evenson and co-workers (1968, 1971; Evenson and Mizushima, 1970; Curl et al., 1972; Mizushima et al., 1972; Radford et al., 1974; Houghton et al., 1975; see also Davies et al., 1975) with electrically pumped lasers (H_2O, HCN, CH_3OH and their deuterated analogs) have encountered definite limitations in terms of the number of species detected. Atoms, in particular, are more simply and sensitively detected by atomic resonance in the vacuum ultraviolet. Recent use of optically pumped (with a cw CO_2 laser) laser lines by Radford and co-workers (Radford and Litvak, 1975; Radford, 1975; Davies et al., 1976; Radford and Russell, 1977) appears promising in the development of LMR as a routine detection technique.

Most of the investigations with LMR have so far been directed toward the spectroscopy of free radicals, especially the hydrides. Nevertheless, LMR has also been applied to kinetic studies on OH (Howard and Evenson, 1974, 1976a,b,c) and more recently HO_2 (Burrows et al., 1977; Howard and Evenson, 1977). The application to HO_2 kinetics is important in directly monitoring this radical that has been made. There is no doubt that many more radical reactions will be studied with LMR following acquisition of a more extensive spectroscopic data base on the relevant transitions. Although weak magnetic dipole absorptions of Cl (Dagenais et al., 1976) and O (Davies and Thrush, 1959) have been observed, the prospect for very sensitive detection of atoms by LMR is not expected to be as good as for radicals because of the generally small magnetic dipole transition probabilities of atoms.

D. Chemiluminescence

The subject of chemiluminescence from elementary gas phase reactions has been reviewed (Thrush, 1968; Clyne, 1973; also Cormier et al., 1973). Here we will only comment on those chemiluminescent reactions which can be applied to the detection of atoms and radicals.

The most commonly used chemiluminescent reaction is the "air afterglow emission" for the detection of $O(^3P_J)$ atoms (Kaufman, 1961)

$$O + NO + M \longrightarrow NO_2^* + M,$$

$$NO_2^* \longrightarrow NO_2 + h\nu.$$

The spectral range of this yellowish-green emission extends from 397.5 nm to the infrared with a maximum near 630 nm (Broida *et al.*, 1961; Fontijn and Schiff, 1961; Fontijn *et al.*, 1964; Golde *et al.*, 1973; Woolsey *et al.*, 1977). Although some details of the process are not fully understood, it is generally agreed that the excited state is the $^2B_1$ state of NO_2. Also, the intensity of chemiluminescence I_a is related simply to the concentrations of $O(^3P_J)$ atoms and of $NO: I_a = I_{oa}[NO][O]$; where I_{oa} is a constant (between 1 and 10 Torr total pressure), due to cancellation of pressure-dependent terms (Kaufman, 1973). Below 1 Torr total pressure, the chemiluminescent intensity shows a pressure dependence. Although the absolute magnitude of I_{oa} has been determined several times to be $\sim 8 \times 10^{-17}$ cm^3 sec^{-1} (M = Ar) at 298 K (Baulch *et al.*, 1973; Vanpee *et al.*, 1971; Fontijn *et al.*, 1964; Clyne and Thrush, 1962a; Woolsey *et al.*, 1977), it is necessary to calibrate individual monitoring systems for the proportionality constant. This can easily be done by adding small, known increments of [NO] to a stream of N atoms produced by a microwave discharge of N_2 or N_2 + Ar(He) mixtures. Before the end point, the intensity of chemiluminescence (as detected by a photomultiplier with a suitable filter) will be dominated by the NO α-, β-, and γ-bands (the nitric oxide afterglow) emitted from excited NO^*: $N + O + M \rightarrow NO^* + M$ (Young and Sharpless, 1963). The intensity decreases as the end point is approached until $[NO]_{added} = [N]_o$, when the chemiluminescence vanishes completely. Beyond the end point, $[N] = 0$ and $[O] = [N]_o$; the chemiluminescence will then be dominated by the air afterglow. The quantity I_{oa} is determined from the slope of a plot of the emission intensity I_a versus $([NO] - [NO]_{endpoint})$. Any subsequent determination of [O] is then made by measuring the chemiluminescent intensity with a known addition of NO. Wavelength selection is usually not necessary for this detection method (e.g., Smith and Brown, 1975; Molina *et al.*, 1977; Owens and Roscoe, 1976) except in cases where other chemiluminescent processes have to be eliminated. Sensitivity can be quite high, with quoted detectability limits for [O] of 1×10^{11} cm^{-3}. A disadvantage of the method is the necessity to introduce a potentially reactive species into the reaction system; however, the addition of NO often is acceptable.

The reaction $O + CO + M \rightarrow CO_2^* + M$, in principle, can be used as an indicator for oxygen atoms. The chemiluminescence from the $^1B_2$ state of CO_2 extends from below 300 nm to above 580 nm, with a maximum near 400 nm (Clyne and Thrush, 1962a; Dixon, 1963). The O + CO chemiluminescent reaction has very low quantum efficiency because it involves a spin-forbidden transition. Clyne and Thrush (1962a) verified that $I_c = I_{oc}[O][CO]$, with I_{oc}(M = Ar) about 5000 times less than I_{oa} under similar conditions at 298 K. Unlike $O + NO + M \rightarrow NO_2^* + M$, however, the CO_2 afterglow intensity has a positive temperature dependence and it has found application in the detection of oxygen atoms at high temperatures, e.g., in shock tube experiments (Dean and Steiner, 1977; Schott et al., 1974; Gutman et al., 1967).

The H + NO + M → HNO* + M reaction, which emits radiation from 600 nm to the infrared, with a maximum near 770 nm (Clyne and Thrush, 1962b; Clement and Ramsay, 1961) corresponding to the $^1A''-^1A'$ transition, is a useful chemiluminescent indicator for $H(^2S)$. As for the O + NO + M reaction, the chemiluminescent intensity I_h from H + NO + M is given by $I_h = I_{oh}[H][NO]$, with I_{oh} (when M = Ar) about 100 times less than I_{oa} under similar conditions at 298 K (Clyne and Thrush, 1962b). Because of its low intensity and interference from the O + NO chemiluminescence due to traces of O atoms, the H + NO reaction has not been used so extensively as the O + NO emission (Clyne and Thrush, 1961a, 1963; Fair and Thrush, 1969b,c).

The sulphur afterglow ($B^2\Sigma_u^- - X^3\Sigma_g^-$) has been studied by Fair and Thrush (1969c). The intensity of this uv chemiluminescence is reported to fit the relationship $I = I_o[S]^2$ between 1 and 3 Torr pressure. I_o has the form $2.76 \times 10^{-33}[M]/(1 + 8.3 \times 10^{-18}[M])$ cm^3 sec^{-1} at 298 K. Although it is not used normally for monitoring $S(^3P_J)$ atoms in flow systems, the S_2 chemiluminescence has been used to monitor S atom concentrations in shock tube experiments (Bowman and Dodge, 1977; Bott and Jacobs, 1970).

The reaction F + NO + M → FNO* + M has been reported by Jones and co-workers (Skolnik et al., 1975; Pollack and Jones, 1973). Although the nature of the FNO* emitter is incompletely characterized, the FNO* emission is a simple means of monitoring atomic fluorine. The FNO* emission, first observed by Johnston and Bertin (1959), is continuous from 510 nm to at least 640 nm, with a maximum near 610 nm. Although the relationship $I_f = I_{of}[NO][F]$, with I_{of} a function of M, has not been fully verified, this detection method for $F(^2P_J)$ gives fair agreement of rate constants with those measured using alternative means of monitoring $F(^2P_J)$ atoms (Pollack and Jones, 1973). We note that analogous emitters for the other halogens have not yet been found.

The excited $B^3\Pi(O_u^+)Cl_2^*$ and $B^3\Pi(O_u^+)$ and $A^3\Pi(1_u)Br_2^*$ states from Cl + Cl + M and Br + Br + M, respectively, are useful indicators for monitoring Cl and Br concentrations. The recombination reactions are complex and the functional dependence of the chemiluminescent intensity on [X] (X=Cl, Br) varies with wavelength (Clyne and Stedman, 1968a; Clyne et al., 1972a). However, at low Cl atom concentrations ($<4 \times 10^{13}$ cm^{-3}) the Cl_2^* emission intensity becomes first-order in [Cl], due to elimination of quenching of Cl_2^* by Cl atoms (Clyne and Smith, 1979).

Although chemiluminescence is routinely used for monitoring atom concentrations, its application to molecular radical concentrations is uncommon. The SO radical is the only example to have been studied systematically using chemiluminescence detection. Recent studies of the $O + SO$ afterglow spectrum (Kley and Broida, 1976) have confirmed that the emitter is SO_2^*. The $O + SO$ emission extends from 218.5 to about 510 nm, with maxima near 270 and 450 nm. Thrush and co-workers (Clyne et al., 1966; Halstead and Thrush, 1966a) have demonstrated that $I_s = I_{os}[O][SO]$ with $I_{os} = 2.5 \times 10^{-16}$ cm^3 sec^{-1} at 298 K and with total pressures between 0.2 and 1.6 Torr of argon. It is clearly necessary to calibrate both O and SO concentrations. Provided the S atom concentration is small, the sum of $[SO] + [O]$ can be estimated by titration with NO_2; and knowing $[O]$ from $O + NO$ emission intensity measurements, the absolute concentration of SO can be determined (Clyne et al., 1966; Clyne and Whitefield, 1979). An alternative method for the determination of SO is the use of the reaction

$$SO(^3\Sigma^-) + O_3(^1A_1) \longrightarrow SO_2^*(^1B \text{ or } ^3B) + O_2(^2\Sigma_g^-).$$

The emission, first identified and studied by Halstead and Thrush (1966b), extends from 283.5 to 480 nm, with a maximum of the $^1B-^1A$ transition at 360 nm and band structure of the $^3B-^1A$ between 374.9 and 469.9 nm. The intensity follows the relationship, $I = I_0[SO][O_3]/[M]$. This can be compared with the chemiluminescent reaction $NO + O_3 \rightarrow NO_2^* + O_2$, which involves the stable radical NO: $I = I_0[NO][O_3]/[M]$ (Clyne et al., 1964; Clough and Thrush, 1969).

REFERENCES

Ambidge, P. F., Bradley, J. N., and Whytock, D. A. (1976). J. Chem. Soc., Faraday Trans. I 72, 1157.
Anderson, G. K., and Bauer, S. H. (1977). J. Chem. Phys. 81, 1146.
Anderson, J. B., Andres, R. P., and Fenn, J. B. (1965). Adv. At. Mol. Phys. 1, 345.
Anderson, J. G. (1975). Geophys. Res. Lett. 2, 231.
Anderson, J. G. (1976). Geophys. Res. Lett. 3, 165.
Anderson, J. G., and Kaufman, F. (1972). Chem. Phys. Lett. 16, 375.
Anderson, J. G., Margitan, J. J., and Kaufman, F. (1974). J. Chem. Phys. 60, 3310.
Anderson, J. G., Margitan, J. J., and Stedman, D. H. (1977). Science 198, 501.
Appelman, E. H., and Clyne, M. A. A. (1975). J. Chem. Soc., Faraday Trans. I 71, 2072.
Appelman, E. H., and Clyne, M. A. A. (1978). Am. Chem. Soc., Symp. Ser. 66.
Badachhape, R. B., Kamarchik, P., Conroy, A. P., Glass, G. P., and Margrave, J. L. (1976). Int. J. Chem. Kinet. 8, 23.
Bader, L. W., and Ogryzlo, E. A. (1964). J. Chem. Phys. 41, 2926.
Barker, J. R., and Michael, J. V. (1968). J. Chem. Phys. 58, 1615.
Baulch, D. L., Drysdale, D. D., and Horne, D. G. (1968). "High Temperature Reaction Rate Data," No. 2, University of Leeds.
Baulch, D. L., Drysdale, D. D., and Horne, D. G. (1973). "Evaluation Data for High Temperature Reactions," Vol. II. Butterworth, London.
Becker, K. H., and Haaks, D. (1973). Z. Naturforsch., Teil A 28, 248.
Bemand, P. P. (1974). Ph.D.Thesis, Queen Mary College, University of London, England.

Bemand, P. P., and Clyne, M. A. A. (1972). *J. Chem. Soc., Faraday Trans. 2* **68**, 1758.

Bemand, P. P., and Clyne, M. A. A. (1973a). *J. Chem. Soc., Faraday Trans. 2* **69**, 1643.

Bemand, P. P., and Clyne, M. A. A. (1973b). *Chem. Phys. Lett.* **21**, 555.

Bemand, P. P., and Clyne, M. A. A. (1975). *J. Chem. Soc., Faraday Trans. 2* **71**, 1132.

Bemand, P. P., and Clyne, M. A. A. (1976). *J. Chem. Soc., Faraday Trans. 2* **72**, 191.

Bemand, P. P., and Clyne, M. A. A. (1977). *J. Chem. Soc., Faraday Trans. 2* **73**, 394.

Bemand, P. P., Clyne, M. A. A., and Watson, R. T. (1973). *J. Chem. Soc., Faraday Trans. 1* **69**, 1356.

Bemand, P. P., Clyne, M. A. A., and Watson, R. T. (1974). *J. Chem. Soc., Faraday Trans. 2* **70**, 564.

Bennett, R. G., and Dalby, F. W. (1964). *J. Chem. Phys.* **40**, 1414.

Bergmann, K., and Moore, C. B. (1975). *J. Chem. Phys.* **63**, 643.

Birely, J. H., Kasper, J. V. V., Hai, F., and Darnton, L. A. (1975). *Chem. Phys. Lett.* **31**, 220.

Birks, J. W., Shoemaker, B., Leck, T. J., Borders, R. A., and Hart, L. J. (1977). *J. Chem. Phys.* **66**, 4591.

Blumenberg, B., Hoyermann, K., and Sievert, R. (1977). *Symp. (Int.) Combust. [Proc.]* **16**, 841.

Boden, J. C., and Thrush, B. A. (1968). *Proc. R. Soc. London, Ser. A* **305**, 107.

Boettcher, J., Dransfeld, K., and Renk, K. F. (1968). *Physl Lett. A* **26**, 146.

Bott, J. F., and Jacobs, T. A. (1970). *J. Chem. Phys.* **52**, 3545.

Bowman, C. T., and Dodge, L. G. (1977). *Symp. (Int.) Combust. [Proc.]* **16**, 971.

Bozzelli, J. W., and Kaufman, M. (1973). *J. Chem. Phys.* **77**, 1748.

Braun, W., and Carrington, T. (1969). *J. Quant. Spectrosc. & Radiat. Transfer* **9**, 1133.

Braun, W., Bass, A. M., Davis, D. D., and Simmons, J. D. (1969). *Proc. R. Soc. London, Ser. A* **312**, 417.

Braun, W., Bass, A. M., and Davis, D. D. (1970). *J. Opt. Soc. Am.* **60**, 166.

Brocklehurst, B., and Jennings, K. R. (1967). *Prog. React. Kinet.* **4**, 1.

Broida, H. P., Schiff, H. I., and Sugden, T. M. (1961). *Trans. Faraday Soc.* **57**, 259.

Brown, J. M., and Thrush, B. A. (1967). *Trans. Faraday Soc.* **63**, 630.

Brown, J. M., Coates, P. B., and Thrush, B. A. (1966). *Chem. Commun.* p. 843.

Browne, R. J., and Ogryzlo, E. A. (1970). *J. Chem. Phys.* **52**, 5774.

Burak, I., and Eyal, M. (1977). *Chem. Phys. Lett.* **52**, 534.

Burrows, J. P., Harris, G. W., and Thrush, B. A. (1977). *Nature (London)* **267**, 233.

Campell, I. M., and Thrush, B. A. (1965). *Annu. Rep. Chem. Soc.* **67**, 17.

Campell, I. M., and Thrush, B. A. (1968). *Trans. Faraday Soc.* **64**, 1275.

Chang, J. S., and Kaufman, F. (1977). *J. Chem. Phys.* **66**, 4989.

Cheah, C. T., Clyne, M. A. A., and Whitefield, P. D. (1979). *J. Chem. Soc., Faraday Trans. 2.* Submitted.

Clark, T. C., and Clyne, M. A. A. (1969). *Trans. Faraday Soc.* **65**, 2994.

Clark, T. C., and Clyne, M. A. A. (1970a). *Trans. Faraday Soc.* **66**, 372.

Clark, T. C., and Clyne, M. A. A. (1970b). *Trans. Faraday Soc.* **66**, 877.

Clement, M. J. Y., and Ramsay, D. A. (1961). *Can. J. Phys.* **39**, 205.

Clough, P. N., and Thrush, B. A. (1969). *Trans. Faraday Soc.* **65**, 23.

Clough, P. N., Curran, A. H., and Thrush, B. A. (1970). *Chem. Phys. Lett.* **7**, 86.

Clough, P. N., Curran, A. H., and Thrush, B. A. (1971). *Proc. R. Soc. London, Ser. A* **323**, 541.

Clyne, M. A. A. (1973). *In* "Physical Chemistry of Fast Reactions" (B. P. Levitt, ed.), Vol. 1, p. 245. Plenum, New York.

Clyne, M. A. A., and Connor, J. (1972). *J. Chem. Soc., Faraday Trans. 2* **68**, 1220.

Clyne, M. A. A., and Coxon, J. A. (1966). *Trans. Faraday Soc.* **62**, 2175.

Clyne, M. A. A., and Coxon, J. A. (1967). *Proc. R. Soc. London, Ser. A* **298**, 424.

Clyne, M. A. A., and Coxon, J. A. (1968a). *Proc. R. Soc. London, Ser. A* **303**, 207.

Clyne, M. A. A., and Coxon, J. A. (1968a). *J. Mol. Spectrosc.* **23**, 258.

Clyne, M. A. A., and Cruse, H. W. (1970a). *Trans. Faraday Soc.*, **66**, 2214.

Clyne, M. A. A., and Cruse, H. W. (1970b). *Trans. Faraday Soc.* **66**, 2227.

Clyne, M. A. A., and Cruse, H. W. (1971). *Trans. Faraday Soc.* **67**, 2869.

Clyne, M. A. A., and Cruse, H. W. (1972a). *J. Chem. Soc., Faraday Trans. 2* **68**, 1281.

Clyne, M. A. A., and Cruse, H. W. (1972b). *J. Chem. Soc., Faraday Trans. 2* **68**, 1377.

Clyne, M. A. A., and Holt, P. M. (1979). *J. Chem. Soc., Faraday Trans. 2* **75**, 569, 582.

Clyne, M. A. A., and MacRobert, A. J. (1979). *Int. J. Chem. Kin.* Submitted.

Clyne, M. A. A., and McDermid, I. S. (1975). *J. Chem. Soc., Faraday Trans. 1* **71**, 2189.

Clyne, M. A. A., and McDermid, I. S. (1976). *J. Chem. Soc., Faraday Trans. 2* **72**, 2252.

Clyne, M. A. A., and McDermid, I. S. (1977). *J. Chem. Soc., Faraday Trans. 2* **73**, 1094.

Clyne, M. A. A., and McDermid, I. S. (1978a). *J. Chem. Soc., Faraday Trans. 2* **74**, 644.

Clyne, M. A. A., and McDermid, I. S. (1978b). *J. Chem. Soc., Faraday Trans. 2* **74**, 664.

Clyne, M. A. A., and McDermid, I. S. (1978c). *J. Chem. Soc., Faraday Trans. 2* **74**, 798.

Clyne, M. A. A., and McDermid, I. S. (1978d). *J. Chem. Soc., Faraday Trans. 2* **74**, 807.

Clyne, M. A. A., and Monkhouse, P. B. (1976). *12th Informal Meet. Photochem.*

Clyne, M. A. A., and Monkhouse, P. B. (1977). *J. Chem. Soc., Faraday Trans. 2* **73**, 298.

Clyne, M. A. A., and Nip, W. S. (1976a). *J. Chem. Soc., Faraday Trans. 2* **72**, 838.

Clyne, M. A. A., and Nip, W. S. (1976b). *J. Chem. Soc., Faraday Trans. 1* **72**, 2211.

Clyne, M. A. A., and Nip, W. S. (1977a). *J. Chem. Soc., Faraday Trans. 2* **73**, 161.

Clyne, M. A. A., and Nip, W. S. (1977b). *J. Chem. Soc., Faraday Trans. 2* **73**, 1308.

Clyne, M. A. A., and Nip, W. S. (1978). *Int. J. Chem. Kinet.* **10**, 367.

Clyne, M. A. A., and Piper, L. G. (1976). *J. Chem. Soc., Faraday Trans. 2* **72**, 2178.

Clyne, M. A. A., and Smith, D. J. (1978). *J. Chem. Soc., Faraday Trans. 2* **74**, 263.

Clyne, M. A. A., and Smith, D. J. (1979). *J. Chem. Soc., Faraday Trans. 2* **75**, 704.

Clyne, M. A. A., and Stedman, D. H. (1966). *Trans. Faraday Soc.* **62**, 2164.

Clyne, M. A. A., and Stedman, D. H. (1968a). *Trans. Faraday Soc.* **64**, 1816.

Clyne, M. A. A., and Stedman, D. H. (1968b). *Trans. Faraday Soc.* **64**, 2698.

Clyne, M. A. A., and Thrush, B. A. (1961a). *Proc. R. Soc. London, Ser. A* **261**, 259.

Clyne, M. A. A., and Thrush, B. A. (1962a). *Faraday Discuss. Chem. Soc.* **33**, 139.

Clyne, M. A. A., and Thrush, B. A. (1962b). *Proc. R. Soc. London, Ser. A* **269**, 404.

Clyne, M. A. A., and Thrush, B. A. (1963). *Proc. R. Soc. London, Ser. A* **275**, 544.

Clyne, M. A. A., and Townsend, L. W. (1974). *J. Chem. Soc., Faraday Trans. 2* **70**, 1863.

Clyne, M. A. A., and Townsend, L. W. (1975). *Int. J. Chem. Kinet. Symp.* **1**, 73.

Clyne, M. A. A., and Watson, R. T. (1974a). *J. Chem. Soc., Faraday Trans. 1* **70**, 1109.

Clyne, M. A. A., and Watson, R. T. (1974b). *J. Chem. Soc., Faraday Trans. 1* **70**, 2250.

Clyne, M. A. A., and Watson, R. T. (1975). *J. Chem. Soc., Faraday Trans. 1* **70**, 336.

Clyne, M. A. A., and Watson, R. T. (1977). *J. Chem. Soc., Faraday Trans. 1* **73**, 1169.

Clyne, M. A. A., and White, I. F. (1970). *Chem. Phys. Lett.* **6**, 465.

Clyne, M. A. A., and White, I. F. (1971). *Trans. Faraday Soc.* **67**, 2068.

Clyne, M. A. A., and Whitefield, P. D. (1979). *J. Chem. Soc., Faraday Trans. 2* **75**, 1327.

Clyne, M. A. A., and Williamson, T. D. A. (1979). To be published.

Clyne, M. A. A., Thrush, B. A., and Wayne, R. P. (1964). *Trans. Faraday Soc.* **60**, 359.

Clyne, M. A. A., McKenney, D. J., and Thrush, B. A. (1965). *Trans. Faraday Soc.* **61**, 2701.

Clyne, M. A. A., Halstead, C. J., and Thrush, B. A. (1966). *Proc. R. Soc. London, Ser. A* **295**, 355.

Clyne, M. A. A., Coxon, J. A., and Woon-Fat, A. R. (1971). *Trans. Faraday Soc.* **67**, 3155.

Clyne, M. A. A., Coxon, J. A., and Woon-Fat, A. R. (1972a). *J. Chem. Soc., Faraday Discuss.* **53**, 82.

Clyne, M. A. A., Cruse, H. W., and Watson, R. T. (1972b). *J. Chem. Soc., Faraday Trans. 2* **68**, 153.

Clyne, M. A. A., McKenney, D. J., and Walker, R. F. (1973). *Can. J. Chem.* **51**, 3596.

Clyne, M. A. A., Monkhouse, P. B., and Townsend, L. W. (1976a). *Int. J. Chem. Kinet.* **8**, 425.

Clyne, M. A. A., McDermid, I. S., and Curran, A. H. (1976b). *J. Photochem.* **5**, 201.

Combourieu, J., LeBras, G., Poulet, G., and Jourdain, J. L. (1977). *Symp. (Int.) Combust. [Proc.]* **16**, 863.

Connor, T. R., and Biondi, M. A. (1965). *Phys. Rev. A* **140**, 778.

Cool, T. A., and Stephens, R. R. (1969). *J. Chem. Phys.* **51**, 5175.

Cool, T. A., and Stephens, R. R. (1970). *Appl. Phys. Lett.* **16**, 55.

Coon, J. B., De Wames, R. E., and Lloyd, C. W. (1962). *J. Mol. Spectrosc.* **8**, 285.

Cormier, M. J., Hercules, D. M., and Lee, J. (1973). "Chemiluminescence and Bioluminescence." Plenum, New York.

Cupitt, L. T., and Glass, G. P. (1970). *Trans. Faraday Soc.* **66**, 3007.

Cupitt, L. T., and Glass, G. P. (1975). *Int. J. Chem. Kinet. Symp.* **1**, 39.

Curl, R. F., Evenson, K. M., and Wells, J. S. (1972). *J. Chem. Phys.* **56**, 5143.

Dagenais, M., Johns, J. W. C., and McKellar, A. R. W. (1976). *Can J. Phys.* **54**, 1438.

Davies, P. B. (1977). *Abstr., Int. Symp. Gas Kinet., 5th*, University of Manchester, Institute of Science and Technology.

Davies, P. B., and Thrush, B. A. (1979). To be published.

Davies, P. B., Russell, D. K., Thrush, B. A., and Wayne, F. D. (1975). *J. Chem. Phys.* **62**, 3739.

Davies, P. B., Russell, D. K., Thrush, B. A., and Radford, H. E. (1976). *Chem. Phys. Lett.* **43**, 35.

Davis, D. D., and Braun, W. (1968). *Appl. Opt.* **7**, 2071.

Davis, D. D., Braun, W., and Bass, A. M. (1970). *Int. J. Chem. Kinet.* **2**, 101.

Deaking, J. J., and Husain, D. (1972). *J. Chem. Soc., Faraday Trans. 2* **68**, 1603.

Dean, A. M., and Steiner, D. C. (1977). *J. Chem. Phys.* **66**, 598.

Debely, P. E. (1970). *Rev. Sci. Instrum.* **41**, 1290.

Del Greco, F. P., and Kaufman, F. (1962). *Faraday Discuss. Chem. Soc.* **33**, 128.

Dickenson, P. H. G., Bolden, R. C., and Young, R. A. (1974). *Nature (London)* **252**, 289.

Dieke, G. H., and Cunningham, S. P. (1952). *J. Opt. Soc. Am.* **42**, 187.

Dixon, R. N. (1963). *Proc. R. Soc. London, Ser. A* **275**, 431.

Dixon-Lewis, G., Wilson, W. E., and Westenberg, A. A. (1966). *J. Chem. Phys.* **44**, 2877.

Donovan, R. J., Husain, D., and Kirsch, L. J. (1970). *Trans. Faraday Soc.* **66**, 2551.

Dunn, M. R., Sutton, M. M., Freeman, C. G., McEwan, M. J., and Phillips, L. F. (1971). *J. Phys. Chem.* **75**, 722.

Eltenton, G. C. (1947). *J. Chem. Phys.* **15**, 455.

Evans, H. G. V., and Winkler, C. A. (1956). *Can. J. Chem.* **34**, 1217.

Evenson, K. M., and Mizushima, M. (1970). *Phys. Rev. Lett.* **25**, 199.

Evenson, K. M., Broida, H. P., Wells, J. S., Miller, R. J., and Mizushima, M. (1968). *Phys. Rev. Lett.* **21**, 1038.

Evenson, K. M., Radford, H. E., and Moran, M. M. (1971). *Appl. Phys. Lett.* **18**, 426.

Fair, R. W., and Thrush, B. A. (1969a). *Trans. Faraday Soc.* **65**, 1208.

Fair, R. W., and Thrush, B. A. (1969b). *Trans. Faraday Soc.* **65**, 1550.

Fair, R. W., and Thrush, B. A. (1969c). *Trans. Faraday Soc.* **65**, 1557.

Fehsenfeld, F. C., Evenson, K. M., and Broida, H. P. (1965). *Rev. Sci. Instrum.* **36**, 294.

Felder, W., and Young, R. A. (1972). *J. Chem. Phys.* **57**, 572.

Ferguson, E. E., Fehsenfeld, F. C., and Schmeltekopf, A. L. (1969). *Adv. At. Mol. Phys.* **5**, 1.

Fersht, E., and Back, R. A. (1965). *Can. J. Chem.* **43**, 1899.

Foner, S. N. (1966). *Adv. At. Mol. Phys.* **2**, 385.

Foner, S. N., and Hudson, R. L. (1953a). *J. Chem. Phys.* **21**, 1374.

Foner, S. N., and Hudson, R. L. (1953b). *J. Chem. Phys.* **21**, 1608.

Foner, S. N., and Hudson, R. L. (1958). *J. Chem. Phys.* **28**, 719.

Foner, S. N., and Hudson, R. L. (1962). *J. Chem. Phys.* **37**, 1662.

Foner, S. N., and Hudson, R. L. (1970). *J. Chem. Phys.* **53**, 4377.

Fontijn, A., and Johnson, S. E. (1973). *J. Chem. Phys.* **59**, 6193.

Fontijn, A., and Schiff, H. I. (1961). In "Chemistry of the Lower and Upper Atmosphere," 239. Wiley (Interscience), New York.

Fontijn, A., Meyer, C. B., and Schiff, H. I. (1964). *J. Chem. Phys.* **40**, 64.

Foon, R., and Kaufman, M. (1975). *Prog. React. Kinet.* **8**, 81.

Ganguli, P. S., and Kaufman, M. (1974). *Chem. Phys. Lett.* **25**, 221.

Gering, M., Hoyermann, K., Wagner, H. Gg., and Wolfrum, J. (1969). *Ber. Bunsenges. Phys. Chem.* **73**, 956.

German, K. R., and Zare, R. N. (1969). *Phys. Rev. Lett.* **23**, 1207.

Gilbert, J. R., Slagle, I. R., Graham, R. E., and Gutman, D. (1976). *J. Phys. Chem.* **80**, 14.

Goldberg, I. B., and Schneider, G. R. (1976). *J. Chem. Phys.* **65**, 147.

Golde, M. F., Roche, A. E., and Kaufman, F. (1973). *J. Chem. Phys.* **59**, 3953.

Golden, D. M., Del Greco, F. P., and Kaufman, F. (1963). *J. Chem. Phys.* **39**, 3034.

Graham, and Gutman, D. (1977).

Gutman, D., Hardwidge, E. A., Dougherty, F. A., and Lutz, R. W. (1967). *J. Chem. Phys.* **47**, 4400.

Halstead, C. J., and Thrush, B. A. (1966a). *Proc. R. Soc. London, Ser. A* **295**, 363.

Halstead, C. J., and Thrush, B. A. (1966b). *Proc. R. Soc. London, Ser. A* **295**, 380.

Harteck, P., Reeves, R. R., and Thompson, B. A. (1964). *Z. Naturforsch., Teil A* **19**, 2.

Heidner, R. F., and Kasper, J. V. V. (1972). *Chem. Phys. Lett.* **7**, 179.

Herron, J. T. (1965). *J. Res. Natl. Bur. Stand., Sect. A* **69**, 287.

Herron, J. T., and Penzhorn, R. D. (1969). *J. Phys. Chem.* **73**, 191.

Homann, K. H., Krome, G., and Wagner, H. Gg. (1968). *Ber. Bunsenges. Phys. Chem.* **72**, 998.

Homann, K. H., Soloman, W. C., Warnatz, J., Wagner, H. Gg., and Zetzsch, C. (1970). *Ber. Bunsenges. Phys. Chem.* **74**, 585.

Homann, K. H., Lange, W., and Wagner, H. Gg. (1971). *Ber. Bunsenges. Phys. Chem.* **75**, 121.

Houghton, J. T., Radford, H. E., Evenson, K. M., and Howard, C. J. (1975). *J. Mol. Spectrosc.* **56**, 210.

Howard, C. J., and Evenson, K. M. (1974). *J. Chem. Phys.* **61**, 1943.

Howard, C. J., and Evenson, K. M. (1976a). *J. Chem. Phys.* **64**, 197.

Howard, C. J., and Evenson, K. M. (1976b). *J. Chem. Phys.* **64**, 4303.

Howard, C. J., and Evenson, K. M. (1976c). *J. Chem. Phys.* **65**, 4771.

Howard, C. J., and Evenson, K. M. (1977). *Geophys. Res. Lett.* **4**, 437.

Husain, D., and Kirsch, L. T. (1971). *Trans. Faraday Soc.* **67**, 2025.

Husain, D., and Norris, P. E. (1977). *J. Chem. Soc., Faraday Trans. 2* **73**, 1107.

Hutton, E., and Wright, M. (1965). *Trans. Faraday Soc.* **61**, 78.

Ingold, K. U., and Bryce, W. A. (1956). *J. Chem. Phys.* **24**, 360.

Jackson, D. S., and Schiff, H. I. (1953). *J. Chem. Phys.* **21**, 2233.

Johnson, D. R., and Lovas, F. J. (1972). *Chem. Phys. Lett.* **15**, 65.

Johnson, S. E., and Fontijn, A. (1973). *Chem. Phys. Lett.* **23**, 252.

Johnston, H. S., and Bertin, H. J. (1959). *J. Mol. Spectrosc.* **3**, 683.

Jones, I. T. N., and Bayes, K. D. (1973). *Symp. (Int.) Combust. [Proc.]* **14**, 277.

Jones, W. M., and Davidson, N. (1962). *J. Am. Chem. Soc.* **84**, 2868.

Kanofsky, J. R., Lucas, D., and Gutman, D. (1973). *Symp. (Int.) Combust. [Proc.]* **14**, 285.

Kaufman, F. (1961). *Prog. React. Kinet.* **1**, 1.

Kaufman, F. (1964). *Ann. Geophys.* **20**, 106.

Kaufman, F. (1969). *Adv. Chem. Ser.* **80**.

Kaufman, F. (1973). *In* "Chemiluminescence and Bioluminescence" (M. J. Cormier, D. H., Hercules, and J. Lee, eds.), Plenum, New York.

Kaufman, F., and Kelso, J. R. (1958). *J. Chem. Phys.* **28**, 510.

Kaufman, F., and Kelso, J. R. (1964). *Faraday Discuss. Chem. Soc.* **37**, 26.

Kaufman, F., and Parkes, D. A. (1970). *Trans. Faraday Soc.* **66**, 1579.

Kikuchi, T. (1971). *Appl. Opt.* **10**, 1288.

Kim, P., MacLean, D. I., and Valence, W. G. (1972). *Abstr., 164th Meet., Am. Chem. Soc.*

King, D. L., and Setser, D. W. (1976). *Annu. Rev. Phys. Chem.* **27**, 407.

Kistiakowsky, G. B., and Volpi, G. G. (1957). *J. Chem. Phys.* **27**, 1141.

Klein, F. S., and Herron, J. T. (1964). *J. Chem. Phys.* **41**, 1285.

Kley, D., and Broida, H. P. (1976). *J. Photochem.* **6**, 241.

Kley, D., Washida, N., Becker, K. H., and Groth, W. (1972a). *Z. Phys. Chem.* (*Frankfurt ans Main*) [N.S.] **82**, 109.

Kley, D., Washida, N., Becker, K. H., and Groth, W. (1972b). *Chem. Phys. Lett.* **15**, 45.

Kolb, C E., and Kaufman, M. (1972). *J. Phys. Chem.* **76**, 947.

Krongelb, S., and Strandberg, M. W. P. (1959). *J. Chem. Phys.* **31**, 1196.

Kurylo, M. J., Peterson, N. C., and Braun, W. (1971). *J. Chem. Phys.* **54**, 943.

Larkin, F. S., and Thrush, B. A. (1964). *Faraday Discuss. Chem. Soc.* **37**, 112.

Lee, J. H., Michael, J. V., Payne, W. A., and Stief, L. J. (1978). *J. Chem. Phys.* **69**, 3069.

Leu, M. T., and De More, W. B. (1976). *Chem. Phys. Lett.* **41**, 121.

Leu, M. T., and De More, W. B. (1977a). *J. Phys. Chem.* **81**, 190.

Leu, M. T. and De More, W. B. (1977b). *Chem. Phys. Lett.* **48**, 317.

Leu, M. T., Lin, C. L., and D. More, W. B. (1977). *J. Phys. Chem.* **81**, 190.

Lin, C. L., and Kaufman, F. (1971). *J. Chem. Phys.* **55**, 3760.

Lin, C. L., Parkes, D. A., and Kaufamn, F. (1970). *J. Chem. Phys.* **53**, 3896.

Lossing, F. P., and Tickner, A. W. (1952). *J. Chem. Phys.* **25**, 1031.

Lundell, O. R., Ketcheson, R. D., and Schiff, H. I. (1969). 12th *Symp.* (*Int.*) *Combus.* [*Proc.*] **12**, 307.

Lynch, K. P., Schwab, T. C., and Michael, J. V. (1976). *Int. J. Chem. Kinet.* **8**, 651.

McAfee, K. B., and Hozack, R. S. (1976). *J. Chem. Phys.* **64**, 2491.

McKenzie, A., Mulcahy, M. F. R., and Steven, J. R. (1974). *J. Chem. Soc., Faraday Trans. 1* **70**, 549.

McNesby, J. R., and Okabe, H. (1964). *Adv. Photochem.* **3**, 157.

Mathias, A., and Schiff, H. I. (1964). *Faraday Discuss. Chem. Soc.* **37**, 39.

Mei, C.-C., and Moore, C. B. (1977). *J. Chem. Phys.* **67**, 3936.

Michael, J. V., and Lee, J. H. (1977). *Chem. Phys. Lett.* **51**, 303.

Michael, J. V., and Weston, R. E. (1966). *J. Chem. Phys.* **45**, 3632.

Miller, T. A. (1976). *Annu. Rev. Phys. Chem.* **27**, 127.

Mitchell, A. C. G., and Zemansky, M. W. (1971). "Resonance Radiation and Excited Atoms." Cambridge Univ. Press, London and New York.

Mitchell, D. N., and Le Roy, D. J. (1977). *J. Chem. Phys.* **67**, 1042.

Mizushima, M., Evenson, K. M., and Wells, J. S. (1972). *Phys. Rev. A* **5**, 2276.

Molina, L. T., Spencer, J. E., and Molina, M. J. (1977). *Chem. Phys. Lett.* **45**, 158.

Monkhouse, P. B. (1976). Ph.D. Thesis, Queen Mary College, University of London, England.

Morse, F. A., and Kaufman, F. (1965). *J. Chem. Phys.* **42**, 1785.

Mulcahy, M. F. R., and Pethard, M. R. (1963). *Aust. J. Chem.* **16**, 527.

Mulcahy, M. F. R., and Williams, M. J. (1961). *Aust. J. Chem.* **14**, 534.

Mulcahy, M. F. R., Steven, J. R., and Ward, J. C. (1967). *J. Phys. Chem.* **71**, 2124.

Mulcahy, M. F. R., Steven, J. R., Ward, J. C., and Williams, D. J. (1969). *Symp.* (*Int.*) *Combust.* [*Proc.*] **12**, 323.

Myerson, A. L., and Watt, W. S. (1968). *J. Chem. Phys.* **49**, 425.

Nordine, P. C. (1974). *J. Chem. Phys.* **61**, 224.

Nordine, P. C., and Rosner, D. E. (1976). *J. Chem. Soc., Faraday Trans. 1* **72**, 1526.

Norrish, R. G. W., and Thrush, B. A. (1956). *Q. Rev., Chem. Soc.* **10**, 149.

Ogryzlo, E. A. (1961). *Can. J. Chem.* **39**, 2556.

Ogryzlo, E. A., Reiley, J. P., and Thrush, B. A. (1973). *Chem. Phys. Lett.* **23**, 37.

Okabe, H. (1964). *J. Opt. Soc. Am.* **54**, 478.

Okabe, H. (1969). *Natl. Bur. Stand.* (*U.S.*), *Tech. Note* **496**.

Otvos, J. W., and Stevenson, D. P. (1956). *J. Am. Chem. Soc.* **78**, 546.

Owens, C. M., and Roscoe, J. M. (1976). *Can. J. Chem.* **54**, 984.

Parkes, D. A., and Kaufman, F. (1970). *Trans. Faraday Soc.* **66**, 1579.

Parkes, D. A., Keyser, L. F., and Kaufman, F. (1967). *Astrophys. J.* **149**, 217.

Parson, J. M., and Lee, Y. T. (1972). *J. Chem. Phys.* **56**, 4658.

Passchier, A. A., Christian, J. D., and Gregory, N. W. (1967). *J. Phys. Chem.* **71**, 937.

Perner, D., and Franken, T. (1969). *Ber. Bunsenges. Phys. Chem.* **73**, 897.

Persky, A. (1972). *J. Chem. Phys.* **59**, 3612.

Phillips, L. F. (1973). *J. Photochem.* **2**, 255.

Phillips, L. F. (1975). *J. Photochem.* **4**, 407.

Phillips, L. F. (1976a). *Chem. Phys. Lett.* **37**, 421.

Phillips, L. F. (1976b). *J. Photochem.* **5**, 241.

Phillips, L. F. (1976c). *J. Photochem.* **5**, 277.

Phillips, L. F., and Schiff, H. I. (1962a). *J. Chem. Phys.* **36**, 1509.

Phillips, L. F., and Schiff, H. I. (1962b). *J. Chem. Phys.* **37**, 1233.

Piper, L. G., Richardson, W. C., Taylor, G. W., and Setser, D. W. (1972). *Faraday Discuss. Chem. Soc.* **53**, 100.

Piper, L. G., Velazco, J. E., and Setser, D. W. (1973). *J. Chem. Phys.* **59**, 3323.

Poirier, R. V., and Carr, R. W. (1971). *J. Phys. Chem.* **75**, 1593.

Polanyi, J. C., and Woodall, K. B. (1972). *J. Chem. Phys.* **57**, 1574.

Pollack, T. L., and Jones, W. E. (1973). *Can. J. Chem.* **51**, 2041.

Poulet, G., Le Bras, G., and Combourieu, J. (1974). *J. Chem. Phys.* **71**, 101.

Pruss, F. J., Slagle, I. R., and Gutman, D. (1974). *J. Phys. Chem.* **78**, 683.

Radford, H. E. (1975). *IEEE J. Quantum Electron.* **QE-11**, 213.

Radford, H. E., and Litvak, M. M. (1975). *Chem. Phys. Lett.* **34**, 561.

Radford, H. E., and Russell, D. K. (1977) .*J. Chem. Phys.* **66**, 2222.

Radford, H. E., Evenson, K. M., and Howard, C. J. (1974). *J. Chem. Phys.* **60**, 3178.

Rapp, D., and Johnston, H. S. (1960). *J. Chem. Phys.* **33**, 695.

Roger, W. A., and Biondi, M. A. (1964). *Phys. Rev. A* **134**, 1215.

Rosner, D. E., and Allendorf, H. H. (1971). *J. Phys. Chem.* **75**, 308.

Safrany, D. R. (1971). *Prog. React. Kinet.* **6**, 1.

Samson, J. A. R. (1967). "Techniques of Vaccuum Ultraviolet Spectroscopy." Wiley, New York.

Schatz, G., and Kaufman, M. (1972). *J. Phys. Chem.* **76**, 3586.

Schmatjko, K. J., and Wolfrum, J. (1977). *Symp. (Int.) Combust. [Proc.]* **16**, 819.

Schott, G. L., Getzinger, R. W., and Seitz, W. A. (1974). *Int. J. Chem. Kinet.* **6**, 921.

Seery, D. J., and Britton, D. (1964). *J. Phys. Chem.* **74**, 2238.

Setser, D. W., and Stedman, D. H. (1970). *J. Phys. Chem.* **74**, 2238.

Skolnik, E. G., Veysey, S. W., Ahmed, M. G., and Jones, W. E. (1975). *Can. J. Chem.* **53**, 3188.

Slagle, I. R., Gilbert, J. R., Graham, R. E., and Gutman, D. (1975). *Int. J. Chem. Kinet. Symp.* **1**, 317.

Slemř, S., and Warneck, P. (1977). *Int. J. Chem. Kinet.* **9**, 267.

Smith, I. W. M., and Brown, R. D. (1975). *Int. J. Chem. Kinet.* **7**, 301.

Spencer, J. E., and Glass, G. P. (1977a). *Int. J. Chem. Kinet.* **9**, 97.

Spencer, J. E., and Glass, G. P. (1977b). *Int. J. Chem. Kinet.* **9**, 111.

Spencer, J. E., Endo, A., and Glass, G. P. (1977). *Symp. (Int.) Combust. [Proc.]* **16**, 829.

Stedman, D. H., and Setser, D. W. (1971). *Prog. React. Kinet.* **6**, 4.

Stedman, D. H., Steffenson, D., and Niki, H. (1970). *Chem. Phys. Lett.* **7**, 173.

Strattan, L. W., and Kaufman, M. (1977). *J. Chem. Phys.* **66**, 4963.

Sung, J. P., and Setser, D. W. (1977). *Chem. Phys. Lett.* **48**, 413.

Takacs, G. A., and Glass, G. P. (1973a). *J. Phys. Chem.* **77**, 1060.

Takacs, G. A., and Glass, G. P. (1973b.) *J. Phys. Chem.* **77**, 1182.

Takacs, G. A., and Glass, G. P. (1973c). *J. Phys. Chem.* **77**, 1948.

Tellinghuisen, J., and Clyne, M. A. A. (1976). *J. Chem. Soc., Faraday Trans. 2* **72**, 783.

Thompson, B. A., Reeves, R. R., and Harteck, P. (1965). *J. Phys. Chem.* **69**, 3964.

Thrush, B. A. (1968). *Aunu. Rev. Phys. Chem.* **19**, 371.

Trainor, D. W., Ham, D. O., and Kaufman, F. (1973). *J. Chem. Phys.* **58**, 4599.

Ultee, C. J., and Bonczyk, P. A. (1977). *Chem. Phys. Lett.* **46**, 576.

Vanpee, M., Hill, K. D., and Kineyko, W. R. (1971). *AIAA J.* **9**, 135.

Van Volkenburgh, G., and Carrington, T. (1971). *J. Quant. Spectrosc. & Radiat. Transfer* **11**, 1181.

Verbeke, G. S., and Winkler, C. A. (1960). *J. Phys. Chem.* **64**, 319.

Wagner, H. Gg., and Zellner, R. (1972). *Ber. Bunsenges. Phys. Chem.* **76**, 518.

Wagner, H. Gg., Zetzsch, C., and Warnatz, J. (1972). *Ber. Bunsenges. Phys. Chem.* **76**, 526.

Wagner, H. Gg., Welzbacher, U., and Zellner, R. (1976a). *Ber. Bunsenges. Phys. Chem.* **80**, 902.

Wagner, H. Gg., Welzbacher, U., and Zellner, R. (1976b). *Ber. Bunsenges. Phys. Chem.* **80**, 1023.

Walker, R. W. (1961). *Phys. Fluids* **4**, 1211.

Walther, H. (1976). "Laser Spectroscopy of Atoms and Molecules." Springer-Verlag, Berlin and New York.

Wang, C.-C., and Davis, L. I. (1974). *Phys. Rev. Lett.* **32**, 349.

Warnatz, J., Wagner, H. Gg., and Zetzsch, C. (1971a). *Angew. Chem., Int. Ed. Engl.* **10**, 564.

Warnatz, J., Wagner, H. Gg., and Zetzsch, C. (1971b). *Ber. Bunsenges. Phys. Chem.* **75**, 119.

Warnatz, J., Wagner, H. Gg., and Zetzsch, C. (1972). Report T-0240/92410/01017 to the Frauhofer Gessellschaft, quoted in Foon and Kaufman (1975).

Washida, N., and Bayes, K. D. (1976). *Int. J. Chem. Kinet.* **8**, 777.

Washida, N., Akimoto, H., and Tanaka, I. (1970). *Appl. Opt.* **9**, 1711.

Watson, R. T. (1974a). Ph.D. Thesis, Queen Mary College, University of London, England.

Watson, R. T. (1974b). "Chemical Kinetic Data Survey VIII," NBSIR 74-516. 2nd ed. Natl. Bur. Stand., Washington, D. C.

Wayne, R. P. (1969). *Ad. Photochem.* **7**, 311.

Westenberg, A. A. (1965). *J. Chem. Phys.* **43**, 1544.

Westenberg, A. A. (1973). *Prog. React. Kinet.* **7**, 23.

Westenberg, A. A., and de Haas, N. (1965). *J. Chem. Phys.* **43**, 1550.

Westenberg, A. A., and de Haas, N. (1967). *J. Chem. Phys.* **46**, 490.

Westenberg, A. A., and Wilson, W. E. (1966). *J. Chem. Phys.* **45**, 338.

Wiese, W. L., Smith, M. W., and Glennon, B. M. (1966). "Atomic Transition Probabilities," Vol. I, NSRDS-NBS4. Natl. Bur. Stand., Washington.

Wiese, W. L., Smith, M. W., and Miles, B. M. (1979). "Atomic Transition Probabilities," Vol. II, NSRDS-NBS22. Natl. Bur. Stand., Washington.

Wong, E. L., and Potter, A. E. (1963). *J. Chem. Phys.* **39**, 2211.

Woolsey, G. A., Lee, P. H., and Slafer, W. D. (1977). *J. Chem. Phys.* **67**, 1220.

Wu, C. H., Wang, C. C., Jaspar, S. M., Davis, L. I., Hanabusa, M., Killinger, D., Niki, H., and Weinstock, B. (1976). *Int. J. Chem. Kinet.* **8**, 765.

Young, R. A., and St. John, G. (1968). *J. Chem. Phys.* **48**, 895.

Young, R. A., and Sharpless, R. L. (1963). *J. Chem. Phys.* **39**, 1071.

Young, R. A., Sharpless, R. L., and Stringham, R. (1964). *J. Chem. Phys.* **40**, 117.

Zahniser, M. S., and Kaufman, F. (1977). *J. Chem. Phys.* **66**, 3673.

Zahniser, M. S., Kaufman, F., and Anderson, J. G. (1976). *Chem. Phys. Lett.* **37**, 226.

Zellner, R., Erler, K., and Field, D. (1977). *Symp. (Int.) Combust. [Proc.]* **16**, 939.

2

High Temperature Flow Tubes. Generation and Measurement of Refractory Species

ARTHUR FONTIJN and WILLIAM FELDER

AeroChem Research Laboratories, Inc.
Princeton, New Jersey

I. INTRODUCTION

The first topic discussed in this chapter is an extension of the flow tube technique to temperatures up to ≈ 2000 K. The apparatus for such studies has been given the acronym HTFFR for high temperature, fast-flow reactor. The extension of flow tube methods to higher temperatures requires special consideration of construction materials and species introduction methods and somewhat complicates the flow analysis, but high temperatures also alleviate some of the problems of species concentration measurements by atomic absorption. The reasons which led to the development of the technique, discussed in detail in Section II, are twofold. First, it allows measurement of atoms of refractory species and corresponding free radicals, for which virtually no kinetic data were available prior to this development. Second, in many cases little is known of the relation between "high" temperature kinetic data (e.g., obtained in flame and shock tube studies) and "low" temperature data, such as discussed in the preceding chapter, since Arrhenius-law behavior is often not obeyed over such large temperature ranges.‡ Thus, systematic variations among high and low temperature measurement techniques could unduly influence the apparent T dependences. It is, therefore, advantageous to use a single technique which can span both temperature regimes.

The word refractory like many "well-understood" common terms is somewhat vague and requires pragmatic considerations in delineating the scope of this chapter. One can equate refractory species to high temperature species, which Gole (1976) has recently defined as species, which, in general, require for their study techniques operative at temperatures in excess of 500 K. Accepting this definition as a guide, we consider here as refractory elements, in the first place, all metals (and will use the term metal atoms in this general sense), though a purist can rightfully deem a metal such as Hg too volatile to be called refractory. Second, while most of the discussions concern metal atoms, such elements as B, C, and Si, are without doubt refractory and require techniques similar to the metals for their study; they are therefore also considered.§ Free radicals, containing atoms of the refractory elements, are also discussed here.

HTFFRs, though uniquely suited to obtaining quantitative kinetic data on "metal atoms" over large temperature ranges, are by no means the only technique for studying their reactions. The other major techniques are discussed in more summary form in Section III.¶ Much of the detailed discussion of HTFFR methods is, *mutatis mutandis*, also applicable to the generation and measurement methods involved in these other techniques or further amplifies some of the discussion of Chapter 1. The material to be covered in Section III requires some restrictions. In particular, the bulk of,

---

‡ Basically, the Arrhenius law is an empirical expression which, especially in the form $k(T) = AT^{1/2} \exp(-\Delta E/RT)$, has often provided a surprisingly good procedure for presenting kinetic data. The uncertainty due to breakdown in Arrhenius behavior becomes especially bothersome in cases where either only "low" or "high" temperature data are available which may then require uncertain extrapolations. Such extrapolations are most uncertain from regimes where reactions display small activation energies. There are many discussions of this problem (see, e.g., Gardiner, 1977; Benson, 1975; Singleton and Cvetanovic, 1976; Hulett, 1964; Menzinger and Wolfgang, 1969; Perlmutter-Hayman, 1976).

§ We could also, very practically, have defined as refractory those species which can, in general, not be generated cleanly by the techniques discussed in the preceding chapter and on that basis would have arrived at the same elements for consideration.

¶ The literature covered here is, in general, that which was available to us by January 1978.

especially the early, crossed molecular beam studies concerned the dynamics of alkali metal atom reactions (Gowenlock *et al.*, 1976), but the subject of such scattering observations is outside the scope of the present volume [it has frequently been reviewed, see, e.g., Gowenlock *et al.* (1976), Kinsey (1972), Fite and Datz (1963), Greene and Ross (1968), and Ross (1966)]. However, there is a special class of molecular beam studies which has covered a wide variety of refractory species and employs methods very similar to the others of this chapter. These are chemiluminescence and laser-induced fluorescence studies, in which the "single collision" condition of true crossed molecular beams is in most cases only approximately obtained. Such beam studies are discussed in Section III,A. Section III contains tables of representative studies from which further details on the generation and measurement methods involved can be obtained and which allow comparison of results among the various types of techniques discussed. A similar table summarizing HTFFR data is given in Section II,B. As a further guide to the literature on refractory species, the reader is referred to Hastie (1975), Margrave (1967), the journal "High Temperature Chemistry," and the "Advances in High Temperature Chemistry" book series.

II. HTFFR (HIGH TEMPERATURE, FAST FLOW REACTOR) TECHNIQUES

A. General Description and Procedures

HTFFRs are used for measurements of overall kinetics $k(T)$, photon yields $\Phi(T)$ (and, incidentally, atomic transition oscillator strengths f), rate coefficients for light emission $k_{h\nu}(T)$, and quenching rate coefficients $k_Q(T)$. Here we describe the principles of these measurements and the basic apparatus functions. Details of the apparatus and metal atom generation are discussed in Section II,B; Sections II,C–E are concerned with measurement procedures requiring further detail. Results are summarized in Table I.

1. Overall Kinetics

Schematics of an HTFFR are given in Figs. 1 and 2. This HTFFR consists of a 2.5-cm-i.d. alumina reactor wound with Pt–40% Rh resistance wire in three contiguous, independently powered, zones. The metal to be studied is vaporized in the upstream zone and entrained in a stream of inert carrier gas (Ar or N_2). The useful length of the reactor for kinetic measurements is the uniform temperature zone extending about 23 cm upstream from the observation window. The axially movable oxidant inlet can be traversed over this distance.

Table I

HTFFR Measurements[a]

Reaction	Rate coefficient (mliter molecule^{-1} sec^{-1})	T range (K)	Remarks	Reference
Group IIA elements				
Ba + N$_2$O → BaO(X$^1\Sigma$, A$^1\Sigma$, A$'^1\pi$) + N$_2$	≈ 5 × 10^{-11}	≈1000	Rate coefficients k_O for quenching of BaO chemiluminescence are ≤3 × 10^{-13} (Ar, 600–1000 K); ≤3 × 10^{-12} (He, 460 K); (4.8 ± 2.0) × 10^{-13} (N$_2$, 600 K); (4.2 ± 1.0) × 10^{-10} (N$_2$O, 600 K); (4.8 ± 3.2) × 10^{-10} (O$_2$, 600 K). Collisional interconversion among precursor and radiating states has k_T ≥ 1.5 × 10^{-11} (Ar, 600 K). Atomic excitation observed with excess Ba.	b, c
Ba + O$_2$ → BaO + O	≈ 5 × 10^{-11}	≈1000		b, c
Group IIIA elements				
Al + O$_2$ → AlO + O	(3.4 ± 2.2) × 10^{-11}	300–1700	Temperature independent rate coefficient.	d, e, f
Al + CO$_2$ → AlO + CO	2.5 × 10^{-13} $T^{1/2}$ exp(−1030/T) + 1.4 × 10^{-9} $T^{1/2}$ exp(−14000/T)	300–1900	Non-Arrhenius temperature dependence as given by fitting expression; at 310, 490, 750, 1500, and 1800 K, rate coefficient is (1.5 ± 0.6) × 10^{-13}, (6.9 ± 2.7) × 10^{-13}, (1.6 ± 0.7) × 10^{-12}, (9.0 ± 3.8) × 10^{-12} and (3.8 ± 1.5) × 10^{-11}, respectively. 300–750 K data indicate D(Al-O) ≥ 122 kcal mole^{-1}.	g
Al + NO + Ar → AlNO + Ar	(2.5 ± 1.2) × 10^{-31}	600 (1800)	k in mliter^2 molecule^{-2} sec^{-1}. Reaction gave evidence of shift to bimolecular	h

62

Reaction	Rate coefficient	T (K)	Comments	Ref
$AlO(v = 0, 1) + O_2 \rightarrow AlO_2 + O$	$(7.2 \pm 3.6) \times 10^{-...}$ $(3.1 \pm 1.7) \times 10^{-13}$	300–1400	300 K measurement pertains to $v = 0$ only. Measurements suggest small negative E_A. Within experimental uncertainty, rate coefficient can be expressed: $k(300\text{–}1400\text{ K}) = (4.8 \pm 3.1) \times 10^{-13}$.	j,k

Group IVA elements

Reaction	Rate coefficient	T (K)	Comments	Ref
$Ge(^4{}^3P_{0,1,2}) + N_2O \rightarrow GeO(X^1\Sigma^+,$ $a^3\Sigma, b^3\pi, A^1\pi) + N_2$	$k_{J=0} = (1.7 \pm 0.3)$ $\times 10^{-11} \exp[-(490 \pm 50)/T]$ $k_{J=1} = (8.4 \pm 1.3)$ $\times 10^{-12} \exp[+(100 \pm 40)/T)]$ $k_{J=2} = (1.2 \pm 0.3)$ $\times 10^{-12} \exp[-(170 \pm 100)/T]$	450–900		j,k
$Sn(5^3P_{0,1}) + N_2O \rightarrow SnO(X^1\Sigma^+,$ $a^3\Sigma^+(1), b^3\pi) + N_2$	$k_{J=0} = (8.9 \pm 4.0)$ $\times 10^{-13} \exp[-(2260 \pm 180)]/T$ $k_{J=1} \approx 9 \times 10^{-13}$	500–950 320	$\Phi(a^3\Sigma) = 0.53 \pm 0.26$ independent of T; $\Phi(b^3\pi) = (0.59 \pm 0.29) \exp[-(1200 \pm 200)/T]$. $k_Q(a^3\Sigma)$ are (900 K): $\leq 2.3 \times 10^{-16}$ (N_2); $\leq 4.0 \times 10^{-16}$ (Ar); $\leq 4.0 \times 10^{-14}$ (N_2O); $\leq 4.0 \times 10^{-12}$ (Sn) based on τ_{rad} (a) $\geq 2.5 \times 10^{-14}$ sec. The $\tau_{rad} \times k_Q(b^3\pi)$ values are: 4.8×10^{-20} (N_2); $\leq 2.0 \times 10^{-20}$ (Ar); $\leq 1.0 \times 10^{-17}$ (N_2O); $\leq 1.0 \times 10^{-15}$ (Sn) in units of mliter molecule^{-1}.	k,l,m
$Sn(5^3P_0) + NO_2 \rightarrow SnO + NO$	$\approx 4 \times 10^{-13}$ $\approx 5 \times 10^{-10}$	320 1000	Used as Sn titration reaction at ≥ 900 K.	l,n
$Sn(5^3P_J) + O \rightarrow SnO(X^1\Sigma^+,$ $a^3\Sigma^+(1), b^3\pi, c, A^1\pi)$		360	Band system intensities: a–X $\gg$ b–X, c–X, A–X.	j
$Sn(5^3P_J) + O/O_2 \rightarrow SnO(X^1\Sigma^+,$ $b^3\pi, c, A^1\pi) + \cdots$		360	No a–X emission, except at low $[O_2]$.	j,l

(continued)

Table I (*continued*)

Reaction	Rate coefficient (mliter molecule^{-1} sec^{-1})	T range (K)	Remarks	Reference
$Sn(5^3P_J) + O_2 \rightarrow SnO(X^1\Sigma^+, b^3\pi, c, A^1\pi) + O$		360–1300	Chemiexcited products indicate multistep mechanism excitation; spectral distribution at 360 K similar to Sn/O/O$_2$.	j
	$k_{J=0} \approx 2 \times 10^{-11}$	310		l
	$k_{J=1} \approx 2 \times 10^{-10}$	310		
Group VIII elements				
$Fe + O_2 \rightarrow FeO + O$	$(3.6 \pm 1.4) \times 10^{-13}$	1600	Reaction is 20 kcal mole^{-1} endothermic. Hence preexponential of $k(T)$ must be close to gas kinetic (3×10^{-10}).	o, p

[a] In some chemiluminescence reaction studies some overall (Me disappearance) rate coefficients are based on a few measurements only; such rate coefficients are indicated here by $\approx$. Rate coefficients for spin–orbit excited states—$J = 1, 2$ levels of electronic ground state Ge(4^{3}P) and Sn(5^{3}P)— do not distinguish between physical quenching and chemical reaction.
[b] Felder et al. (1977).   [c] Felder et al. (1976).   [d] Fontijn et al. (1975).
[e] Fontijn et al. (1977).   [f] Fontijn et al. (1974).   [g] Fontijn and Felder (1977a).
[h] Fontijn (1977).   [i] Felder and Fontijn (1976).   [j] A. Fontijn (unpublished results).
[k] Fontijn and Felder (1977c).   [l] Felder and Fontijn (1978).   [m] Felder and Fontijn (1975).
[n] Fontijn and Felder (1977b).   [o] Fontijn et al. (1973).   [p] Fontijn and Kurzius (1972).

64

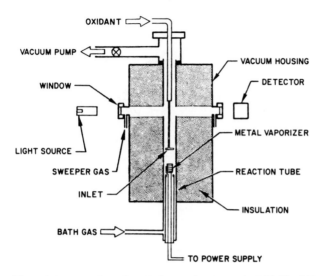

Fig. 1. Schematic of an HTFFR. [From Fontijn *et al.* (1973, Fig. 1).]

Overall rate coefficient measurements require *relative* atom concentrations, $[Me]_{rel} = [Me]/[Me]_i$,‡ as a function of oxidizer concentration $[OX]$, reaction time t, bath gas pressure P, and gas temperature T. $[Me]_i$ denotes the initial $[Me]$ in the absence of oxidizer and $[Me]$ the concentration at time t. $[Me]_{rel}$ is measured using the chopped radiation of a hollow-cathode lamp of the element of interest in absorption and/or fluorescence at the observation plane, Fig. 2. Typical $[Me]$ and $[MeO]$ used for overall kinetic measurements are on the order of 10^9–10^{12} mliter^{-1} and higher $[OX]$ are generally used to allow pseudo-first-order measurements to be made. The actual $[OX]$ used are determined by the desirability to observe about a factor of ten decrease in $[Me]_{rel}$ as a result of the reaction and hence depend on the value of $k(T)$, cf. Clyne and Nip (Chapter 1). $[OX]$ up to ≈ 0.1 $[M]$ (where $[M]$ is the total gas density) can be used; higher $[OX]$ creates flow disturbances and hence k measurement errors. Typically, the $[Me]_i$ used corresponds to 50% absorption of a given line for absorption experiments and 2–20% absorption for fluorescence experiments, thus allowing at least a factor of ten difference in $[Me]_i$ in a set of experiments. (Use of atomic transitions of different oscillator strengths allows an even wider variation in $[Me]_i$.) The choice $[OX] \gg [Me]$ rather than $[OX] \ll [Me]$ is an obvious one on practical grounds. The variation in $[Me]_{rel}$ over the reaction zone is readily obtained by the methods given here. Measurement of $[Me]$ is far more complicated; it is required for photon yield measurements and discussed in Section II,C,1. $[OX]$ is obtained

‡ The symbol [ ] without "rel" subscript denotes absolute concentrations.

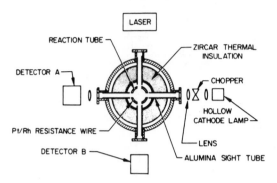

Fig. 2. Cross section of the observation plane of an HTFFR. Detector A for atomic absorption and free-radical laser-induced fluorescence measurements. Detector B for atomic fluorescence measurements. [From Felder and Fontijn (1976, Fig. 1).]

from the reactor pressure and the OX volume flow rate divided by the total volume flow rate (Clyne and Nip, Chapter 1) and is considered constant. For the metal-in-excess case, measurement of changes in [OX], at least by the optical techniques employed, would be very difficult, probably impossible, at the concentrations of interest. Last, high [OX] is readily achieved, whereas high [Me] is not.

The ability to use both absorption and fluorescence techniques in a set of measurements provides a guard against flow measurement errors in that absorption measurements integrate over the full width of the reaction zone, while fluorescence observations yield [Me]$_{rel}$ in the center of the reactor. Detector A in Fig. 2 is a photomultiplier tube (PMT) monochromator combination, i.e., a spectrometer. Since hollow-cathode fluorescence signals are considerably weaker than absorption signals or laser-induced fluorescence signals, detector B, used for atomic fluorescence measurement, is an interference filter/PMT combination. The signals of the PMTs are fed to lock-in amplifiers. We have found that, at the highest temperatures used, the radiation from the hot reactor walls is sufficiently intense to preclude atomic fluorescence measurements. The actual highest useful temperature for fluorescence measurements depends, of course, on the λ of the atomic line, since the intensity of the background radiation in the uv/visible region of interest decreases with λ in accordance with the emissively corrected blackbody radiation laws. For example, for the 309.3 nm Al resonance line, 1500 K is found to be an upper useful limit in our work when a 1 nm FWHN interference filter is used.

In most of the work, pseudo-first-order rate coefficients k_{ps_1} are obtained by traversing the OX inlet nozzle. The first-order kinetics equation is used in the form

$$k_{ps_1} = -d \ln[Me]_{rel}/dt \tag{1}$$

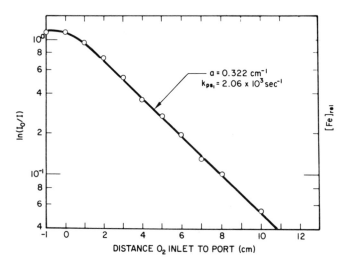

Fig. 3. Fe-atom concentration profile (absorption). $\overline{T} = 1590$ K, $\bar{v} = 49$ m sec^{-1}, $[O_2] = 8.9 \times 10^{14}$ mliter^{-1}, $\lambda = 372.0$ nm, $P = 15$ Torr. [From Fontijn et al. (1973, Fig. 2).]

From the local slope

$$a = -d \ln[Me]_{rel}/dx \qquad (2)$$

of the data plots (e.g., Fig. 3) k_{ps_1} is obtained via the equation

$$k_{ps_1} = \eta a\bar{v}(1 + aD_{Me}\bar{v}^{-1}) \qquad (3)$$

In Eq. (3), η is a factor equal to 1 for plug flow, and approximately equal to 1.6 (Ferguson et al., 1969) for parabolic flow, $\bar{v}$ is the mean bulk linear-gas velocity, D_{Me} the diffusivity of Me atoms, and $(1 + aD_{Me}\bar{v}^{-1})$ the correction factor for the effect of axial diffusion (Kaufman, 1961) which is small ($\leq 10\%$, and usually $\leq 1\%$, for the conditions of our experiments). For our calculations, we take η to be 1.3, which choice is further discussed in Sections II,D and II,E.

The rate coefficient k_{ps_1} incorporates the summation of heterogeneous and homogeneous contributions to the rate, i.e.,

$$k_{ps_1} = k_w + k_5[OX] + k_6[OX][M] \qquad (4)$$

where k_w is the wall contribution and k_5 and k_6 correspond to binary and ternary mechanisms, i.e., to the reaction types

$$Me + OX \xrightarrow{k_5} MeO + X \qquad (5)$$

and

$$Me + OX + M \xrightarrow{k_6} MeOX + M \qquad (6)$$

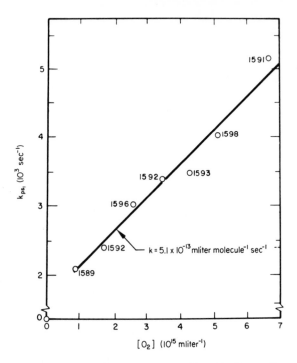

Fig. 4. Fe/O_2 reaction rate coefficient at 15 Torr. $\overline{T} = 1590$ K, $\bar{v} = 49$ m sec^{-1}, [M] $=$ 0.91×10^{17} mliter^{-1}, $\lambda = 372.0$ nm. Numbers beside each individual data point indicate the $\overline{T}$ at which it was obtained. [From Fontijn *et al.* (1973, Fig. 3).]

A series of measurements of k_{ps_1} at constant temperature, as a function of [OX] at various constant values of [M], identifies, cf. Fig. 4, the dominant homogeneous reaction and its rate coefficient. The intercepts of plots like Fig. 4 yield the wall oxidation coefficients, which may be seen to be substantial (γ, the probability for a metal atom to be oxidized on the reactor walls, is usually in the range $10^{-3}-10^{-1}$). Under some circumstances, results are obtained somewhat faster by using the OX inlet nozzle in one position, in which case the time-integrated equation

$$-\ln[\text{Me}]_{\text{rel}} = k_w t + k_5[\text{OX}]t + k_6[\text{OX}][\text{M}]t \qquad (7)$$

yields the desired rate coefficients from plots of $[\text{Me}]_{\text{rel}}$ versus [OX], e.g., Fig. 5. Typically, such measurements are made at two or three nozzle positions and if good agreement is obtained the rate coefficient measurements for the particular flow condition are considered to be completed. (It may be added that in the extreme case of repeating these stationary nozzle measurements at as many distances as used in a k_{ps_1} determination, Fig. 3, the traversing and

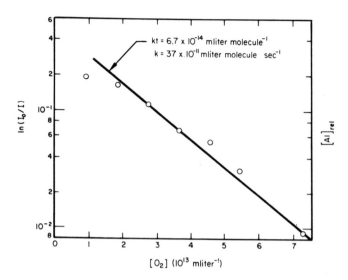

Fig. 5. Al/O$_2$ rate coefficient from fixed nozzle position data (absorption). $\bar{T} = 1490$ K, $\bar{v} = 51$ m sec^{-1}, $\lambda = 309.3$ nm, $P = 37$ Torr, $t = 1.8 \times 10^{-3}$ sec. [From Fontijn *et al.* (1975, Fig. 4).]

stationary nozzle measurements become essentially identical.) In a typical set of measurements enough individual rate coefficient determinations are made to cover at least a factor of ten in pressure P (3–60 Torr is the range typically used), average gas velocities $\bar{v}$ (typically, 10–100 m sec^{-1}), and [Me]$_i$. It may be noted that both these P and $\bar{v}$ ranges are an order of magnitude higher than those used in conventional electrical discharge flow tube studies of non-refractory species (Clyne and Nip, Chapter 1). The reason for this high $\bar{v}$ is that at lower $\bar{v}$ adequate metal evaporation/transport can usually not be achieved. The relatively rapid wall oxidation, noted above, requires $P \geq 3$ Torr. Typically, P and $\bar{v}$ ranges used are the same over the 300–1900 K range (Fontijn *et al.*, 1975, 1977; Felder and Fontijn, 1976; Fontijn and Felder, 1977a) and reactions having a rate coefficient $\gtrsim 1 \times 10^{-15}$ mliter molecule^{-1} sec^{-1} can be measured.

Thus far, we have discussed the methods for Me overall kinetics measurements. Basically, MeO rate coefficients can be obtained in a similar manner, laser-induced fluorescence being the [MeO]$_{rel}$ measurement method used. The principles of laser-induced fluorescence measurements are discussed in the chapter by Lin and McDonald (Chapter 4) and do not need amplification here. Thus far, we have only studied AlO($v = 0, 1$) kinetics, which species can readily be measured with either an Ar$^+$ laser (λ pump 488 nm) (Felder and Fontijn, 1976) or a N$_2$ laser-pumped dye laser (7-diethylamino-4

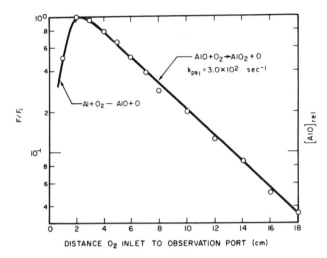

Fig. 6. Normalized AlO($v = 0$) concentration profile. $\overline{T} = 310$ K, $\bar{v} = 10$ m sec^{-1}, $[O_2] = 4.6 \times 10^{13}$ mliter^{-1}, $P = 3.0$ Torr. [From Fontijn *et al.* (1977, Fig. 2).]

methylcoumarin at 465 and 485 nm) (Fontijn *et al.*, 1977). The AlO could be generated readily in this case, since $k(T)$ for Al/O$_2$ is about 100 times that for AlO($v = 0, 1$)/O$_2$ and, by using a sufficient excess $[O_2]$, AlO formation is essentially completed near the O$_2$ inlet nozzle, cf. Fig. 6.

In closing this section a few general remarks are in order. It may be noticed that the observation region, Figs. 1 and 2, differs in two respects from the reaction region upstream from it since the reactor here has holes (1-cm diam) facing the windows of the vacuum housing. As a result, the observation region generally is at a somewhat lower temperature and the flow may be locally disturbed. However, as has been extensively discussed by Westenberg and de Haas (1967), when the inlet nozzle is traversed under pseudo-first-order reaction conditions, at a given observation point one measures the result of changes in the limiting reagent (Me) concentration in the region upstream from this point and the variations at the observation point cancel out. That the window effect is not serious also follows from the frequent observation that when results from the stationary nozzle mode, Eq. (7) (where the measurements do not exclude the window region), are compared to the traversing nozzle mode [Eqs. (5) and (6)], the results are indistinguishable.

Whereas the reactor can be used up to about 2000 K, the maximum operating temperature for a given reaction is determined by the stability of the oxidizer. Thus, O$_2$ can be used over the full temperature range, but N$_2$O only up to 950 K. A first idea of the maximum temperature for a given oxidizer can be obtained from thermal equilibrium calculations taking the dissociation products into consideration. The latter are important since, if a dissociation product cannot react with Me as rapidly as the parent oxidizer, somewhat more dissociation could be acceptable than when it reacts, or could react on the basis of a priori considerations, more rapidly than this parent (Fontijn and Felder, 1977a). In practice, since dissociation is a kinetic process and the time available is on the order of milliseconds, such calculations are conservative and somewhat higher temperatures can usually be used, based on experimental criteria. For instance, in the case of N_2O we found that the Sn/N_2O reaction led to emission only from $SnO(a^3\Sigma, b^3\Pi)$, whereas the Sn/O_2 reaction additionally produces $SnO(c$ and $A^1\Pi)$ (Felder and Fontijn, 1978; Fontijn and Felder, 1977b; A. Fontijn, unpublished results). Onset of emission from the latter states with increasing temperature thus indicates N_2O decomposition. The 950 K upper limit for N_2O was arrived at on this basis; it was also shown that above this temperature the scatter in k measurements increased (Felder and Fontijn, 1978). The observed N_2O decomposition must result from heterogeneous processes since the rate coefficient for homogeneous N_2O decomposition is known and can be used to show that this process is far too slow to explain the observed degree of decomposition (Felder and Fontijn, 1978). It may therefore be possible to increase the maximum temperature for a given oxidant by using a cooled inlet design, but this has not yet been tried.

2. Photon Yields, Rate Coefficients for Light Emission, and Atomic Oscillator Strengths

Many metal atom oxidation reactions are high yield chemiluminescent processes, leading to intense radiation in the visible and ultraviolet. To study their kinetics, HTFFRs offer advantages similar to those for the overall (all product states) oxidation reactions, i.e., those of independent control of individual reactant and quencher concentrations, pressure, temperature, and reaction time interval observed. The reactions involved can be written symbolically

$$Me + OX \longrightarrow MeO^* + X \tag{8}$$

$$Me + OX \longrightarrow MeO + X \tag{9}$$

$$MeO^* \longrightarrow MeO + h\nu \tag{10}$$

$$MeO^* + Q \longrightarrow MeO + Q \tag{11}$$

For such reactions the photon yield Φ may be defined as the fraction of reaction events leading to emission of a photon. In the extreme case of negligible quenching [Eq. (11)], the photon yield is identical to the branching ratio

$k_8/(k_8 + k_9)$. A quantity sometimes confused with Φ is the rate coefficient for light emission k_{hv}, which is defined by

$$I = d[hv]/dt = k_{hv}[\text{Me}][\text{OX}] \tag{12}$$

where I is the rate of light emission per unit volume. Hence,

$$k_{hv} = \Phi k_{\text{overall}} \tag{13}$$

It follows that the measurement of Φ requires the measurement of absolute intensity and absolute [Me]. Measurement of the latter is discussed in Section II,C. The measurement of the former is achieved by comparison to the O/NO standard chemiluminescent reaction observed under identical geometry with the HTFFR at room temperature. For this reaction

$$\text{O} + \text{NO} \xrightarrow{\hspace{1.5cm}} \text{NO}_2 + hv \tag{14}$$

the true (photons mliter^{-1} sec^{-1}) spectral distribution $I_S(\lambda)$ versus λ and hence also the integrated intensity I_S in these same units have been determined (Fontijn et al., 1964; Golde et al., 1973). To determine the intensity I_X of the Me/OX reaction, the following equation can now be used (Fontijn et al., 1964):

$$I_X = I_S \int_{\lambda_1}^{\lambda_2} R_X(\lambda)\, d\lambda \bigg/ \int_{\lambda_1}^{\lambda_2} R_S(\lambda)\, d\lambda \tag{15}$$

where R is the output of the spectrometer PMT (Detector A of Fig. 2). The O/NO reaction is suitable for emission at $\lambda \geq 390$ nm; for near uv wavelengths we have found it practical to use the O/SO reaction and/or a quartz iodine lamp (A. Fontijn, unpublished results).

For many reactions, determination of I_X and the reactant concentrations at a given point in the reaction zone combined with Eq. (12) suffices to determine k_{hv}. If k_{overall} is known, Φ can then be obtained from Eq. (13). However, in many instances Φ is the primary quantity of interest and use of Eqs. (12) and (13) leads to an uncertainty in Φ combining the inaccuracy of both k_{hv} and k_{overall}. A preferred procedure for Φ determination is therefore to volume integrate I_X and divide this by the flux of limiting reagent (the reagent present in the smallest concentration) consumed. In the HTFFR, volume integration of I_X is achieved by traversing the nozzle in steps of 2 cm (equalling the viewing aperture of the spectrometer used as the detector), i.e., by viewing the chemiluminescent zone in these steps. Ideally, the limiting reagent is essentially completely consumed in the maximum observed reaction distance (time), in which case the flux of this reagent consumed simply equals its flux entering the reaction zone; if this cannot be achieved extrapolation procedures can be used (Felder and Fontijn, 1978). For Me as the limiting reagent, the decrease in [Me] could be measured directly. However, when Me is the limiting reagent

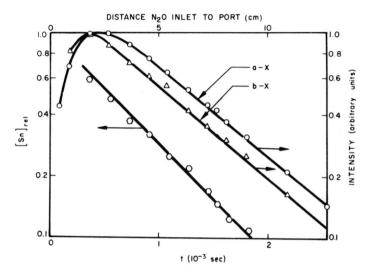

Fig. 7. Normalized time–(distance) dependence of chemiluminescence intensity from the a–X and b–X systems of SnO and of $[Sn]_{rel}$. $\bar{v} = 50$ m sec^{-1}, $[Sn] = 2 \times 10^{11}$ mliter^{-1}, $[N_2O]$ $= 5 \times 10^{11}$ mliter^{-1}, $P = 10$ Torr. [From Felder and Fontijn (1975, Fig. 2). Copyright of North–Holland Publishing Co.]

the reaction can usually be made to go to completion in the observable reaction distance, cf. Fig. 7, since there is little limitation in the [OX] which can be used, but $[Me] \gtrsim 1 \times 10^{14}$ mliter^{-1} is often hard to achieve.

It should be cautioned that the direct use of Eqs. (12) and (13) to determine k_{hv} and/or I without volume integration may only be practiced if the emitters MeO radiate in essentially the same volume element in which they are formed, i.e., if the radiative lifetimes are short compared to transport times. Because of the faster flow velocities in HTFFRs as compared to regular flow-tube reactors, this condition is more often not obeyed in refractory species studies than in nonrefractory species studies (Clyne and Nip, Chapter 1), though the general procedures outlined here are equally applicable to the latter. Φ measurements in HTFFRs have thus far concerned emitters SnO (Felder and Fontijn, 1978) and GeO (A. Fontijn, unpublished results) with radiative lifetimes on the order of 10^{-3} sec which is not short compared to transport times; $k_{hv}(T)$ can be obtained in these cases by taking the product $k(T)\Phi(T)$.

In the study (Felder and Fontijn, 1978) of

$$Sn + N_2O \longrightarrow SnO(X^1\Sigma, a^3\Sigma, b^3\Pi) + N_2 \tag{16}$$

both conditions $[Sn] \ll [N_2O]$ and $[Sn] \gg [N_2O]$ were employed in the Φ determinations. In the former case $2 \times 10^{10} \leq [Sn] \leq 8 \times 10^{11}$ mliter^{-1} was used as determined from Sn absorption and f numbers, cf. Section II,C,1;

in the latter case $1 \times 10^{13} \leq [\text{Sn}] \leq 1 \times 10^{14}$ mliter^{-1} and [Sn] were determined by a chemiluminescent titration, cf. Section II,C,4. The agreement between the Φ measurement for these two conditions was well within the accuracy limits of the combined experiments. The best agreement was obtained using $f(\text{Sn}^3\text{P}_0, 286.4 \text{ nm}) = 0.20$ and $f(\text{Sn}^3\text{P}_1, 300.9 \text{ nm}) = 0.052$. Literature values (deZafra and Marshall, 1968; Penkin and Slavenas, 1963; Lawrence *et al.*, 1965) for these numbers range for the former from 0.186 to 0.230, for the latter from 0.033 to 0.059. It follows that such comparisons allow determinations of atomic f numbers. Such use of chemiluminescent reactions of refractory metal species appears to be an interesting development; f numbers are poorly known for a large number of refractory elements because of the difficulties involved in accurately determining concentrations in the high temperature cells often used for this purpose (Wiese, 1970).

A more direct approach to f number determination in flow tubes would be to directly measure a given atom concentration via titration (see also Clyne and Nip, Chapter 1) and in absorption. Such an approach was followed successfully by Lin *et al.* (1970) and Morse and Kaufman (1965) for O, N, and H atoms for which much longer observation times for the titration reaction are available, and hence lower atom concentrations could be measured in titration. [Since the concentration of the titrant at the end point equals the original atom concentration, the time required for a titration reaction to go to completion is determined by the product $(k_5[\text{Me}])^{-1}$.] The lowest [Sn] that could be measured in the HTFFR in titration was 10^{13} mliter^{-1}, at which concentration the system is optically thick for the resonance lines and hence no quantitative absorption measurement is possible.

3. Quenching Rate Coefficients

Of the above reaction scheme for chemiluminescent reactions Eq. (11), quenching, remains to be discussed. To determine k_Q, it is again necessary to observe the complete luminescent reaction region, which changes in length with the concentration of the quencher Q, which can be a nonreactive gas or either reactant. Basically, the Stern–Volmer equation in the form

$$I_0/I = \Phi_0/\Phi = 1 + \tau_{\text{rad}}k_Q[\text{Q}] \tag{17}$$

which follows from a steady treatment of Eqs. (8), (10), and (11), can be used to determine k_Q. In its original form the Stern–Volmer treatment pertained to photoexcited species and I_0 is the fluorescence intensity in the absence of the quencher. However, when k_Q for quenching by a reactant in a chemiluminescent reaction has to be determined, this definition clearly cannot be used and we define I_0, or Φ_0, instead as maximum intensity or photon yield as a function of the concentration of that reactant.

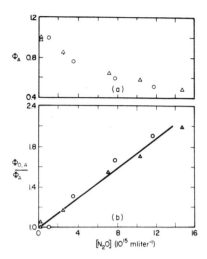

Fig. 8. Influence of $[N_2O]$ on BaO $(A^1\Sigma-X^1\Sigma)$ integrated intensity. (a) Relative photon yield. (b) Quenching data obtained from (a). 3×10^9 mliter$^{-1} \leq [Ba] \leq 3 \times 10^{10}$ mliter^{-1}; $T = 600$ K; $\bar{t} = 30$ m sec^{-1}. Key: $\triangle$, $[Ar] = 2.4 \times 10^{17}$ mliter^{-1}; $\bigcirc$, $[Ar] = 4.8 \times 10^{17}$ mliter^{-1}. [From Felder et al. (1977, Fig. 5).]

The basic procedures were first established in a study of the Ba/N_2O chemiluminescent reaction (Felder et al., 1977). In that reaction the emitting $BaO(A^1\Sigma)$ molecules are populated via collision-induced transitions of BaO in a precursor state and the complete equations become rather complicated, though they can be shown (Felder and Fontijn, 1978; Felder et al., 1977) to be of the basic Stern–Volmer form; a representative plot is shown in Fig. 8.

Ba/N_2O reaction occurs at approximately every collision and special procedures had to be employed to ascertain that the observations were not falsified by diffusion/reactant mixing effects or light blockage near the inlet. By contrast for k_Q measurements on SnO(a, b) from the Sn/N_2O reaction [O], which has a $k(T)$ several orders of magnitude lower (Felder and Fontijn, 1978), the normal (Section II,B,1) OX inlet configuration could be used. For quenching of $BaO(A^1\Sigma)$ by the nonreactive quenchers Ar, He, and N_2, the optimum N_2O inlet configuration was found to be an L-shaped tube perforated with three small holes near the reaction tube axis. Under the conditions of these experiments, $[Ba] \ll [N_2O] \ll [Q]$, this configuration gave rise to a cone-shaped uniform (not streaky) glow, with the apex at the inlet, which extended for distances up to 15 cm downstream. At constant flow velocity, the length of the glow increased with increasing pressure, indicating that reactant mixing determined the length of the reaction zone. To reduce the integrating times needed, a coarse screen was attached to the N_2O delivery tube a short distance (2–6 cm) downstream from the inlet. The screen creates a planar flow disturbance across which mixing is intense and the Ba/N_2O reaction abruptly approaches completion. Pressures up to 120 Torr were covered in this work. Above 70 Torr some of the glow was observed to "leak" through the screen; however, measurements downstream of the screen indicated the escaping fraction to be $< 10\%$ and essentially identical results were obtained with the screen removed and the far more time-consuming integration procedure from measurements with the nozzle moved in steps of 2 cm. For the Sn/N_2O reaction, which is kinetically controlled, the latter procedure was the only possible one.

The large $[N_2O]$, see Fig. 8, required to make N_2O quenching measurements makes it necessary to change the inlet configuration to prevent N_2O back diffusion and nonuniform glows. The L-shaped, three-hole inlet was replaced by a multiperforated, axially symmetric distributor ring similar to those used in HTFFR overall kinetic measurements. Hole sizes in the distributor ring were small and a number of closely spaced holes was used to maximize the dispersion and exit velocity of the N_2O, thus improving the mixing of N_2O with the Ba laden bath gas stream. Holes were placed on the inner circumference of the ring so that N_2O entered the bath gas flow at angles between $45°$ and $90°$ from the downstream direction. Just upstream of this inlet, a flow constraining washer was attached to the inlet tube in order to channel the bath gas flow to the inner circumference of the inlet. The washer is 2.3-cm o.d., 0.9-cm i.d. This inlet configuration results in the formation of a conical, uniform glow with its base attached to the inlet ring. The visually observable glow did not contact the reaction tube walls. For all $[N_2O]$ investigated, the chemiluminescence intensity upstream of the inlet (due to N_2O back diffusion) was $< 5\%$ of the total intensity.

Since obtaining complete spectra at every nozzle position and every $[Q]$ can be a very time-consuming procedure, it usually suffices to take measurements of one band, or one band per band system, combined with some well-selected measurements of the changes in spectral distribution as functions of t and $[Q]$.

B. The Reactors and Their Operation; Metal Evaporation

1. Single Unit HTFFR

The essential features of the most recent version of this HTFFR are shown in Fig. 9. [This version represents a considerable improvement in design and operation over the original one discussed in Fontijn et al. (1972).] The vacuum chamber is a brass cylinder (15-cm i.d., 80-cm long) closed at both ends by demountable flanges. The reaction tube (2.5-cm i.d., 3.2-cm o.d., 84-cm long) is made of McDanel 998 alumina tubing (99.8 % Al_2O_3). This tube is grooved on the outside at 1.2 turns per cm; the grooves are 0.13-cm deep and 0.18-cm wide and are wound with 0.13-cm-diameter, Pt–40% Rh resistance wire. The windings are protected with three thin coats of alumina-base cement (Aremco Ceramabond 503) and are thermally shielded by a 998 alumina tube (4.4-cm i.d., 5.0-cm o.d., 75-cm long). Zircar fibrous zirconia cylinders (type ZYC, Zircar Products, Inc., Florida, New York) are used as thermal insulation between the shield tube and the water-cooled vacuum chamber wall. The reaction tube rests on the base flange T. To allow for thermal expansion an asbestos fiber collar V provides a flexible seal between the top (downstream end) of the reactor and the brass tube providing the connection between the reactor and the mechanical vacuum pump (3.7 or 7.3 m^3 min^{-1} capacity).

The Pt–40% Rh resistance wire is arranged to provide three independently controlled, ≤20-cm long, heating zones. The reaction zone is covered by the downstream 8 cm of the middle zone and the upstream 15 cm of the downstream zone. Each heating zone is separately powered by a variable auto-transformer (2.4 kVA) and 36 A stepdown transformer and has a 0–30 A current meter. Connecting wires run between the reaction and shield tubes to the feedthroughs R. This arrangement provides a uniform temperature over the reaction zone and allows the region around the metal source S to be at a radically different temperature from the reaction zone. The maximum useful temperature attainable with the apparatus as presently constructed is fixed by the 2040 and 2220 K upper recommended operating temperatures of the Pt–40% Rh wire and of the alumina tubing, respectively (Campbell and Sherwood, 1967). [For the properties of Pt–40% Rh wire see, e.g., Matthey Bishop, Inc. (1968); for those of alumina see, e.g., McDanel Refractory Porcelain Co. (1970).]

Optical observations are made through four 2.5-cm-diameter quartz windows flange mounted on the vacuum jacket. Alumina sight tubes (1.6-cm i.d., 2.5-cm o.d.) passing through 2.5 cm holes in the Zircar insulation and shield tube J and four open ports (0.95-cm diameter) in the reaction tube are aligned with the windows. The optical path external to the reaction tube can be swept with a small inert gas flow introduced at the inlets D. This window sweeper volume flow is typically ≤0.2 cc (STP) cc^{-1} [as compared to the 20–300 cc (STP) sec^{-1} bath gas flow through the reactor] and is used mainly during reactor conditioning, though its presence has been found not to influence results in an actual experiment.

The metal vaporizer source support tubes (1.4-cm-i.d., 1.9-cm-o.d. AV30 alumina) can readily be withdrawn through the 3.2-cm-i.d. ball valve O and sliding seal B_2, thus permitting the replacement of the metal atom sources without drastically altering flow conditions. Details of the 75-cm-long source support tube are shown in Fig. 10. The metal atom source is placed atop the support tube where it is heated by the reaction tube. Additional resistance

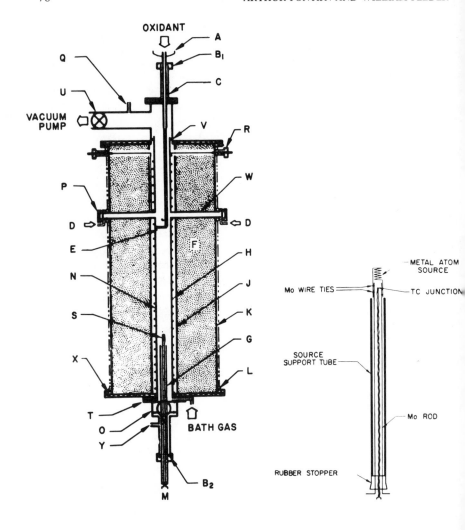

Fig. 9. **Fig. 10.** Source assembly.

Fig. 9. Single unit HTFFR. A, thermocouple; B, sliding O-ring seal; C, sleeve; D. window sweeper gas; E, oxidant inlet and thermocouple well; F, Zircar insulation; G, source support tube; H, Pt–40 % Rh resistance wire; J, alumina heat shield; K, brass vacuum housing; L, cooling coil; M, to power supply; N, alumina reaction tube; O, ball valve; P, window in annular flange; Q, pressure tap; R, high current vacuum feedthrough; S, metal atom source; T, base flange; U, throttling valve; V, asbestos fiber collar; W, sight tube; X, O-ring; Y, auxiliary vacuum outlet.

heating power can be supplied through the 0.27-cm-diameter Mo rods which hold the source. These rods, which are fed through stacked double-bore mullite insulator tubes (0.3-cm i.d., 0.9-cm o.d.), are connected to a 30 A external power supply. For most operating conditions, even for quite refractory materials such as Al or Sn, this power supply can be left in the off position as sufficient heat is supplied by the upstream HTFFR heating zone. The actual placement of the source within this region provides a good means for controlling atom fluxes. To aid in this positioning, a thermocouple (TC) is provided just below the source to register the temperature of the bath gas impinging on the metal atom source.

On the top (downstream side) of the reactor is a 0.95-cm-i.d. stainless steel sleeve C (Fig. 9), through which the oxidant and thermocouple inlet tube are fed by means of a 0.6-cm sliding O-ring seal B_1. This sleeve is offset 0.6 cm from the centerline of the reaction tube, so that the inlet system, shown in detail in Fig. 11, does not block the optical path, Fig. 2. The upper, cooler, part of the inlet system consists of a thin wall stainless steel tube (0.6-cm o.d., 40-cm long) to which are sealed, with epoxy cement, the double bore 998 alumina thermocouple well (0.24-cm o.d., 55-cm long) and the single bore 998 alumina oxidant tube (0.16-cm i.d., 0.24-cm o.d., 55-cm long). To uniformly distribute the oxidant over the reaction tube cross section, a multihole ring

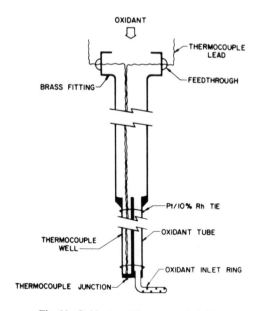

Fig. 11. Oxidant and thermocouple inlet.

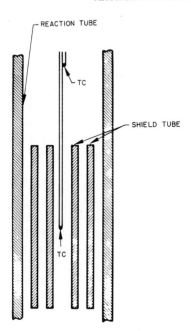

Fig. 12. Representation (to scale) of experimental setup for comparison of unshielded and doubly shielded thermocouples. [Reprinted with permission from A. Fontijn and W. Felder, *J. Phys. Chem.* (1978). Copyright by the American Chemical Society.]

distributor is attached with alumina cement to the tip of the oxidant inlet tube. The ring is formed of Pt–10% Rh tubing (0.13-cm i.d., 0.16-cm o.d.) bent to a 1.2-cm-i.d. circle. Twelve ≈0.08-cm-diameter inlet holes are situated in the plane perpendicular to the reactor axis on offset hexagonal patterns along the inner and outer circumference of the ring. Alumina cement is used to cover the TC junction and close the thermocouple well. Pt/Pt–10% Rh TCs are used, except for $T \gtrsim 1600$ K where Pt–20% Rh/Pt–40% Rh is preferred for longer life expectancy (Slaughter and Margrave, 1967).

As discussed in Section II,D,3, since the TC of Fig. 11 is essentially un-shielded, its reading is subject to a small radiation correction. To determine this correction for each individual reaction condition, the assembly shown in Fig. 12 is used by traversing the reaction zone with it before or after a kinetic measurement. This assembly consists of an unshielded TC situated at the reaction tube centerline which is compared to a doubly shielded TC placed 5 cm downstream from it in the same radial position. Shielding for the doubly shielded TC is provided by 0.65-cm-i.d., 0.95-cm-o.d. and 1.6-cm-i.d., 1.9-cm-o.d. alumina tubes, each of which extends 2.5 cm upstream and 2.5 cm downstream of the TC.

2. *Vaporization Sources*

A first concern in vaporization source materials is compatibility with the
Me species at and below the temperature at which a sufficient flux of Me atoms
is obtained for kinetic studies. Experience in this area is limited, and a certain
amount of trial and error is involved. In addition to our experience described
here, the reader is referred to the discussions and tables of Section III and the
original literature quoted there for guidance. A good starting point is often
provided by the thin film literature (e.g., Glang, 1970), and manufacturers'
information on vaporization sources (e.g., Mathis, 1973). (Both of these
references give tables summarizing the available information.) Another useful
compendium of information on high temperature materials is Campbell and
Sherwood (1967). It should be remembered, though, that in thin film and
beam work (Section III,A), unlike in the HTFFR and other bulk methods, no
carrier gas is used.

To establish the required vaporization temperature, data on the vapor
pressure of the elements are needed as a guide. At HTFFR entrainment conditions (3–75 Torr, 10–80 m sec^{-1}), to achieve atom concentrations on the
order of 10^{10}–10^{12} mliter^{-1} (i.e., 10^{-6}–10^{-4} Torr), we have found that
temperatures giving Me pressures of roughly 10^{-2}–10^{-1} Torr at equilibrium
are needed [i.e., the condition given in vapor pressure tables and graphs such
as are given in Honig (1967)]. Of course, these conditions depend somewhat
on the "container" used; the larger the surface area of metal exposed to the
bath gas, the lower the temperature that is required. This criterion, "maximize
the surface area within the constraints set by the reaction tube," is the golden
rule of HTFFR source design.

A simple source design with a large surface area for a material which is a
solid at its vaporization temperature, e.g., Fe, is the following (Fontijn *et al.*,
1972, 1973). An Fe "beaker" (1.0-cm i.d., 1.2-cm o.d., 4.1-cm long) is inverted
over a resistance heater consisting of W wire wound on a 0.2-cm-i.d., 0.5-cm-
o.d. alumina tube. The heater wire is attached to the Mo rods, Fig. 10. The
W-wound ceramic fits snugly into an alumina liner (0.70-cm i.d., 0.95-cm o.d.)
placed into the Fe beaker. Another method for exposing a large surface area of
a solid material is to suspend chunks of it inside a W or Mo coil.

For materials liquid at their vaporization temperature, crucibles (e.g.,
1.0-cm i.d., 1.3-cm o.d., 2–4-cm deep; boron nitride for Al, alumina for Ge, Sn)
contained inside a coil of 0.075-cm-diameter W wire are suitable. The coils
again connected to the Mo rods, Fig. 10, support the crucible and can be used
for supplying additional heat when needed. However, if the material being
evaporated can wet a refractory metal wire, the use of such wires, which provide better contact with the bath gas, is preferable over that of crucibles. Thus,
e.g., Mathis Co. (1973) three-stranded helical (0.075-cm-diameter strands) W

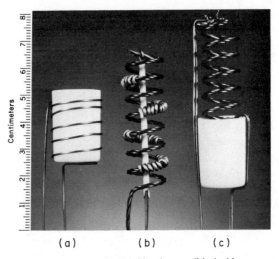

Fig. 13. Various evaporation sources. (a) Alumina crucible, inside a tungsten wire coil, used e.g., for $[Sn] \approx 1 \times 10^{10} - 1 \times 10^{12}$ mliter^{-1}. (b) Three-stranded tungsten coil with Al strands. The Al melts inside the reactor, producing a fairly evenly Al-wetted coil for Al studies. (c) Alumina crucible with three-stranded tungsten coil "wick," used, e.g., for $[Sn] \approx 1 \times 10^{13} - 1 \times 10^{14}$ mliter^{-1}.

coils are ideal for vaporizing Al. Loops of Al are bent around the coil which becomes fairly evenly coated when the Al melts once inside the reactor. (1.5 g Al usually suffices for 12–24 hr of operation.) For high Sn fluxes and Ge, a combination of the crucible and coil method has been found useful. A W wire coil is placed inside an alumina crucible and the crucible is filled with the metal; in this mode the metal inside the crucible serves as a reservoir and the coil acts as a wick to which current is always supplied to assist evaporation. In this way $[Sn]$ at 7×10^{13} mliter^{-1} could be maintained for several hours for titration (Section II,C,4) and quenching (Section II,A,3) experiments (Felder and Fontijn, 1978). A few of these source designs are shown in Fig. 13. The original references quoted in Table I can be consulted for additional details.

3. Modular HTFFRs

The single unit HTFFR as described above can, in principle, be used down to room temperature for the study of reactions of volatile species or permanent gases [in which case it is used as an ordinary flow reactor (Clyne and Nip, Chapter 1)]. However, when refractory species are vaporized, sufficient heat is generated by the source that higher reaction zone temperatures result (e.g.,

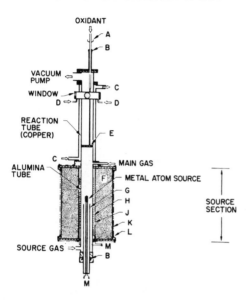

Fig. 14. Schematic of a cooled modular HTFFR. A, thermocouple leads; B, sliding O-ring seal; C, coolant; D, window sweeper gas; E, oxidant inlet and thermocouple well; F, Zircar insulation; G, source support tube; H, Pt–40% Rh resistance wire; J, alumina heat shield; K, brass vacuum housing; L, cooling coil; M, to power supply. [From Fontijn *et al.* (1977, Fig. 1).]

700–800 K for Al, Sn) despite no additional heating of the reaction tube. For such species, a modular reactor was constructed to obtain temperatures near 300 K. This reactor, shown in Fig. 14, consists of a 27-cm-long, heated source section followed by a 27-cm-long (to the observation plane) water-cooled copper reaction tube. The source section differs from the usual HTFFR in that it has only one heating zone. The alumina tube in the source and the reaction tube are each 2.5-cm i.d. Bath gas for the reaction tube is introduced in part through the source tube, and in part through a "main gas" inlet, located just downstream of the source section. Dividing the flow in this manner keeps part of the bath gas from being subjected to heating in the source section; it is this aspect of the reactor which, in addition to the cooling water, allows the achievement of a uniform (± 10 K) reaction tube temperature of ≈ 300 K. The proportion of the Ar introduced at the source gas inlet is typically varied from 3% at the highest reaction tube volume flow rates (experiments at 20 Torr, 50 m sec^{-1}) to 100% at the lowest volume flows through the reactor (3 Torr, 10 m sec^{-1}) (Fontijn *et al.*, 1977; Fontijn and Felder, 1977a). Other details are similar to the single unit HTFFR.

For work at temperatures between 300 K and the lowest achievable reaction zone temperature in single unit HTFFRs, a third modification,

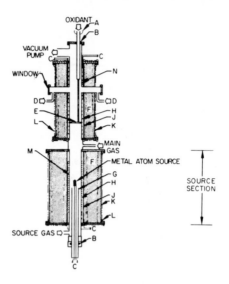

Fig. 15. Schematic of a modular HTFFR. A, alumina tube; B, sliding O-ring seal; C, to power supply; D, window sweeper gas; E, oxidant inlet and thermocouple well; F, Zircar insulation; G, source support tube; H, Pt–40% Rh resistance; J, alumina heat shield; K, brass vacuum housing; L, cooling coil; M, alumina source tube; N, alumina reaction tube. [From Fontijn and Felder (1977a, Fig. 1).]

shown in Fig. 15, is used. The source section here is of identical design to that of Fig. 14; it is followed by a short (≈ 5 cm) uninsulated copper section where room-temperature main gas may be introduced. The combined gas flow emanates from this section at $T \lesssim 400$ K and flows into the independently heated reaction tube section. This section, 20-cm long to the observation tube windows, has an alumina reaction tube and is similar in design to the single unit HTFFRs except that it has only two independently controlled heating zones. This, the latest of the HTFFR designs, has been used at temperatures from about 400 to 1400 K (Fontijn and Felder, 1977a; Felder and Fontijn, 1978). It basically represents the most versatile of the HTFFRs and there appears no reason why with suitable modifications its usefulness cannot be extended to ≈ 2000 K.

C. Atomic Concentration Measurements

1. Measurement of Absolute [Me] by Resonance Absorption Spectrophotometry

Absolute [Me] is obtained from the observed attenuation of hollow-cathode lamp Me resonance lines, using a calculation method similar to that described by Clyne and Nip (Chapter 1). At room temperature and in the low pressure operating regime of "regular" flow tubes, the utility of such calculations for absolute concentration measurements depends critically on line shape matching between the lamp emission and absorbing atoms. However, the elevated temperature and pressure of HTFFRs result in considerable absorber line broadening and this allows simplified treatment or neglect of many of these line shape matching problems such as hyperfine structure effects. In this section the line shapes and operating characteristics of the hollow cathode source are discussed first and then the calculation procedure for elevated temperature and pressure [Me] determinations is outlined.

Hollow-cathode lamps are commercially available since they are used extensively for analytical chemistry applications. Typically, a cup-shaped cathode is fabricated from the element(s) whose emission lines are desired or such element(s) are placed on the inner surfaces of an, e.g., Al cup. The lamps are sealed with an inert gas at a pressure of ≈ 1 Torr [typically the fill gas is Ne, although special applications may require other fill gases or higher pressure (Bruce and Hannaford, 1971)]. In operation, a normal glow discharge is established between the cathode and a ring- or pin-shaped anode. The lamps can be run dc or modulated. Emission arises mainly in the negative glow region of the discharge (Bruce and Hannaford, 1971). For several lamps which have been directly investigated using interferometric techniques, the emission lines can be characterized by a Doppler temperature between 350 and 700 K (Bruce and Hannaford, 1971; Wagenaar and deGalan, 1973; West and Human, 1976) when run at low (5–25 mA) dc currents with some dependence of this width upon the element excited and the lamp's manufacturer. At higher currents, broadening due to self-absorption and increased lamp Doppler temperature is noted. In the low current dc regime, $k_o m$ [see Clyne and Nip (Chapter 1) for a discussion of this notation] values for the self-absorbing layer in a Ca lamp were found to be in the 0.1–1.5 range (Bruce and Hannaford, 1971; Wagenaar and deGalan, 1973), increasing with increasing operating current. Modulation of the lamp current ($\approx 50\%$ duty cycle) results in narrower, less self-reversed lines than does dc operation; self-reversal and line width increase with "on" time during the modulation and decrease with modulation frequency (Piepmeier and deGalan, 1975). Direct cooling (Vidale, 1960) and the use of small i.d. (<0.5 cm) hollow cathodes (Bruce and Hannaford, 1971; Piepmeier and deGalan, 1975) also reduce the emission line widths apparently by allowing a more rapid rate of condensation of absorbers on the cathode surface.

Hyperfine structure due to nuclear spin and isotope shifts has a decisive effect on the emission profiles in the low current regime because line broadening effects are small (Bruce and Hannaford, 1971; Wagenaar and deGalan, 1973; Piepmeier and deGalan, 1975; West and Human, 1976; Vidale, 1960;

Tellinghuisen and Clyne, 1976). The complete spectrum of an emitted line can be described by a combination of the individual hyperfine components (Mitchell and Zemansky, 1934), but lack of information on splittings of many transitions prevents such detailed treatment in most cases. Large numbers of closely spaced and therefore unresolved hyperfine components make the detailed treatment difficult even in cases where the hyperfine structure is known (Clyne and Nip, Chapter 1). In such cases, the low current lamp emission can be adequately characterized by grouping those hyperfine components whose splittings are less than the Doppler half-width and describing the lamp output profiles as a combination of these groups using a Gaussian (Doppler) type function (Bruce and Hannaford, 1971; Vidale, 1960).

The HTFFR measurements are carried out using the low lamp current regime (typically 5–15 mA). A Doppler temperature of ≈ 600 K is assumed for the lamp emission. This is somewhat on the high side of that suggested by the measurements of isolated lines to allow for the hyperfine broadening and small self-reversal effects (Clyne and Nip, Chapter 1). The bandpass of the detection system (> 0.2 nm) is sufficiently wide so that these groups of lines are completely passed and not resolved.

The absorption line contains the same structures as the emission line; however, the high T and moderate P of many of the experiments make Doppler and Lorentz (pressure) broadening dominant effects. A Voigt profile (Mitchell and Zemansky, 1934) is used to describe the absorber line shape

$$k_v l = (k_o l) \frac{a_0}{\Pi} \int_{-\infty}^{\infty} \frac{e^{-y^2}}{a_0^2 + (\omega - y)^2} \, dy \tag{18}$$

The symbols here have the same meaning as in Chapter 1; the corresponding equation there is the more simple one giving the Gaussian profile:

$$k_v = k_o \exp[-\omega^2] \tag{19}$$

The parameter a_0 is a measure of the Lorentz broadening effect

$$a_0 = \Delta v_L / \Delta v_D (\ln 2)^{1/2} \tag{20}$$

where

Doppler broadened half-width $= \Delta v_D = 7.2 \times 10^{-7} v_0 (T/M)^{1/2} \tag{21}$

Lorentz broadened half-width $= \Delta v_L = \dfrac{2}{\Pi} [M] \left(2\Pi R T \left(\dfrac{1}{M} + \dfrac{1}{M_B} \right) \right)^{1/2} \sigma_L$

$$\tag{22}$$

and v_0 is the line center frequency (sec^{-1}), T is the reaction gas temperature, M, M_B are the molecular weights of Me and bath gas, respectively, R is the universal gas constant, and σ_L is the pressure broadening cross section. With this form of the absorption and the emission line shape discussed above, the fractional absorption

$$A = 1 - \frac{I}{I_0} = \frac{\int_{-\infty}^{\infty} \exp[-(\omega/\alpha)^2][1 - \exp(-k_v l)]\, d\omega}{\int_{-\infty}^{\infty} \exp[-(\omega/\alpha)^2]\, d\omega} \qquad (23)$$

where k_v is given by Eq. (18), $\alpha = (T_l/T)^{1/2}$, and $T_l = 600$ K, the lamp temperature. This integral is evaluated numerically using k_v values for $0 \le a_0 \le 0.5$ taken from Hummer (1965). Results are obtained as values of A for input values of $k_0 l$.

As an example of the results we will use the study of Sn reactions in N_2. Figure 16 shows the effect of temperature on the absorption at low pressure ($a_0 = 0$, $P \le 10$ Torr), while Fig. 17 shows the effect of pressure broadening on the absorption at 1000 K (midrange HTFFR temperatures) for $0 \le a_0 \le 0.5$. Using $\sigma_L \approx 100 \times 10^{-16}$ cm^2 (Lovett and Parsons, 1977), the pressure range covered by these values of a_0 is $0 \le P_{N_2} \le 140$ Torr at 1000 K, easily encompassing that used in HTFFR experiments. The upper scales in Figs. 16 and 17 have been labeled to show values of $[Sn(^3P_0)]$ corresponding to I/I_0 and percent absorption at the 286.4 nm resonance line; total [Sn] is obtained from the assumption of thermal equilibrium among the spin–orbit states of

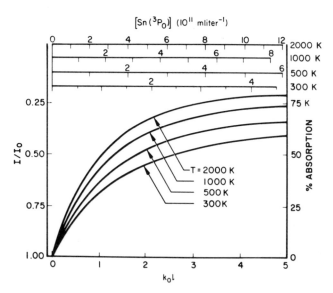

Fig. 16. Effect of reaction zone temperature on absorption at low pressure ($a_0 = 0$; $P \le$ 10 Torr). The upper abscissa $[Sn^3P_0]$ is given at the T indicated; $T_l = 600$ K. [Reprinted from A. Fontijn and W. Felder, *J. Phys. Chem.* (1979). Copyright by the American Chemical Society.]

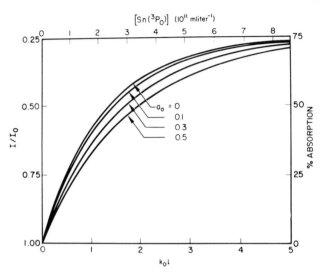

Fig. 17. Effect of bath gas pressure on absorption at 1000 K. The values of a_0 are those for the pressure range $0 \leq P \leq 140$ Torr for Sn in N_2 bath gas; $T_l = 600$ K. [Reprinted with permission from A. Fontijn and W. Felder, *J. Phys. Chem.* (1979). Copyright by the American Chemical Society.]

the Sn electronic ground state [$^3P_0$, $^3P_1$ (1692 cm^{-1}), and $^3P_2$ (3428 cm^{-1}) (Moore, 1958)] and the relationship (Mitchell and Zemansky, 1934)

$$k_o l = \left(\frac{4 \ln 2}{\Pi}\right)^{1/2} \frac{\Pi e^2}{M_e c \Delta \nu_D} [Sn]_{J=0} f_{J=0} l$$

$$= 1.15 \times 10^{-11} [Sn]_{J=0} f_{J=0} l \qquad \text{(at 1000 K)} \qquad (24)$$

where the symbols are defined as in Chapter 1.

This method for determining absolute [Me] is particularly suitable for the high temperature, moderate pressure regime covered by HTFFRs. The large Doppler broadening effect of the high temperatures and the somewhat smaller pressure broadening are quite forgiving of the vastly simplified treatment used to characterize lamp emission. Thus, the curves presented in Figs. 16 and 17, pertaining to $T > T_l$, are expected to be most accurate, while those for lower temperatures are the least accurate.

2. Resonance Line Absorption Measurements of [Me]$_{rel}$

The determination of [Me]$_{rel}$ for HTFFR kinetic measurements carried out in absorption is based on the assumption that the Lambert–Beer law is obeyed, i.e.,

$$\ln(I/I_0) = C[Me]_{rel} \qquad (25)$$

where I and I_0 are measured hollow-cathode lamp intensities in the presence and absence of Me in the absorption path, respectively, and $C = \ln(I/I_0)_i$. [Note that in plots of kinetic data such as Figs. 3 and 5, $\ln(I_0/I)$ is used simply to obtain a negative slope indicating the consumption of Me.] It is recognized that Eq. (25) is valid under limited conditions, and the range of its application and corrections to it in near room-temperature kinetic studies have been the subject of a number of publications (e.g., Clyne and Nip, Chapter 1; Davis and McFarlane, 1977; Phillips, 1976; Bemand and Clyne, 1973; Donovan *et al.*, 1975; Husain and Norris, 1977b; Foo *et al.*, 1975). As pointed out in Section II,C,1, however, the higher temperatures encountered in HTFFR work result in quite broad (mostly Doppler broadened) absorber line shapes which extend the range of [Me] over which Eq. (25) is valid. This moderating effect of temperature can be seen in Fig. 18, derived using Eq. (18) and taking $T_1 = 600$ K, $a_0 = 0$. The plots show that $\ln(I/I_0)$ has an increasing linear range as

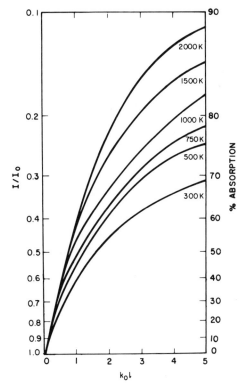

Fig. 18. Effect of reaction zones temperature on relative absorption showing that deviations from Lambert–Beer behavior become smaller with increasing T. For this plot $a_0 = 0$ ($0 \leq P \leq 10$ Torr) and $T_i = 600$ K.

temperature increases and that $[Me]_i \propto k_o l \lesssim 1$ (corresponding to 40–60%
absorption) presents quite linear relationships for kinetic measurements for
$T > T_l$; thus reasonable precision can be expected.

There is currently a controversy in the kinetics literature concerning one
particular modified form of the Lambert–Beer law proposed and used by
Donovan et al. (1975), Husain and Norris (1977b), and Foo et al. (1975) to
analyze near room-temperature kinetic results obtained in flash-photolysis
resonance absorption studies. These workers employ an empirically derived
constant, γ (≤ 1), to modify Eq. (25) as follows:

$$\ln(I/I_0) = C'[Me]_{rel}^{\gamma} \qquad (26)$$

If Eq. (26) held, rate coefficients obtained by resonance absorption would have
to be multiplied by $1/\gamma$ to obtain "true" values. However, Eq. (26) must be an
erroneous formulation (Davis and McFarlane, 1977; Phillips, 1976; Bemand
and Clyne, 1973) since it does not reduce to the simple Lambert–Beer law for
small values of absorption (unless, of course, $\gamma \approx 1$, which is a fact of numerous
experimental investigations). Consequently, curve-of-growth calculations
which have proven adequate over the years to describe the dependence of
absorption on atom density cannot be matched to the predictions of Eq. (26)
(Phillips, 1976; Bemand and Clyne, 1973; Donovan et al., 1975). Davis and
McFarlane (1977) have performed calculations to investigate the validity of
Eq. (26) for a wide variety of source and absorber line shape combinations and
conclude that in cases where significant line broadening effects operate (such
as in the elevated T regime), γ is constant and nearly unity, i.e., the Lambert–
Beer law is obeyed. When broadening effects are small, they find that γ varies
widely with experimental conditions of P, T, and $[Me]$ and that a constant
single value of γ less than unity cannot be applied.

Despite the inappropriateness of Eq. (26), the necessity for caution in
applying Eq. (25) near room temperature and at low pressure still remains. In
this context some form of validation of the experimental applicability of
Eq. (25) is always desirable. For a few nonrefractory species (cf. Chapter 1),
resonance absorption lamps can be calibrated against chemical titrations in
situ, but for most atoms, and especially for refractory species, such titrations
are not yet available (see, however, Section II,C,4). The use of equilibrium
vapor pressure furnaces for calibrating absorption lamps of refractory species
is not a desirable alternative to titration since the T and P conditions accessible with such devices are not generally those of the experiment. One
solution to this dilemma lies in a more extensive use of resonance fluorescence
(cf. Section II,C,5) as a complementary kinetic tool. Such measurements will
not completely supplant absorption measurements because of their limitation
by background emission at high temperature (cf. Section II,A,1) and because
of the experimental simplicity of absorption measurements. However, since

at low [Me] (a condition easily achievable!), fluorescence intensity is linearly proportional to [Me] and this linearity is unaffected by lamp parameters (Chapter 1), agreement between kinetic results obtained using absorption and fluorescence techniques can provide the desired validation of absorption measurements. The use of fluorescence has the additional benefit of extending the range of [Me] covered in a set of measurements (cf. Section II,A,1). In HTFFR studies, rate coefficients obtained using both resonance absorption and resonance fluorescence for the determination of $[Al]_{rel}$ at similar flow conditions were indistinguishable at all reaction zone temperatures where both techniques could readily be used, i.e., from ≈ 500 to ≈ 1500 K (Fontijn et al., 1977; Fontijn and Felder, 1977a).

3. Blended Line Absorption Measurements

In some cases, two or more closely spaced multiplet lines in hollow-cathode lamp emission make it difficult to spectrally isolate a single fine structure line for use in kinetic measurements, but these lines are too broadly spaced to be merged by broadening effects as is the case for the hyperfine structure lines. Furthermore, the fine structure lines are characterized by distinct transition oscillator strengths rather than by a single transition oscillator strength distributed over hyperfine components. While it is usually better to try to separate the lines, it is of some importance to consider the implications of making blended line measurements. Two types of such measurements can be considered: $[Me]_{rel}$ during reaction and [Me] in the absence of reaction. The latter measurement is often practical, the former only sometimes.

The Lambert–Beer law has to be modified for blended fine structure lines. In order to obtain meaningful kinetic data using such lines for $[Me]_{rel}$ measurements, extension of the simple absorption treatment is necessary. The blended emission intensity of two lines is

$$I_0 = I_{0,1} + I_{0,2} \tag{27}$$

In the presence of absorber, the attenuated intensity is

$$I = I_1 + I_2 \tag{28}$$

and the density is

$$[Me] = [Me]_1 + [Me]_2 \tag{29}$$

where the subscripts 1 and 2 refer to the fine structure levels. It can be shown that the observed attenuation can be written as

$$\ln(I/I_0) = \ln\{(I_1/I_{0,1})(1 + \underline{C})^{-1} + (I_2/I_{0,2})(1 + 1/\underline{C})^{-1}\} \tag{30}$$

where $\underline{C} = I_{0,2}/I_{0,1}$. From the Lambert–Beer law,‡

$$[Me]_1 = C_1 \ln(I_1/I_{0,1}), \qquad [Me]_2 = C_2 \ln(I_2/I_{0,2}),$$

i.e.,

$$[Me] = C_1 \ln(I_1/I_{0,1}) + C_2 \ln(I_2/I_{0,2}) \tag{31}$$

To solve these equations to obtain [Me] from a single measurement of the blended lines, a relationship between $[Me]_1$ and $[Me]_2$ has to be found. In cases where the energy spacing between the fine structure levels is small compared to kT, it is reasonable to assume that these levels are in thermal equilibrium and that the Boltzmann equation holds, i.e.,

$$[Me]_2 = [Me]_1(g_2/g_1) \exp(-\Delta\varepsilon_{12}/kT) \tag{32}$$

which may be rewritten (using the Lambert–Beer law):

$$\ln(I_2/I_{0,2}) = C_3 \ln(I_1/I_{0,1}) \tag{33}$$

where

$$C_3 = (C_1/C_2)(g_2/g_1) \exp(-\Delta\varepsilon_{12}/kT)$$

and where $\Delta\varepsilon_{12}$ is the energy spacing between the fine structure levels. Using Eq. (33) we can rewrite Eqs. (30) and (31) as

$$\ln(I/I_0) = \ln\{(I_1/I_{0,1})(1 + \underline{C})^{-1} + (1 + 1/\underline{C})^{-1} \exp[C_3 \ln(I_{0,1}/I_1)]\} \tag{34}$$

$$[Me] = C_1 \ln(I_1/I_{0,1})\{1 + (C_2 C_3/C_1)\} \tag{35}$$

and

$$[Me]_{rel} \propto \ln(I_1/I_{0,1}) \tag{36}$$

To obtain $[Me]_{rel}$, the measured blended line absorption I/I_0 is used and Eq. (34) is numerically solved to obtain a value of $I_1/I_{0,3}$ which is then used in (36). The values of C_1 and C_2 depend upon the oscillator strengths of the transitions, while C_3 depends on the generally more accurately known value of the ratio of the oscillator strengths. Thus, in determining $[Me]_{rel}$, only this ratio is needed. This method depends critically upon the assumption of thermal equilibrium for its validity. If there are great differences in reactivity among the fine structure levels, then collisional repopulation of the reaction-depleted level must be rapid compared to the reaction in order to maintain equilibrium. If such rapid repopulation occurs, then valid $[Me]_{rel}$ and hence valid overall rate coefficient measurements will be obtained using blended lines. However, if the fine structure levels are widely spaced in energy, such a

‡ No loss of generality in this method is incurred by using the Lambert–Beer law here for algebraic simplicity. Equation (23) of Section II,C,1 could equally well be used to express $[Me]_1$ and $[Me]_2$ in terms of absorption of the individual fine structure lines.

repopulation may not easily be effected, especially at low reaction zone pressures, and incorrect rate coefficient values will be obtained. Because of this potential problem and because even an accurate blended line measurement can lead only to an overall rate coefficient reflecting a blend of the reactivities of the fine structure levels measured, this is not, in general, a preferred method for rate coefficient measurement.

Under HTFFR operating conditions in the absence of oxidizer, equilibrium between Me fine structure lines is often found to be maintained at the observation window. The absorption of blended lines can then be conveniently used in absolute initial (i.e., in the absence of reactions) concentration measurements. To do this, one simple method is to prepare a graph of blended lines absorption against [Me] in advance by solving Eqs. (34) and (35) for selected values of I/I_0. Then, experimental values of I/I_0 can be used to read off [Me] from the graph. Such a plot is shown for Ge at 900 K and low pressure ($a_0 = 0$) in Fig. 19 using the full absorption equation [Eq. (23)] given in Section II,C,1 to obtain the analogs of Eq. (31). The absorption is of

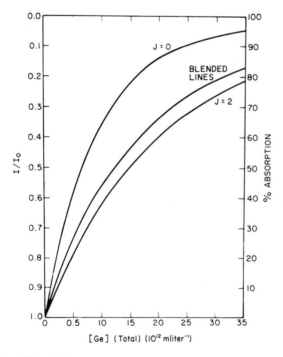

Fig. 19. Use of blended lines absorption for determining absolute atom density of Ge at $T = 900$ K and $P = 4$ Torr Ar ($a_0 = 0$). The lines observed are Ge ($4^3P_0 \rightarrow 5^3P_1^0$ and $4^3P_2 \rightarrow 5^3P_2^0$) at 265.2 nm. [Ge] (total) is the atom population in all fine structure levels, $4^3P_{0,1,2}$ at 900 K. $T_l = 600$ K.

the blended lines near 265.2 nm. For comparison, the absorptions of the individual fine structure lines are also shown. The abscissa of the figure is [Ge] (total), i.e., the density of Ge in all populated fine structure levels $^3P_{0,1,2}$ as calculated based on equilibrium.

4. Absolute Concentrations via Chemical Titrations

As discussed in Sections II,A,2 and II,A,3, photon yield and k_Q^{Me} measurements may require considerably higher [Me] than can be measured in absorption. For such [Me], chemiluminescence titration methods can be used. They are discussed more extensively in Chapter 1 and require a fast titration reaction and a second, usually slower, indicator reaction. For Sn the following procedure has been developed. The reaction

$$Sn + NO_2 \longrightarrow SnO + NO \qquad (37)$$

in the $\approx 900-1100$ K temperature range was found to be adequately rapid; $k_{37} \approx 1 \times 10^{-10}$ mliter molecule^{-1} sec^{-1}. At ≈ 300 K this reaction has a rate coefficient some 10^3 times smaller; thus the method presented here is not suitable for that temperature. The Sn/NO_2 reaction is apparently not chemiluminescent; however, the Sn/N_2O reaction produces an intense chemiluminescence. Thus, by introducing NO_2 through the movable inlet, Fig. 1, and by introducing some N_2O at one of the observation windows a chemiluminescent titration technique is obtained (Fontijn and Felder, 1977b). A typical titration plot is shown in Fig. 20. The end point was determined to be independent ($\pm 5\%$) of NO_2 inlet-to-observation port distance (reaction time) and SnO emission band system observed (Fontijn and Felder, 1977b).

As noted briefly in Section II,A,2, there is a lower limit to the concentration of the atoms which can be titrated; in the present case, where flow velocities ≥ 10 m sec^{-1} (and hence reaction times $\leq 2 \times 10^{-2}$ sec) are required for Sn transport even for low [Sn], this lower limit was found to be [Sn] $\approx 1 \times 10^{13}$ mliter^{-1}. The limitations imposed on the rate coefficients of titration reactions by these transport conditions are more stringent than those encountered at similar atom concentrations in conventional electrical discharge flow system titrations, where lower average gas velocities and hence longer reaction times can be used. For a flow velocity of 60 m sec^{-1}, this is illustrated by the calculated titration plots shown in Fig. 20. These plots were calculated using the time-integrated form of the bimolecular rate equation and the measured $[Sn]_0 = 4.4 \times 10^{13}$ mliter^{-1}, for $k_{37} = 3 \times 10^{-10}$ mliter molecule^{-1} sec^{-1}, i.e. approximately gas kinetic, and 3×10^{-11} mliter molecule^{-1} sec^{-1}. Even the latter, still very large, rate coefficient value (approximately equal to that for the N/NO titration reaction) would cause deviations from linearity at such low $[NO_2]_0$ as to make the titration at best a qualitative procedure. The observation, Fig. 20, that the experimental

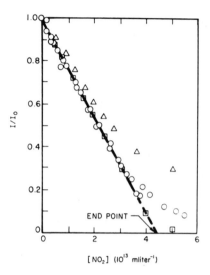

Fig. 20. Titration of Sn with NO_2 using SnO(a–X) chemiluminescence at 579 nm as indicator. $T = 895$ K; $[Ar] = 9.7 \times 10^{16}$ mliter^{-1} (9 Torr); $\bar{v} = 62$ m sec^{-1}, $t = 1.6$ msec. End point indicates $[Sn]_0 = (4.4 \pm 0.3) \times 10^{13}$ mliter^{-1}. ○ Experimental data, □ Calculated using $k_{37} = 3 \times 10^{-10}$ mliter molecule^{-1} sec^{-1}, △ Calculated using $k_{37} = 3 \times 10^{-11}$ mliter molecule^{-1} sec^{-1}. [From Fontijn and Felder (1977b, Fig. 1). Copyright of North Holland Publishing Co.]

titration plot falls between the two calculated extremes confirms the approximate value of k_{37} of 1×10^{-10} mliter molecule^{-1} sec^{-1}.

5. Resonance Fluorescence Measurements of $[Me]_{rel}$

The measurement of $[Me]_{rel}$ by resonance fluorescence is entirely similar to the procedures discussed by Clyne and Nip in Chapter 1. The measurement is aided at elevated temperatures by the thermal broadening effects on the absorber lineshape. Lamp broadening effects such as self-absorption decrease intensity at the line center, while increasing it in the wings of the line. This intensity, which may not be useful for exciting fluorescence if the lamp line width is broader than the absorber line width, will be recovered as the absorber line width broadens. It can also be noted that the analysis of results in the case of blended fluorescence lines is much simpler than in the case of absorption. This is so because of the linear (rather than logarithmic) relationship between fluorescence intensity from the individual lines and [Me]. This in turn results in a linear proportionality between the blended fluorescence intensity and [Me]. As with absorption, however, blended line measurements yield overall rate coefficients, and care must be taken to insure equilibration of fine structure levels under reaction conditions.

D. Flow Considerations

1. Reaction Time and Flow Profile

As discussed in Section II,A,1, the determination of overall rate coefficients requires knowledge of the time–distance relationship in the flow tube, i.e., of the flow profile. Such considerations have been extensively investigated for "low" temperature flow tubes (e.g., Clyne and Nip, Chapter 1; Kolts and Setser, Chapter 3; Walker, 1961; Westenberg, 1973; Ferguson *et al.*, 1969; Ogren, 1975; Bolden *et al.*, 1970; Huggins and Cahn, 1967; Farragher, 1970; Kaufman, 1961; Hoyermann, 1975) and have been recently summarized by Hoyermann (1975). Basically, two extreme cases have been treated, namely, plug flow (Walker, 1961; Westenberg, 1973; Ferguson *et al.*, 1969; Kaufman, 1961; Hoyermann, 1975) and fully developed laminar (parabolic) flow (Ferguson *et al.*, 1969; Ogren, 1975; Bolden *et al.*, 1970; Huggins and Cahn, 1967; Farragher, 1970; Kaufman, 1961; Hoyermann, 1975); in both cases, the combined effects of axial and radial diffusion with wall losses have been studied. A convenient expression given by Ferguson *et al.* (1969) shows the corrections to the simple, uniformly mixed, constant flow velocity plug flow condition derived from realistic gas dynamic models to be applied to measurements of $[Me]_{rel}$ in order to calculate rate coefficients. In terms of rate (i.e., k_{ps_1}) measurements, this expression can be written as

$$k_{ps_1} = \beta a \bar{v} \left(1 + \sum_{i=1} \alpha_i \right) \qquad (38)$$

where β is a correction factor dealing with pressure measurement which for the moment will be neglected, but is discussed separately in Section II,D,2; the correction factors, α_i, are α_1, for the velocity profile; α_2 for axial diffusion; and α_3, for reagent inlet effects. The factor α_1 is zero for plug flow and approximately 0.6 for fully developed laminar flow (Ferguson *et al.*, 1969). Axial diffusion corrections are to a first approximation given by (Kaufman, 1961)

$$\alpha_2 = (1 + \alpha_1) a D_{Me} \bar{v}^{-1}$$

For the multiperforated distribution ring inlet used in HTFFR kinetic work, α_3 is taken to be zero, cf. Farragher's (1970) arguments for this inlet configuration, i.e., nearly instantaneous planar mixing of reagents is assumed to occur. Inserting these values into Eq. (38) we recover Eq. (3)

$$k_{ps_1} = \eta a \bar{v}(1 + a D_{Me} \bar{v}^{-1}) \qquad (3)$$

where $\eta = 1 + \alpha_1$ is the flow velocity correction factor. The parenthetical term in Eq. (3) represents a straightforward correction ($\leq 10\%$) which usually can be neglected; however, the η factor needs further consideration. In the

HTFFR neither plug nor parabolic profiles apply and an intermediate situation exists (leading us to the discussed choice of using $\eta = 1.3$) as the following arguments show.

Plug flow (Chapter 1) represents the beginning of fully developed laminar flow, i.e., plug flow is the form of the gas flow near the entrance to a duct such as a flow tube. It is often referred to as "entrance flow" (Ferguson et al., 1969; Eckert and Drake, 1959; Brokaw, 1960). Because of the growth of a boundary layer along the flow tube walls, entrance flow will smoothly change over to fully developed (parabolic velocity profile) laminar flow. The tube length required for this complete development is (Ferguson et al., 1969; Dushman and Lafferty, 1962)‡

$$L_e = 0.114 \, l \, \text{Re} \tag{39}$$

where Re is the Reynolds number of the flow:

$$\text{Re} = 5.67 \times 10^{-5}(MQ_M/l\bar{\eta})$$

i.e.,

$$L_e = 6.5 \times 10^{-6}(MQ_M/\bar{\eta})\text{cm} \tag{40}$$

where Q_M is the bath gas flow in mliter (STP) sec^{-1} and $\bar{\eta}$ is the bath gas viscosity in poise. For typical HTFFR conditions: $50 \leq Q_M \leq 500$ mliter (STP) sec^{-1} at $T = 1000$ K with Ar [$M = 39.94$ g mole^{-1}, $\bar{\eta}$ (1000 K) $= 540 \, \mu$ poise (Brokaw, 1960; Hirschfelder et al., 1963; Reid and Sherwood, 1966)], we obtain $82.6 \leq \text{Re} \leq 826$ and $24 \leq L_e \leq 240$ cm.

This entry length may be compared with the physical dimensions of the HTFFRs discussed in Section II,B. For the full-sized HTFFR, Fig. 9, it should be kept in mind that in the region of the source support tube, gas flow is constrained to an annulus of inner diameter $0.75 \, l$ and outer diameter l, i.e., to an area $\frac{7}{16}$ the area of the unobstructed tube. This constriction causes an increase in the gas flow velocity above that for the unobstructed tube by the inverse of the ratio of the areas, i.e., a factor of approximately two. At the downstream end of the source support, i.e., at the metal atom source, the annulus of higher velocity gas begins to relax to fill the tube. In the first instance, there is a velocity defect on the flow tube centerline, with the gas near the walls moving faster than the centerline gas. Swirling around the source, the gas flow assumes a plug profile and a boundary layer begins to grow at the walls. We may thus take the "entrance" to the HTFFR to be at the downstream end of the source tube. Hence, the distance from the flow tube "entrance" to the reaction zone boundary is decreased by the length of insertion of the source tube into the

‡ The uncommon symbol l is used here for diameter since l is optical path length as defined in Section II,C,1 in accord with optical practice and this path length in the present case equals the tube diameter.

HTFFR, typically ≈ 20 cm, and the effective length for flow development of the full-sized HTFFR to the upstream end of the reaction zone is ≈ 30 cm. For the modular designs, Figs. 14 and 15, this distance is either ≥ 15 or 10 cm depending on whether additional bath gas is introduced downstream of the vaporizer section. It is clear from comparison of these dimensions with L_e that, for most of the range of HTFFR operations, only partially developed laminar flow can be realized and that development will be ongoing throughout the length of the reaction zone (20–23 cm).‡

At the other extreme, the criteria for plug flow in a tubular reactor have been developed by Walker (1961) and Westenberg (1973) for homogeneous and heterogeneous first-order (or pseudo-first-order) kinetics in terms of the dimensionless parameters;

$$u = \bar{v}l/2D, \qquad B^2 = k_{ps_1}l^2/4D, \qquad \delta = 8D/k_w l^2 \qquad (41)$$

With a series of numerical comparisons, Walker concludes that plug flow analysis is valid (to within $\approx 5\%$) for $\delta > 1, B^2 < 1$, and $u < 20$. The values of u and δ are more important (Walker, 1961) in determining this validity than is the value of B^2. Again, we consider a typical HTFFR experiment to compare with these criteria. From our study of the Al/CO$_2$ reaction (Fontijn and Felder, 1977a) carried out in Ar bath, the parameters in the above at 1000 K are

$$\bar{v} = 2500 \text{ cm sec}^{-1}, \qquad D \approx 40 \text{ cm}^2 \text{ sec}^{-1} \qquad [P = 25 \text{ Torr (Hirschfelder}$$
$$\textit{et al.}, 1963; \text{Reid and}$$
$$\text{Sherwood}, 1966)],$$

$$k_w \approx 200 \text{ sec}^{-1}, \qquad k_{ps_1} \approx 500 \text{ sec}^{-1}$$

which gives $u \approx 80, B^2 \approx 20$, and $\delta \approx 0.25$. From this example, and since $\bar{v} \geq 1000$ cm sec^{-1} has been found to be a practical lower limit for Me transport, and $k_w < 100$ sec^{-1} is seldom encountered, the combination $u < 20, \delta > 1$ is rare for HTFFR operation. Thus, for most conditions covered in HTFFR studies, the plug flow description is not appropriate either.

That our choice of $\eta = 1.3$ based on all these considerations is a reasonable one§ also follows from an inspection of the individual HTFFR rate coefficient measurements (see the original data in the references quoted in Table I). Such inspection shows that the scatter in rate coefficient values

---

‡ It might be argued that the analysis of kinetic results could be made more exact by simply increasing the distance from the source to the reaction zone in the HTFFR, thus ensuring fully developed laminar flow. It is somewhat doubtful if such a reactor could indeed be used; considerations of materials, metal transport, the height of the laboratory, and excessively high costs have prevented us from looking further into building such a reactor.

§ The effect of this choice on accuracy of the results is discussed in Section II,E,2.

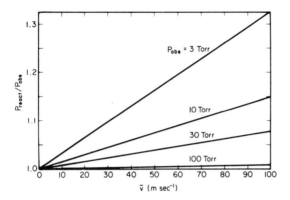

Fig. 21. HTFFR pressure measurement correction chart. P_{react} = average reaction zone pressure; P_{obs} = measured pressure at downstream pressure tap.

obtained under nearly identical flow conditions is comparable to that which pertains at widely different flow conditions, i.e., exact values of η would not reconcile this scatter. Variable wall reactivity due to continuous formation of thin wall deposits‡ may be an additional factor causing this scatter.

2. Pressure

The exigencies of HTFFR construction, i.e., insertion of the source tube through the upstream (bottom) end and movable reagent inlet at the other end, make it most practical to locate the pressure tap downstream from the reactor in the vertical arm leading to the pump. As a result, pressures somewhat lower than the actual reaction pressure are being read. (The largest pressure drop occurs at the tee connecting the reactor and this arm, Fig. 9.) To arrive at the correct reaction zone pressure, the movable oxidant inlet can be replaced with a pressure probe for calibration experiments and a calibration chart can be prepared that gives the reaction pressures corresponding to the pressures read on the permanent pressure gauge. The movable pressure probe consists of a 0.3-cm-diameter alumina tube with a 0.1-cm-diameter hole located 0.4 cm from its blunted upstream end, cf. the recommendations of Perry et al. (1963). The required corrections were found to be, within experimental accuracy, independent of reaction zone temperature and are largest at low pressure and high flow velocity. Figure 21 shows some representative results plotted as the ratio of average reaction zone pressure to the observed downstream pressure against the nominal $\bar{v}$, where $\bar{v}$ is based on the permanent

‡ The surface roughness caused by such deposits also works against development of parabolic flow.

pressure tap reading. This ratio P_{react}/P_{obs} is the inverse of the factor β given in Eq. (38). For ultimate accuracy the pressure drop ΔP over the ≈ 20-cm-long reaction zone should also be taken into account. In practice this factor, which cannot readily be corrected for, may be neglected at the flow conditions used and the overall accuracy attainable. To give an idea of the magnitude of this drop for a midrange velocity (35 m sec^{-1}) for $P = 3, 10, 30, 100$ Torr, $\Delta P = 0.21, 0.30, 0.35, 0.40$ Torr, respectively.

3. Temperature

a. Source Support Tube Influence   The presence of this tube has an important effect on attaining high, uniform, reaction zone temperatures. It nearly fills the upstream end of the reactor and forces the entering room temperature gas into close contact with the hot HTFFR walls. Two additional factors enhance gas heating in the source section: (i) the source support tube itself is hot due to heating from the reactor tube walls and (ii) the constraining effect on the flow caused by this tube results in high Reynolds numbers [Re(source section) $\approx 8 \times$ Re(unobstructed tube section)] and local turbulence in the flow to enhance heat transfer, i.e., boundary layer formation is inhibited and more effective heat transfer results. The relatively large diameter of the source support tube (1.8-cm o.d.) thus is advantageous in this respect. In preliminary experiments it was verified that this tube yields higher reaction zone tempera-tures than did a smaller diameter (0.6-cm) tube. The effect of the source itself is an extension of the effect of the source support tube and when the source is heated directly additional heating of the bulk gas, of course, results.

b. Thermocouple Accuracy   For the flow conditions used in HTFFR studies (Mach numbers ≤ 0.2; Re with respect to reaction tube diameter ≤ 2300 at 300 K, ≤ 1000 at the other temperatures; Re with respect to the coated thermocouple, TC, diameter ≈ 0.13 times those with respect to the reaction tube diameter), the analysis given by Moffat (1962) shows that radiative heat transfer between the TC junction and the reactor walls is the potential major source of uncertainty. As a first step in evaluating the significance of this uncertainty, calculations were made of axial and radial temperature distribu-tions for unobstructed laminar flow of Ar in a uniformly heated HTFFR by solving the Graetz equation (Eckert and Drake, 1959).‡ These calculations showed that (i) it is indeed possible to achieve gas temperatures indicated by the TC within the heated distance available in the HTFFR and (ii) the flow conditions used satisfy the criteria that the calculated mean gas temperature at the upstream boundary of the reaction zone be within 5% of the nominal

---

‡ Transport properties of (Ar) used as input to this calculation were obtained from Brokaw (1960).

wall temperature and that the extremes of radial temperature variation be within 5% of this mean temperature.

Experimentally, this problem was investigated by Fontijn and Felder (1979) by comparing the readings of the shielded and unshielded centerline thermocouples of Fig. 12 for a variety of conditions. It was observed in that work that if the source section of the HTFFR is hotter than the reaction zone, the unshielded TC reads somewhat lower than the shielded, i.e., the gas was hotter than the reaction zone walls and the unshielded TC was subject to the largest cooling effect. The reverse situation obtained when the source section of the HTFFR was cooler than the reaction section. These effects were the larger the higher the volume flow rate. The largest T difference observed between these two TCs was 5%.

To arrive at the approximately correct reaction zone temperature, use is made of the work of Moffat (1962) according to which the radiation error of a TC is reduced by shielding by at least a factor of $n + 1$ where n is the number of shields. Let the temperature of the TC used in the kinetic measurements be T_1, that of the unshielded TC of Fig. 12 T_2, and the shielded one from that figure T_3. If the error in T_2 is S, then the true centerline temperature is $T_3 + S/3$ according to the $(n + 1)$ formula. Thus, the magnitude and sign of S are determined by the measurement $T_3 - T_2 = 2S/3$. Additional experiments with two unshielded TCs, one on the centerline and one 0.3 cm from the walls, as in a kinetic experiment, suggest radial temperature gradients $T_2 - T_1$ of magnitude $S/3$. Taking the average T in a cross section of the reaction tube as that in between the center and the wall, we arrive at $T_3 + S/3 - S/6 = T_3 + S/6$ as the approximate reaction zone temperature. Since these procedures are approximate, we will allow for an uncertainty of $S/3$ ($\leq 2.5\%$) in this temperature estimate.

E. Accuracy

Assessing the accuracy of the various HTFFR results discussed in previous parts of this section involves two steps. The first is to propagate random (statistical) errors through the data reduction chain (schematically shown for kinetic data in Fig. 22), using standard methods (Bevington, 1969; Pugh and Winslow, 1966; Wilson, 1952) to obtain a calculated standard deviation in the result. The second step is to add to this standard deviation an estimate of the systematic uncertainty in the experimental procedures. The latter step is more difficult to quantify than the former because, of course, careful procedures have been adopted to reduce errors of this kind to a minimum. In this section, we will describe these two steps for HTFFR measurements, giving rate coefficient procedures in detail and outlining the corresponding methods for photon yield measurements. A brief discussion of special procedures for

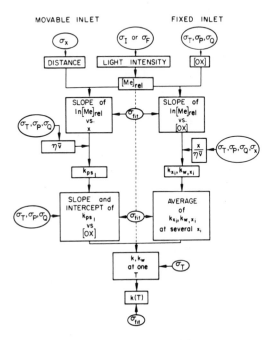

Fig. 22. Data reduction chain for HTFFR rate coefficient measurements. Contributing error sources are identified at each measurement or calculation stage. Errors are propagated through calculations as weighting factors on the input data.

situations in which (i) significant OX decomposition occurs and (ii) the use of pseudo-first-order data treatment is not valid is also given.

1. Propagation of Random Errors

a. Sources and Magnitude of Random Errors   All HTFFR results flow from five primary determinations: (i) gas flow Q; (ii) pressure P; (iii) temperature $\overline{T}$; (iv) distance x; and (v) light intensity I (or F). Uncertainties in these values together with scatter in the kinetic data ultimately determine the precision of reported results. Table II contains a list of the ranges over which these five determinations are made and the usual ranges of their respective uncertainties. These uncertainty values are estimated based on reading errors σ_r defined appropriately [e.g., as $\pm\frac{1}{2} \times$ (smallest scale division for analog meters)] and the standard deviations (scatter) in the calibration data of the measuring devices used. To cover the complete range of any measurement, several similar devices with overlapping spans are usually used and this procedure "randomizes" specific calibration errors which would otherwise cause systematic errors. We stress that scatter in the calibration data for any measuring device

Table II

Random Error Sources and Magnitudes in Primary HTFFR Measurements

Measurement (units)	Device/method used	Range used	Uncertainty range[a]
Gas flow, Q (mliter (STP) sec^{-1})	Volume displacement	0.05–250	0.01–15
	Critical flow orifices	5–500	1.0 –5.0
Pressure, P (Torr)	Closed end manometers	1–1000	0.1 –5.0[b]
	Bourdon gauges	1–50	0.1 –2.5[b]
	U-tube manometer	0.5–20	0.05–1.0[b]
Temperature, T (K)	Pt vs. Pt–10% Rh thermocouple	300–1500	10 –45[c]
	Pt vs. Pt–20% Rh thermocouple	1000–2000	30 –60[c]
Distance, x (cm)	Centimeter scale	0–25	0.05
Light intensity I (mV)	Absorption-monochromator/PMT	10–100	0.5 –2.0
F (mV)	Fluorescence-light filter/PMT	1–20	0.1 –1.0
I_{CL} (pA)	Chemiluminescence-calibrated monochromator/PMT	0.1–500	0.02–50[d]

[a] Values correspond to uncertainties at extremes of range, respectively; units are same as measurement units.

[b] Includes contribution of scatter in pressure correction data ($\pm 4\%$) discussed in Section II,D,2.

[c] Includes contributions of scatter in calibration data ($\pm 1\%$), axial temperature profile ($\pm 1\%$), and uncertainty in TC radiation correction and radial temperature profiles of $\pm 2.5\%$ discussed in Section II,D,3,b.

[d] Includes contribution of scatter in detection system calibration ($\approx \pm 5\%$) used to obtain absolute intensities.

can be validly included in estimates of random error, but that only inter-comparisons among several similar devices can minimize systematic error in the device.

b. Random Uncertainty in Calculated Quantities The primary uncertainties now are used to derive the error in calculated quantities using the propagation of errors method (Bevington, 1969; Pugh and Winslow, 1966; Wilson, 1952). The defining equation is

$$\sigma_\mu = \left[\sum_i \left(\frac{\partial \mu}{\partial \phi_i} \sigma_i \right)^2 \right]^{1/2} \tag{42}$$

Table III

Random Error in Calculated Quantities

Quantity determined	Method and formula	Contributing primary error sources	Fractional uncertainty
$[Me]_{rel}$	Atomic absorption $C[Me]_{rel} = \ln(I/I_0)$	σ_I	$\sigma^A_{[Me]_{rel}}$ $= \{2[\ln(I/I_0)]^{-2}\sigma_I^2\}^{1/2}$
	Fluorescence $[Me]_{rel} = F/F_0$	σ_F	$\sigma^F_{[Me]_{rel}} = (2\sigma_F^2)^{1/2}$
$[Me]$	Atomic absorption, Eq. (23)	$\sigma_I, \sigma_T, \sigma_P$	$\sigma_{[Me]} = \{2(\ln(I/I_0))^{-2}\sigma_I^2$ $+ (1 \times 10^{-6})T^2\sigma_T^2$ $+ (1 \times 10^{-4})P^2\sigma_P^2\}^{1/2}$
$[OX]$	Measure OX fraction X_{OX} of total density: $[OX] = X_{OX} K_0 P/T$ $[K_0$ is the conversion constant from perfect gas law to pressure units used; $X_{OX} = Q_{OX}/$ $(Q_{OX} + Q_M)]$	$\sigma_Q, \sigma_P, \sigma_T$	$\sigma_{[OX]} =$ $\{2[Q_M/(Q_M + Q_{OX})]^2\sigma_Q^2$ $+ \sigma_P^2 + \sigma_T^2\}^{1/2}$
$\bar{v}$	Total gas flow through HTFFR: $\bar{v} = [(Q_M + Q_{OX})/$ $(\pi l^2/4)]K_1 (T/P)$ $(K_1$ is the conversion constant from metering P, T to reactor P, T)	$\sigma_Q, \sigma_P, \sigma_T$	$\sigma_{\bar{v}} = \{2\sigma_Q^2 + \sigma_P^2 + \sigma_T^2\}^{1/2}$

where σ_μ is the fractional uncertainty‡ in the calculated quantity μ, ϕ_i is a primary measurement, and σ_i is its fractional uncertainty. Equation (42) is used at several points in the calculation chain shown in Fig. 22 where the primary measurement uncertainties enter the calculations. Table III shows the fractional uncertainties in the intermediate quantities $[Me]_{rel}$, $[Me]$, $[OX]$, and $\bar{v}$ which arise from using Eq. (42). These are summarized for convenience of notation since it is possible to propagate the primary uncertainties at each calculation step. To obtain an expression for the complex dependence of $\sigma_{[Me]}$ on σ_T and σ_P, Fontijn and Felder (1979) used an empirical

‡ Fractional uncertainty is the magnitude of the ratio (value of measured uncertainty) ÷ (value of measurement).

correlation deduced from Figs. 16 and 17 and additional plots like them covering the ranges $0 \leq a_0 \leq 0.5$ and $300 \text{ K} \leq T \leq 2000 \text{ K}$ to obtain the $\sigma_{[Me]}$ given in Table III. They found that for the large ranges of conditions used in HTFFRs, the [Me] which yields a given I/I_0 increases approximately as $0.001 \, T$ and $0.01 \, P$ for $T \geq T_1$ in the range $0.3 \leq I/I_0 \leq 0.95$ (the lower limit decreases as T increases).

Following the calculation chain in Fig. 22, the kinetic quantities are obtained as discussed in Section II,A,1. The introduction of the primary uncertainties as weighting factors in this chain is shown explicitly in Table IV for the steps of Fig. 22. These weighting factors are derived from Eq. (42) and are given by

$$W_\mu = (1/\sigma_\mu)^2 \tag{43}$$

When uncertainties are involved in the determination of both ordinate and abscissa values in plots leading to the calculated quantities of Table IV, the ordinate uncertainty is combined with the abscissa uncertainty as a vector sum. Weighted least-squares fitting procedures (Bevington, 1969; Pugh and Winslow, 1966; Wilson, 1952; Cvetanovic et al., 1975; Cvetanovic and Singleton, 1977a,b) are used to obtain values of a and $k_{x_i} t$ from measurements of $\ln[Me]_{rel}$ against x or [OX], respectively. Standard deviations of the fit σ_{fit} obtained from these calculations are combined as vector sums with uncertainties introduced at succeeding stages of the kinetic calculations (cf. Table IV). This procedure is repeated at each stage of data reduction until the ultimate value k at one temperature is obtained.

Special note should be taken of a fact emphasized by Cvetanovic et al. (1975) and Cvetanovic and Singleton (1977a,b) when fitting data obtained in linear form to logarithmic functions. They point out that such a change of variable requires that the weighting factors W_μ given in Table IV also have to be changed

$$W'_\mu = W_\mu \mu^2 \tag{44}$$

to correctly apply the fitting procedures in such cases.

In applying these procedures to HTFFR kinetic measurements, it has invariably been found that the largest uncertainties are those arising from fitting the raw data, i.e., from scatter. Similar magnitude errors result in values of k obtained from fitting k_{ps_1} against [OX] or from the weighted average of the $k_{x_i} t$. The source of such scatter is undoubtedly multifold but it is thought to arise in the first place from differences in the variations of wall reactivity from experiment to experiment and within an experiment. Overall, the currently achievable precision obtainable in HTFFR kinetic measurements is in the range of 25–60%.

Table IV

Random Uncertainties in Kinetic Calculations

Quantity calculated	Weighting factor on individual data	Uncertainty
a, slope of $\ln[\text{Me}]_{\text{rel}}$ vs. x (movable inlet kinetic data)	$\{(\sigma_{[\text{Me}]_{\text{rel}}}^{\text{A(or F)}})^2 + \sigma_x^2\}^{-1}$	$\sigma_a = (\sigma_{\text{fit}}^2)_{\text{slope}}^{1/2}$
$k_{x_i}t$, slope of $\ln[\text{Me}]_{\text{rel}}$ vs. $[\text{OX}]$ at one distance, x_i (fixed inlet kinetic data)	$\{(\sigma_{[\text{Me}]_{\text{rel}}}^{\text{A(or F)}})^2 + \sigma_{[\text{OX}]}^2\}^{-1}$	$\sigma_{k_{x_i}t} = (\sigma_{\text{fit}}^2)_{\text{slope}}^{1/2}$
$k_{\text{w},x_i}t$, intercept of $\ln[\text{Me}]_{\text{rel}}$ vs. $[\text{OX}]$ (fixed inlet kinetic data)	$\{(\sigma_{[\text{Me}]_{\text{rel}}}^{\text{A(or F)}})^2 + \sigma_{[\text{OX}]}^2\}^{-1}$	$\sigma_{k_{\text{w},x_i}t} = (\sigma_{\text{fit}}^2)_{\text{int}}^{1/2}$
k_{ps_1}, pseudo-first-order rate coefficient (movable inlet)	—	$\sigma_{k_{\text{ps}_1}}^a = \{\sigma_a^2 + \sigma_{\bar{v}}^2\}^{1/2}$
k_{x_i}, values of rate coefficient at x_i (fixed inlet)	—	$\sigma_{k_{x_i}}^a = \{\sigma_{k_{x_i}t}^2 + \sigma_x^2 + \sigma_{\bar{v}}^2\}^{1/2}$
k_{w,x_i}, value of wall reaction rate coefficient obtained at x_i (fixed inlet)	—	$\sigma_{k_{\text{w},x_i}} = \{\sigma_{k_{\text{w},x_i}t}^2 + \sigma_x^2 + \sigma_{\bar{v}}^2\}^{1/2}$
k, rate coefficient value at one T. Slope of k_{ps_1} vs. $[\text{OX}]$	$\{\sigma_{k_{\text{ps}_1}}^2 + \sigma_{[\text{OX}]}^2\}^{-1}$	$\sigma_k = \left\{\sigma_{\text{fit}}^2 + \left(\dfrac{\partial k}{\partial T}\right)^2 \sigma_T^2\right\}^{1/2}$
k_{w}, wall reaction rate coefficient at one T. Intercept of k_{ps_1} vs. $[\text{OX}]$	$\{\sigma_{k_{\text{ps}_1}}^2 + \sigma_{[\text{OX}]}^2\}^{-1}$	$\sigma_{k_{\text{w}}} = \left\{\sigma_{\text{fit}}^2 + \left(\dfrac{\partial k_{\text{w}}}{\partial T}\right)^2 \sigma_T^2\right\}^{1/2}$
k, rate coefficient value at one T. Average of k_{x_i}	$\sigma_{k_{x_i}}^{-2}$	$\sigma_k = \left\{\sigma_{\text{fit}}^2 + \left(\dfrac{\partial k}{\partial T}\right)^2 \sigma_T^2\right\}^{1/2}$
k_{w}, wall reaction rate coefficient at one T. Average of k_{w,x_i}	$\sigma_{k_{\text{w},x_i}}^{-2}$	$\sigma_{k_{\text{w}}} = \left\{\sigma_{\text{fit}}^2 + \left(\dfrac{\partial k_{\text{w}}}{\partial T}\right)^2 \sigma_T^2\right\}^{1/2}$
$k(T)$, temperature dependence of rate coefficient	σ_k^{-2}	$\sigma_{k(T)} = (\sigma_{\text{fit}}^2)^{1/2}$

^{a} Axial diffusion [cf. Eq. (3)] is neglected since it contributes an uncertainty, $\sigma_{\text{axial}} \leq 0.1$ $(\sigma_a^2 + \sigma_{D\text{Me}}^2)^{1/2}$, where $\sigma_{D\text{Me}} \approx 0.05$ (Reid and Sherwood, 1966); σ_{axial} is small compared to σ_a or $\sigma_{\bar{v}}$.

Table V

Fractional Error in Rate Coefficient Values Introduced by Temperature Uncertainties for Several Common Forms of Temperature-Dependent Rate Coefficients

Rate coefficient temperature dependence	Contribution to σ_k
Simple Arrhenius	
$k(T) = A \exp(-E_A/RT)$	$(E_A/RT)\sigma_T$
Collision theory	
$k(T) = AT^{1/2} \exp(-E_A/RT)$	$\{(E_A/RT)^2 - E_A/RT + \frac{1}{4}\}^{1/2} \times \sigma_T$
Transition state	
$k(T) = AT^n \exp(-E_A/RT)$	$\{(E_A/RT)^2 - (2nE_A/RT) + n^2\}^{1/2} \times \sigma_T$

c. Effect of σ_T on Rate Coefficient Uncertainties In addition to its contribution to σ_k through its contribution to $\sigma_{\bar{v}}$ and $\sigma_{[\text{OX}]}$, temperature uncertainty has another effect on the measured values of k. For a temperature dependent rate coefficient, an individual measurement at one temperature will be uncertain due to its variation within the range σ_T. Thus, using Eq. (42), an additional contribution to σ_k given by $(\partial k/\partial T)\sigma_T$ must be combined with the other uncertainties propagated through the data reduction chain.

The formulas for calculating this additional contribution to σ_k for some common forms of the temperature dependence used in describing kinetic data are shown in Table V. It can be seen from this table that single temperature or small temperature range measurements on activated reactions can suffer gross effects due to small T uncertainties. As an example, consider the results of HTFFR measurements on the Al/CO_2 reaction in the 310–750 K range where the reaction adheres quite well to simple Arrhenius behavior and for which an $E_A = 2.6 \text{ kcal mole}^{-1}$ was found (Fontijn and Felder, 1977a). Within this T range, the contribution of σ_T to the individual k values ranges from $12.7\% (310 \text{ K}) \leq (E_A/RT)\sigma_T \leq 5.2\% (750 \text{ K})$ due solely to the 3% temperature uncertainty. Clearly it is desirable, especially when E_A is large, to reduce the $(\partial k/\partial T)\sigma_T$ contribution to σ_k by making several measurements and expressing k as a weighted average. Even more preferable is to measure $k(T)$ over a very wide T range, which approach is a major strength of the HTFFR technique. The final expression $k(T)$ obtained by such a practice is more precise than any single measurement of k at one T.

The preceding discussion is by no means unique to HTFFR measurements and the effect of σ_T on single temperature rate coefficient measurements often appears to be overlooked. While σ_T will generally be somewhat smaller than the 3% HTFFR value in more conventional experiments, more attention should be paid to this problem in analyzing and presenting kinetic results.

2. Treatment of Systematic Errors

Discussion of systematic (nonrandom) errors conventionally takes the form of an "estimate of experimental accuracy." Typically, reports of "corrected" data such as the pressure and temperature corrections discussed in Sections II,D,2 and II,D,3 represent the results of recognizing and accounting for these kinds of errors. For two interesting cases of unrecognized systematic error in experimental results see Cvetanovic et al. (1975).

A special instance of systematic error in HTFFR work which is not simply amenable to an experimental "fix" is that discussed in Section II,D,1 concerning our choice of $\eta = 1.3$. The question to be addressed here is how accurate that choice might be. We have in the past allowed for a possible 23 % uncertainty in η (Fontijn et al., 1973, 1975, 1977; Felder and Fontijn, 1976; Fontijn and Felder, 1977a), but have also treated it as a random uncertainty, combining it at the end of the calculation chain with σ_k. The reasoning for this was that the wide pressure and flow velocity ranges spanned in the HTFFR measurements would tend to give η values normally distributed about 1.3. However, in view of the fact that neither extreme, $\eta = 1.0$ (plug flow) or $\eta = 1.6$ (fully developed parabolic flow), is adhered to under the usual HTFFR conditions, it seems likely that the original 23 % estimate is overconservative. In view of the range of conditions covered in the HTFFR, it is felt that 10 % is a more reasonable estimate of the possible range η values might have. However, since there is no certainty that this range of 10 % in η is normally distributed about $\eta = 1.3$, we consider that 10 % as a systematic uncertainty and add it directly to the precision values as calculated above. The result of the error analysis procedures given is an overall accuracy for HTFFR kinetic data in the 40–70 % range. This can be compared with the 10–20 % accuracy range accessible in room-temperature flow tube studies of nonrefractory species with small wall-recombination coefficients, cf. Chapter 1.

3. Random and Systematic Uncertainties in Photon Yield Measurements

Following the treatment of Section II,A,2, for $[OX] \gg [Me]$:

$$\Phi = I_{CL}/\eta(\Pi l^2/4)\bar{v}[Me] \tag{45}$$

where I_{CL} (photons sec^{-1}) is the absolute volume and wavelength integrated chemiluminescence intensity and the denominator is the flux of Me (atoms sec^{-1}) consumed in the reaction. When all Me is consumed (as is verified in such studies), $[Me]$ is the initial value in the absence of oxidizer. A single Φ measurement has a random error given by

$$\sigma_\Phi = \{\sigma_{I_{CL}}^2 + \sigma_{\bar{v}}^2 + \sigma_{[Me]}^2\}^{1/2} \tag{46}$$

where quantities in the braces are given in Tables II and III. As in the treatment of kinetic data, the σ_Φ are used as weighting factors in determining the average Φ value from a number of Φ measurements.

Unlike rate coefficient data, Φ measurements have been found to be quite precise (Felder and Fontijn, 1978) and the scatter between measurements is generally smaller than $\sigma_\Phi \approx 0.15$. The largest errors in Φ determinations (and hence also in k_{hv}) are the systematic uncertainties involved in determining [Me] by absorption and the additional 10% contribution of η. The potential major source of error in this [Me] measurement is in the f numbers used to evaluate Eq. (23) which, in general, are not well known for refractory species. Thus, a factor of two uncertainty in values of [Me] calculated would not appear to be an overconservative error estimate. To narrow this uncertainty range, an independent [Me] measurement, such as provided by chemical titration, can be used (cf. Section II,A,2).

4. Special Cases in Kinetic Measurements

a. OX Decomposition Generally, it is advisable to extend measurements only up to the temperatures where OX is stable. Any OX dissociation introduces systematic errors, the degree of which depends upon: (i) the resulting decrease in [OX] and (ii) the reactivity of the decomposition products with respect to Me. As discussed in Section II,A,1, actual dissociation generally occurs to a smaller degree than corresponds to equilibrium. However, one can calculate an upper limit to the systematic errors thus introduced by thermochemical equilibrium calculations and prior knowledge (or upper limit estimates) of rate coefficients of the decomposition products/Me reaction. For example, in the Al/CO_2 study (Fontijn and Felder, 1977a), the effect of possible CO_2 dissociation at 1880 K on the data was considered. The principal dissociation products are CO and O_2, of which only O_2 can react with Al at an appreciable rate. Over the range of [CO_2] used, equilibrium calculations indicated that between 7 and 57% of [CO_2]$_0$ could have dissociated, producing $0.035\,[CO_2]_0 \leq [O_2] \leq 0.285\,[CO_2]_0$. Therefore, the data were analyzed using "corrected" forms for the stationary and traversing kinetic equations

$$-\ln[\text{Al}]_{\text{rel}}^{\text{corr}} = -\ln[\text{Al}]_{\text{rel}}^{\text{obs}} - k_2[O_2]t = k_1[CO_2]t + k_w t \tag{47}$$

$$k_{\text{ps}_1}^{\text{corr}} = k_{\text{ps}_1}^{\text{corr}} - k_2[O_2] = k_1[CO_2] + k_w \tag{48}$$

where k_1 and k_2 are the Al/CO_2 and Al/O_2 rate coefficients, respectively, and $k_2/k_1 = 1.1$. The mean of the k_1 data thus obtained was found to be 6% higher than that calculated ignoring possible dissociation, which uncertainty is an acceptable one considering the other factors affecting accuracy. Some subsequent observations at 1950 K indicated that the actual dissociation at 1880 K was probably negligible (Fontijn and Felder, 1977a).

b. Violation of Pseudo-First-Order Condition   At high reaction rates it may not be possible to use [OX] $\gg$ [Me] and still maintain sufficient span in the data to give reliable kinetic results. In such cases, the full bimolecular rate equation must be employed to analyze the data; consequently both [Me] and [OX] need to be measured. For example, in some measurements of the Al/CO_2 reaction at 1880 K, the condition [Al] $\approx$ [CO_2] was used (Fontijn and Felder, 1977a). [Al] was calculated from Eq. (23) and f values, which in this case can be considered quite accurate because of good agreement among published determinations, using different methods (Smith and Liszt, 1971; Wiese *et al.*, 1969; Froese-Fischer, 1976). The basic reliability of such absolute [Me] calculations at elevated temperatures and conversely of the "usual" HTFFR kinetic procedures was evidenced in this work since it was found that comparisons of rate coefficients obtained where the full bimolecular rate equation was appropriate, agreed, within experimental accuracy ($\approx 40\%$), with those obtained under pseudo-first-order conditions.

III. FURTHER TECHNIQUES FOR KINETIC MEASUREMENTS ON REFRACTORY SPECIES

A. Beam Studies With Luminescence Detection

1. Purpose and Scope

Some aspects of the oxidation reactions of refractory species, namely, their often high exothermicities and efficiencies and, as a consequence, their tendency to lead to excited products and chemiluminescence, have been exploited in a special class of very low pressure studies that use molecular beam techniques coupled with optical detection methods. These studies obtain (i) dynamical information in the form of chemiluminescence or laser-induced fluorescence spectra of the nascent e, v, J distributions of product molecules and (ii) kinetic information in the form of total cross sections, chemiluminescence cross sections, and reaction orders from the concentration dependences of the spectral features and intensities. By contrast to these nascent distributions, in bulk (many collision) work, relaxed product distributions are usually observed. Kinsey (1977) has recently reviewed the laser-induced fluorescence aspects of these beam studies and he, as well as Lin and McDonald in Chapter 4, discuss such fluorescence per se. Since the basic approach of this type of beam work is similar to that used in beam chemiluminescence measurements, we will concentrate on the latter here.

These studies are essentially limited to fast reactions; kinetic data can generally not be obtained for reactions that proceed less often than one in every ten collisions, depending somewhat on the sensitivity of the detectors used. Cross section values are of lower accuracy than are obtainable for rate coefficients in flow tubes; however, the achievable accuracy is steadily improving. Observations are generally free of interference by wall collisions and reactions are observed at pressures in the 10^{-3}–10^{-6} Torr range, though higher pressures have been used in a few instances, usually by adding inert gas to the reaction chamber.

Much of the work has used beam–scattering gas arrangements, though some beam–beam (crossed beams) chemiluminescence experiments have also been performed. The distinction between these two operating modes is in some experiments not a priori clear, and in some of the original, especially early, literature was not clearly drawn. The problem has recently been discussed thoroughly by Gole et al. (1977) who point out that when the radiative lifetime of emitters formed by reactions is short compared to the time between collisions, the two types of experiments become indistinguishable as far as spectral observations are concerned. In the opposite case, $\tau_{coll} \ll \tau_{rad}$, beam–gas experiments can approach (J. L. Gole, private communications) a situation of being essentially bulk experiments, like the other methods discussed in this chapter.

Since a beam is a directed flow (either effusive or nozzle type), a Boltzmann distribution among the x, y, z components of velocity does not exist, and reaction temperatures cannot be defined conventionally. Consequently, kinetic results are expressed as cross sections $\sigma(\bar{v})$ rather than rate coefficients $k(T)$. (Photon yield can still be defined as in Section II,A,2 and is now the ratio σ_{hv}/σ_{tot}.) In the work of Preuss and Gole (1977b), Gole and Preuss (1977), and Yokozeki and Menzinger (1977), it has nonetheless been shown to be possible to obtain Arrhenius activation energies over limited effective temperature ranges ($\Delta T_{eff} = 0.1 T_{eff}$–$0.2 T_{eff}$). The effective temperatures are defined by combining the mass-weighted scattering gas temperature (usually room temperature) and beam temperature (the variable metal vaporizer temperature). Somewhat larger changes in T_{eff} are possible by heating the scattering chamber (Yokozeki and Menzinger, 1977).

The first study of atomic chemiluminescence under single collision conditions was by Moulton and Herschbach (1966) on KBr† + Na → K* + NaBr, followed by the molecular chemiluminescence studies of Ackerman (1967) on NO/O_3 and of Ottinger and Zare (1970) on the reactions of Ba and Ca with NO_2 and N_2O. A major advance in technique was the development of the beam-gas LABSTAR type apparatus in Zare's laboratory at Columbia.

2. Apparatus and Evaporation Methods

A diagram of a beam–gas LABSTAR is shown in Fig. 23. The basic description is due to Jonah et al. (1972); a large number of variations due to general

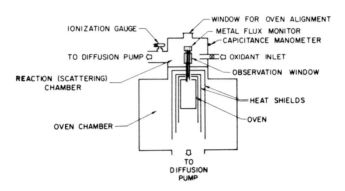

Fig. 23. Schematic of a beam-gas type apparatus.

improvements and the requirements of individual reactant combinations have since been developed. Typically, a ≈ 0.3-cm-diameter beam of Me atoms effuses from an oven source which is surrounded by heat shields and differentially pumped, e.g., by a 750 liter sec^{-1} oil diffusion pump. This beam enters the reaction chamber (pumped, e.g., by a 1500 liter sec^{-1} oil diffusion pump) where it intersects an uncollimated beam of the oxidizer gas that flows into this chamber through a ≈ 0.3-cm-diameter orifice. The chemiluminescence is viewed at right angles to both beams through a quartz window mounted at the side of the reaction chamber.

The oven is heated to sufficiently high temperatures (measured by optical pyrometry or thermocouples) to produce pressures of Me in the oven chamber on the order of 10^{-2}–10^{-1} Torr (Dickson *et al.*, 1977). No carrier gas is used to entrain the Me atoms, which is a major difference with the HTFFR and other bulk techniques. Typical atom supply rates in the reaction region are on the order of 10^{14}–10^{15} atoms sec^{-1} (with a beam diameter of 0.3 cm, this corresponds to fluxes of 10^{15}–10^{16} atoms cm^{-2} sec^{-1}). The oxidant is usually metered with a needle valve and its pressure is measured in the reaction chamber, preferentially (Dickson *et al.*, 1977) with a capacitance manometer.

In the original work on group IIA metals (Jonah *et al.*, 1972), the Me beam was formed by effusion from a molybdenum oven heated with a three-phase molybdenum-wire resistance winding (50–60 A per leg, ≈ 1 kW total power consumption). A molybdenum heat shield and a water-cooled copper heat shield surrounded the oven and provided part of the collimation of the beam. The oven was a closed cylinder 4-cm diameter and 10-cm high. A more recent design, using radiant heating of the crucible, is shown in Fig. 24. The actual construction materials vary, cf. Section II,B,2, with the compatibility of the construction material with the metal being vaporized and the required vaporization temperatures. Table VI provides a guide to the original literature for

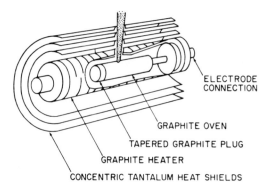

ELECTRODE
CONNECTION

GRAPHITE OVEN

TAPERED GRAPHITE PLUG

GRAPHITE HEATER

CONCENTRIC TANTALUM HEAT SHIELDS

Fig. 24. Metal atom source, designed by R. C. Oldenborg. A few hundred amperes are passed through a graphite cylinder held between two water-cooled copper busbars. The graphite heater has slots cut on its body to increase the resistance. Inside the heater and supported from one end is a graphite crucible containing the metal sample. [From Dickson *et al.* (1977, Fig. 2).]

details of specific reactant combinations; here we will discuss a few interesting designs and methods. Chalek and Gole (1976) in a study of Sc and Y with O_2 used a vaporizer design like that of Fig. 24 with a 1.6-cm-diameter, 10-cm-long tantalum or tungsten crucible surrounded by a 2.5-cm-diameter, 0.05-cm-thick tantalum radiator. Temperatures on the order of 2300 K were obtained with these crucibles and temperatures up to 2700 K were achieved by substituting tungsten carbide crucibles (J. L. Gole, private communications). To vaporize Sc, a quite expensive metal, Chalek and Gole (1976) developed a chemical technique by use of the reaction

$$2Sc_2O_3(s) + 3Th(s) \longrightarrow 3ThO_2(s) + 4Sc(g) \qquad (49)$$

which proceeds readily at temperatures around 2000 K. Stoichiometric mixtures of powdered Sc_2O_3 and Th produced an essentially pure Sc beam. Y can be vaporized by the same technique. Manos and Parson (1975) in a study of these same metals used a tantalum oven heated by bombardment with 2–4 keV electrons. In a later study from the same laboratory, radiant heating was supplied by a resistively heated tungsten mesh (Liu and Parson, 1977). Fite *et al.* (1974) succeeded in producing a beam of U atoms ($\approx 4 \times 10^{12}$ atoms sec^{-1}) in a resistance heated tungsten foil oven at 2000 K. This is a solution to a particularly difficult problem because of the "low" melting point of uranium (1405 K) and the property of liquid U to act as a nearly universal solvent; their beams could be maintained for periods of 2–3 hr. However, these fluxes, while adequate for chemi-ionization studies [see Fontijn (1972, 1974) for reviews of metal atom reactions of this type], appear inadequate for chemiluminescence observations.

Table VI

Some Beam Chemiluminescent Reactions

Reaction	Remarks	Reference
Group IA elements		
$Me_2 + OX \rightarrow MeO^* + MeX$ $\quad Me_2 = K_2, Rb_2, Cs_2$; $\quad OX = Cl_2, Br_2, I_2, IBr,$ $\quad ICl, ClF$	Spectra have a broad unidentified molecular feature, spanning most of the visible region, with superimposed atomic lines. For the mixed halogen reactions the MeX species is the more stable salt molecule that can be formed from the halogens of the OX pair. No molecular emissions observed in reactions of Na_2 and of F_2.	[a]
$\quad Me_2 = K_2, Rb_2, Cs_2$; $\quad OX = Cl_2, Br_2, I_2, ICl$	Same spectra as Ref. *a*. Main reaction path is formation of $Me + MeO + X$.	[b]
$Me_2 + OX \rightarrow Me^*$ $\quad + MeO + X$ $\quad Me_2 = K_2, Rb_2, Cs_2$; $\quad OX = Cl_2, Br_2, I_2, ICl$	Many Me levels excited.	[b]
$Me_2 + X \rightarrow Me^* + MeX$; $\quad Me_2 = Na_2, K_2, Rb_2$; $\quad X = Cl, Br, I$	All energetically allowed Me levels within measurement region observed.	[b]
$Na + KBr(v) \rightarrow K(4^2P)$ $\quad + NaBr$		[c]
Group IIA elements		
$Me + OX \rightarrow MeO^* + X$ $\quad Me = Ca, Sr, Ba$; $\quad OX = ONO, ON_2$	For Ba/NO_2 the emitting state is $(A^1\Sigma)$. The other monoxide emitters not definitely identified. $Mg + NO_2$ gave no detectable emission, $Mg + N_2O$ gave very weak MgO emission only.	[d, e]
$Ba + N_2O \rightarrow BaO^* + N_2$	σ_{hv} appears to increase with N_2O vibrational excitation, probably due to bending mode.	[f]
$Ba + O_3 \rightarrow BaO^* + O_2$	Same emitter as BaO* from Ba/N_2O.	[g]
$Me + ClO_2$ $\quad \searrow MeO(A'^1\pi) + ClO$ $\qquad\qquad\quad MeCl^* + O_2$ $\quad Me = Mg, Ca, Sr, Ba$	σ_{hv} for both reaction paths are comparable. σ_{tot} much larger. MeCl* represents various excited states.	[h] [h]
$Me + Cl_2 \rightarrow MeCl_2^*$ $\quad \rightarrow MeCl_2 + hv$ $\quad Me = Ba, Sr$		[i]
$Ba + OX + M \rightarrow BaOX^* + M$ $\quad OX = Cl_2, Br_2, I_2$	Suggests that this process rather than radiative recombination (Ref. *i*) is the dominant process.	[j]
$Ca + OX$ $\quad \xrightarrow{+(M)} CaOX^*(+M)$ $\qquad\qquad CaO^* + X$ $\quad OX = F_2, Cl_2, Br_2$	CaO* represents a number of Ca monohalide states.	[k]

Table VI (*continued*)

Reaction	Remarks	Reference
$Sr + F_2 \rightarrow SrF(A^2\pi, B^2\Sigma, C^2\pi, D^2\Sigma) + F$		[l]
$Me + S_2Cl_2 \rightarrow S_2(B^3\Sigma_u^-) + MeCl_2$ $Me = Ba, Sr, Ca$	Some MeCl emission and a continuum perhaps due to $MeCl_2$ also observed.	[m]
Group IIIA elements		
$B + N_2O \rightarrow BO^* + N_2$	Pulse of B atoms having from 1 to 4 eV energy.	[n]
$Al + O_3 \rightarrow AlO(B^2\Sigma) + O_2$	Weak emission probably due to polyatomic Al oxide also observed.	[o]
	From non-Boltzmann vibrational distribution concludes that nascent product is $AlO(A^2\pi)$ followed by collision-induced transition to $AlO(B^2\Sigma)$.	[p]
$Al + O_3 \rightarrow AlO(A^2\pi) + O_2$		[q]
Group IIIB elements		
$Me + O_2 \rightarrow MeO(A^2\pi, A'^2\Delta) + O$ $Me = Sc, Y$		[l, r, s]
$La + O_2 \rightarrow LaO(A^2\pi, B^2\Sigma, C^2\pi) + O$		[s, t]
$Me + OX \rightarrow MeO(A^2\pi, B^2\Sigma) + X$ $Me = Sc, Y;$ $OX = ONO, ON_2$		[l]
$La + OX \rightarrow LaO(A^2\pi, B^2\Sigma, C^2\pi, D^2\Sigma) + X$ $OX = ONO, ON_2, O_3$		[l, t]
$La + OCS \rightarrow LaS(C^2\pi) + CO$		[u]
$Me + OX \rightarrow MeO^* + X$ $Me = Sc, Y, La;$ $OX = F_2, Cl_2, Br_2, F_2,$ $SF_6, ClF, ICl, IBr.$	Several of these systems characterized by selective excitation of a $^3\Sigma^+$ monohalide state, which has a low quenching cross section and is formed with a high Φ. Selectivity is most pronounced for ScF formation ($\Phi \geq 8\%$ from $Sc + F_2$) and least for La reactions.	[v, w]
Lanthanide elements		
$Sm + N_2O \rightarrow SmO^* + N_2$	Two excited states are produced; they yield Arrhenius plots of opposite curvature. σ_{hv} is increased by vibrational excitation of N_2O (presumably in the bending mode). Absolute Φ values given in Ref. *e*.	[x]
$Me + OX \rightarrow MeO^* + X$ $Me = Eu, Sm;$ $OX = O_3, N_2O, NO_2,$ and F_2	Of this group Sm/F_2 yields the highest Φ (11.8% according to Ref. *e*).	[y]
$Me + OX \rightarrow MeO^* + X$ $Me = Sm, Yb;$ $OX = N_2O, O_3, F_2, Cl_2$		[z]

(*continued*)

Table VI (continued)

Reaction	Remarks	Reference
$Yb + F_2 \rightarrow YbF(A^2\pi, B) + F$	Emissions show extensive vibrational perturbations. An emission from a probably lower-lying YbF state is also observed.	[aa]
$Yb + ClO_2 \rightarrow YbCl(A^2\pi, B) + O_2$	Contrary to the group IIA/ClO_2 reactions, no YbO* is observed. Emission from a probably lower-lying YbCl state is also observed.	[aa]
$Ho + N_2O \rightarrow HoO^* + N_2$	Burst of Ho atoms having 0.5–3.5 eV energy.	[n]
Group IVA elements		
$Pb + O_3 \rightarrow PbO(a, b) + O_2$		[bb]
$Pb + N_2O \rightarrow PbO (B) + N_2$	Reaction is 0.6 eV endothermic. Observed from threshold up to 3 eV center of mass energy.	[cc]
Group IVB elements		
$Ti + OX \rightarrow TiO(B^3\pi, C^3\Delta, c^1\Phi) + X$ $OX = O_2, ONO, ON_2$	The N_2O reaction additionally produced TiO(D). Emission from Ti/O_2 is mainly due to excited Ti atoms.	[dd]

[a] Oldenborg et al. (1974).   [b] Struve et al. (1975).   [c] Moulton and Herschbach (1966).
[d] Jonah et al. (1972).   [e] Dickson et al. (1977).   [f] Wren and Menzinger (1975).
[g] Schultz and Zare (1974).   [h] Engelke et al. (1976).   [i] Jonah and Zare (1971).
[j] Wren and Menzinger (1974).   [k] Menzinger (1974).   [l] Gole and Preuss (1977).
[m] Engelke and Zare (1977).   [n] Tang et al. (1976).   [o] Gole and Zare (1972).
[p] Lindsay and Gole (1977).   [q] Sayers and Gole (1977).   [r] Chalek and Gole (1976).
[s] Manos and Parson (1975).   [t] Gole and Chalek (1976).   [u] Jones and Gole (1977).
[v] Gole and Chalek (1978).   [w] Gole and Chalek, (1979).   [x] Yokozeki and Menzinger (1977).
[y] Dickson and Zare (1975).   [z] Yokozeki and Menzinger (1976).   [aa] Lee and Zare (1977).
[bb] Oldenborg et al. (1975).   [cc] Wicke et al. (1978).   [dd] Dubois and Gole (1977).

Me atom fluxes into the reaction region are usually measured by a thin film thickness monitor. An improved design Me flux monitor is shown in Fig. 25. With this design (Dickson et al., 1977) the total Me flux is captured, whereas with the film thickness devices not all Me striking the monitor is collected if the sticking coefficient is not unity, which it often is not.

3. Beam–Gas Measurements

Observation of the chemiluminescence intensity I in beam–gas experiments allows measurement of the order of the chemiluminescent reaction, the total cross section for atom removal‡ by OX, σ_{tot}, and of σ_{hv}, the chemilumin-

‡ The observed σ_{tot} can be equated with a total reactive cross section if it is assumed that scattering out of the metal beam can be neglected. With proper precautions, this is reasonable for fast reactions.

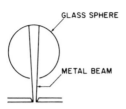

Fig. 25. Me flux determination using a thin, hollow glass sphere. In order to obtain a weighable sample, the orifice of the entrance port to the scattering chamber is enlarged. While the number of Me atoms sec^{-1} that are collected increases, the flux (number of Me atoms cm^{-2} sec^{-1}) is not altered by the small change made. [From Dickson *et al.* (1977, Fig. 5).]

escence cross section as follows (Jonah *et al.*, 1972; Dickson *et al.*, 1977). The measured chemiluminescence intensity at the position viewed by the detector (a spectrometer or filtered PMT) obeys the general equation

$$I \propto F_{Me} P_{OX}^n \tag{50}$$

where F_{Me} is the Me flux at that position, P_{OX} is the OX pressure, and n is the order of the light-producing reaction. Because of the presence of the oxidizer gas, Me will be attenuated on traversing the reaction chamber according to

$$F_{Me} = F_{Me_0} \exp(-\alpha P_{OX}) \tag{51}$$

where F_{Me_0} is the flux at the entry port, and the attenuation parameter

$$\alpha = 1.33 \times 10^{-13} x \sigma_{tot}/kT \tag{52}$$

for P_{OX} in Torr and x the beam path length in the reaction chamber from the port of entry to the observation position in cm. The constant 1.33×10^{-13} has units of dyne $Torr^{-1}$ cm^{-16}. Substituting Eq. (51) into Eq. (50), we obtain

$$I \propto F_{Me_0} P_{OX}^n \exp(-\alpha P_{OX}) \tag{53}$$

By fitting I to expressions of this form for various n, the order of reaction can be determined. In many cases $n = 1$. However, cases of $n = 2$ and $1 < n < 2$ have been observed (Jonah *et al.*, 1972; Gole *et al.*, 1977), indicative of secondary reactions such as collision-induced radiationless transition from metastable states to the emitting states. Further, or corroborating, evidence for such secondary reactions can be obtained from spectroscopic observations; for example, the occurrence of metal atom emissions or of product emission from levels exceeding the known reaction exoergicity (Engelke *et al.*, 1976) [taking into account that even in single collisions some endoergic population can occur with the appropriate activation energy (Fontijn and

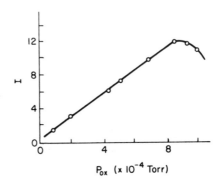

Fig. 26. Determination of σ_{tot} from the pressure maximum for Sm + N_2O. [From Dickson *et al.* (1977, Fig. 6).]

Johnson, 1973)]. The attenuation parameter, and hence σ_{tot}, can be determined in two independent ways:

(1) by studying the chemiluminescence intensity versus oxidant pressure for constant x, and

(2) by studying the chemiluminescence intensity versus x for a constant oxidant pressure.

The first method determines α from the pressure maximum (i.e., from $(dI/dP_{OX} = 0)$, which gives

$$\alpha = n/P_{OX}^{max} \tag{54}$$

from which σ_{tot} may be found by using Eq. (52). A typical plot for the Sm/N_2O reaction is given in Fig. 26. The second method plots $\ln I$ versus x (see Fig. 27); the slope of this plot is given by

$$d(\ln I)/dx = -1.33 \times 10^{-13}(P_{OX}/kT)\sigma_{tot} \tag{55}$$

from which σ_{tot} can be directly determined. The latter method is preferable since higher P_{OX} are avoided and it avoids the necessity of locating a maximum in the data which is often difficult to do accurately. Moreover, at $P \gtrsim 1 \times 10^{-3}$ Torr some of the oxidant gas tends to creep into the oven chamber. In a technique for varying x which has recently been described by Engelke and Zare (1977), the reaction chamber is replaced by a Pyrex or quartz tube and the attenuation is measured with a filtered PMT movable along the length of the tube, a method similar to that used in flow tube kinetic studies (Chapter 1).

From the above treatment and the discussion of k_{hv} in Section II,A,2, it follows that the chemiluminescence cross section can be determined from

$$\sigma_{hv} = I/F_{Me}[OX]^n V_{obs} \tag{56}$$

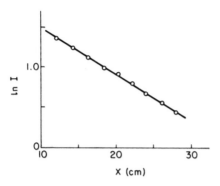

Fig. 27. Plot of the logarithm of the chemiluminescent intensity vs. path length for the reaction Sm + N_2O. [From Dickson *et al.* (1977, Fig. 7).]

where V_{obs} is the observed reaction volume. I has been determined (Dickson *et al.*, 1977) for the Sm/N_2O ($n \approx 1$) reaction by comparison to a standard lamp. For this reaction the accuracy of Φ thus obtained is estimated (Dickson *et al.*, 1977) to be about $\pm 50\%$. The spectrum of the Sm/N_2O reaction shows little structure at low resolution (0.5 nm) and covers a wide wavelength range (450–750 nm). It can serve as a photon flux standard for other single-collision observations (Dickson *et al.*, 1977), similar to the use of the O/NO reaction in flow tubes (Section II,A,2). It should again be cautioned that this method for $\sigma_{h\nu}$ determination is suitable only for relatively short-lived emitters, i.e., those that radiate before diffusing out of the observation zone; whether this condition is fulfilled can often be established by visual observation.

In addition to spectral observations and σ_{tot} and $\sigma_{h\nu}$ experiments, other types of measurements, which will be mentioned briefly here, have been made with the types of apparatus discussed. By adding up to 20 Torr of Ar to the reactive gases in the scattering chamber, quenching information on group IIIB halides has been obtained (J. L. Gole, private communications). For nonchemiluminescent reactions laser-induced fluorescence has been used to obtain nascent product state distributions, radiative lifetimes, and lower limits to bond energies of product molecules, see, e.g., Liu and Parson (1977), Zare and Dagdigian (1974), Dagdigian *et al.* (1975), Pruett and Zare (1975), Kinsey (1977), and Pasternack and Dagdigian (1977). Lower limits to bond energies can also be obtained from chemiluminescence observations. The arguments for both types of determination are rather similar: In fluorescence the highest excitation level observed is that which contains the sum of the internal energy of the product molecule and the known fluorescence pumping energy. In chemiluminescence, if the highest excited level observed corresponded to the reaction exoergicity and the bond energy of the reactant were known, the excitation energy of this level would give a direct measure of the bond energy of the product after appropriate correction for translational energy of the reactants and products and internal energy of the reactants. However, since the translational energy of the products is not known and since chemiexcitation often will not occur up to the highest energetically allowable level, only a lower limit to the bond energy is obtained

$$D(\text{MeO}) \geq D(\text{OX}) + E_{\text{int}}(\text{MeO}) + E_{\text{trans}}(\text{MeO})$$

$$+ E_{\text{trans}}(X) - E_{\text{int}}(\text{OX}) - E_{\text{trans}}(\text{OX}) - E_{\text{trans}}(\text{Me}) \cdots \quad (57)$$

The basic method here was discussed by Jonah et al. (1972). For some recent refinements and further literature references see Preuss and Gole (1977a). Generally, such bond energy limit determinations have given values in good agreement with bond energies obtained by other methods, i.e., the limit values are close to the actual values. Some caution is, however, needed. In some cases, metastable atoms have been produced by electrical discharges in the oven region, Oldenborg et al. (1975), the presence of which would, of course, cause errors in the bond energy limit determinations. A method, from the same laboratory, to guard against such errors is discussed in Section III,A,4. The occurrence of such discharges is now utilized to study reactions of metastable atoms (R. C. Estler and R. N. Zare, private communications).

Friichtenicht (1974) has described a technique for producing beams of Me atoms with energies in the ≈ 0.5–10 eV range. Metal films are plated on a transparent slide and flash evaporated with a Q-switched ruby laser. The resulting burst of atoms enters a reaction chamber filled with a low pressure, e.g., 5×10^{-4} Torr, of the target gas. Chemiluminescence is measured as a function of Me energy (arrival time as a function of Me velocity). Overall kinetics is simultaneously observed with a mass spectrometer. Probably most of the beam consists of ground state atoms, though metastable excited atoms can be present. The technique has been applied to detailed studies of the Pb/N_2O (Wicke et al., 1978) and B, Ho/N_2O (Tang et al., 1976) reactions; a survey of relative chemiluminescence yields of a large number of elements with N_2O and NO_2 has also been reported (Friichtenicht and Tang, 1976). Because of the higher energies involved, results are not directly comparable to those obtained with thermal beams or the other reactors discussed in this chapter.

4. Crossed Beam Studies

To close out the discussion of beam experiments, we discuss here a few techniques from crossed beam studies. In such studies the oxidizer gas is expanded from a third differentially pumped vacuum chamber. By changing the temperature of the nozzle of such a chamber from 280 to 1400 K, Manos and Parson (1975) could observe, in a study of the group IIIB atoms with O_2, the influence of the interaction energy on the observed spectra (hence the branching ratios between the various excited states). Engelke *et al.* (1976) made a crossed beam study of the reactions of alkaline earth metal atoms with ClO_2. They demonstrated the reactions to be of first order in both reactants by showing that at fixed F_{Me} the chemiluminescence intensity is of first order in the ClO_2 source pressure, while at fixed F_{ClO_2} the logarithm of this intensity varies as the inverse of the metal source temperature. Based on the Clausius–Clapeyron equation, $d \ln P/dT^{-1} = -\Delta H_{vap}/R$, this source temperature dependence indicates first-order dependence on the metal oven pressure and hence F_{Me}. The slope of these plots gave the correct value, to within 10%, for the atomic heat of vaporization, suggesting that in this case the reactions are indeed due to the ground state metal atoms, not dimers or excited atoms. Because of the collisional isotropy in beam–gas experiments, similar arguments in such work are more convincing (Preuss and Gole, 1977a,b; Gole and Preuss, 1977).

The influence of vibrational excitation of N_2O on σ_{Cl} in reactions with Ba and Sm was established by Yokozeki and Menzinger (1977) and Wren and Menzinger (1975).‡ They used He-seeded supersonic N_2O beams which were crossed with effusive metal atom beams. The N_2O vibrational temperature T_v was taken to equal the temperature of the heatable nozzle (270–620 K), since vibrational relaxation is slow on the time scale of the nozzle-beam expansion. T_v was varied independently of the translational energy of the supersonic beam, which is determined by the He/N_2O seeding ratio and the nozzle temperature.

B. Atomic Diffusion Flames

Atomic diffusion flames as developed by Polanyi (1932) represent the oldest technique for obtaining kinetic information on free metal atoms.§ They are hardly used any more since for overall kinetic studies, primary and secondary reactions can often not be properly distinguished and the rate coefficients obtained are thus often global and not elementary and moreover are generally not very accurate. Furthermore, concentrations as well as temperature in most of the flames are only approximately known. For refractory species these flames also often suffer from the presence of large quantities of condensed reaction products, so that it becomes rather difficult to distinguish between homogeneous and heterogeneous processes [measuring the distribution of the condensed product over the reaction zone was indeed an

‡ Increases in σ_{CL} with T_v were observed, probably due to increases in the flux of N_2O molecules in the v_2 (bending) mode.
§ For a recent review of results see Gowenlock *et al.* (1976).

essential element of the original investigations (Polanyi, 1932)]. However, in recent years there has been a selective rebirth of the technique because it offers the opportunity to observe high concentrations of luminescing reaction products in a relatively small volume, i.e., to obtain relatively high intensity spectra of species not otherwise conveniently accessible because of their refractory (and often free radical) nature. In this form the atomic diffusion flame is primarily a spectroscopic tool. However, photon yields Φ can often also be determined in such experiments and the extent of the flame gives a qualitative idea of overall rate coefficients (which makes the technique suitable for scanning of a series of reactions). Moreover, observation of changes in spectra as a function of pressure supplies mechanistic information.

The use of the word "flame" in the present context refers to reaction chemiluminescence; the systems here discussed are not true self-sustaining flames, which involve chain reactions and are discussed in Section III,C,1. "Flame" has indeed become even more of a convenient misnomer here now that laser-induced fluorescence is frequently used to observe nonchemiluminescent product states and to obtain spectra of the pumped states. The decay of this fluorescence is frequently used for obtaining radiative lifetimes of these states (e.g., Sakurai et al., 1970; Johnson, 1972) and can yield information on their quenching (Johnson, 1972).

A representative design of recent atomic diffusion flame apparatus (Jones and Broida, 1973, 1974; West et al., 1975) is shown in Fig. 28. Metal atoms are evaporated from a small crucible surrounded by a resistively heated tungsten wire basket. The entrained metal is mixed with the oxidant in a large reaction chamber. Temperatures in the reaction zone are usually somewhat above room temperature, although, depending on furnace design and operating conditions, temperatures of up to about 700 K have also been used. The metal vaporizer design shown in Fig. 28 is that used for relatively volatile metals such as Ba. Various other designs have been described by West et al. (1975); they are similar to those discussed in Section II,B. The metal feed rates are typically on the order of $10^{16}-10^{19}$ atoms sec^{-1} and initial metal atom concentrations in the reaction zone are on the order of $10^{13}-10^{14}$ mliter^{-1} (Jones and Broida, 1973, 1974; Capelle and Brom, 1975; Capelle and Linton, 1976; Eckstrom et al., 1974). The total reaction zone pressure is in the range 0.1–100 Torr. The metal atom feed rate is usually obtained by measuring the weight loss of the metal in the vaporizer over a period of hours. This procedure often leads to fairly good estimates, though losses due to reaction of the bulk metal with the container and the metal vapor in transport can be causes of error. Alternately, the metal feed rate can be obtained by collection on a probe inside the reactor in the absence of oxidizer (Eckstrom et al., 1974). Because sticking coefficients are often less than unity (Section III,A,2), this procedure is not necessarily more reliable.

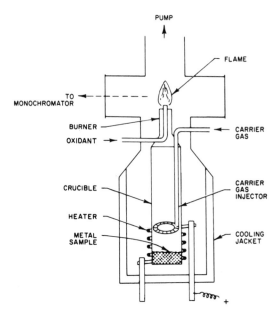

Fig. 28. Schematic diagram of atomic diffusion flame apparatus. Metal atoms are thermally evolved from the sample, entrained in a carrier gas, and reacted with an oxidizer at the burner to form the flame. [From Jones and Broida (1974, Fig. 1).]

For fast reactions with sufficient excess oxidizer, the "closed" flames that can be obtained show that the reaction has gone to completion; hence the metal consumption rate equals the metal feed rate. Under these conditions, dividing the number of photons emitted per second by this feed rate gives the photon yield. The photon emission rate is measured with a spectrometer, calibrated absolutely with the use of a standard lamp at the position of the flame, and geometrically corrected with approximate information on the flame size and shape. The distance of the flame to the spectrometer entrance slit is fairly large [107 cm in the work of Jones and Broida (1973, 1974)]. For the weaker emissions, a lens is employed for which the calibration correction factor is determined by monitoring a bright flame of about the same shape and size with and without the lens. One way to circumvent the problem of accurately determining the metal feed rate, probably useful only for the more volatile metals, is the use of metal-in-excess closed flames, for which, of course, only the easily determinable oxidizer feed (and hence consumption) rate needs to be known (Eckstrom et al., 1975; Palmer et al., 1975; Obenauf et al., 1972b).

The diffusion flame apparatus at the laboratory of Palmer et al. (1975) operates at lower pressures (3×10^{-4}–2×10^{-1} Torr) than in the other work discussed. In effect it fills a gap between observations at lower pressures in most molecular beam studies and the higher pressures of other atomic diffusion flames and HTFFR studies. Having information over a wide range of pressures is especially important for reactions, such as Ba/N_2O, where both Φ and spectral distribution are strongly dependent on pressure, indicative of collision-induced radiationless transitions.

An interesting recent variant in the generation of atomic diffusion flames is the heat pipe oven reactor (HPOR), described by Hesser et al. (1975), Sakurai and Broida (1976), and Luria et al., (1976b,c,d). The principle of heat pipes will be recognized by chemists as that of reflux columns. They provide uniform measurable high densities of Me atoms at a well-defined constant temperature, by virtue of the fact that the metal of interest is actually boiling, and thus the operating pressure provided by an inert gas such as Ar determines the vapor pressure of the metal. Typically, after startup, the metal vapor displaces the inert gas and a region of pure metal vapor is obtained. At the end of this vapor region, Me condenses and is transported back to the heated zone along a wick. The temperature in the region of the pure metal vapor is also quite uniform and equal to the boiling point at the preset pressure. Further external heating does not raise the temperature but rather increases the metal vaporization rate. Physically, this causes the refluxing zone to grow in size.

Figure 29 shows an HPOR. Metal vapor is produced in the 15-cm-long uniformly heated central section of well-controlled (better than ± 1 K) temperature, typically in the 700–800 K range. The reactions are carried out in this section. A spherical diffusion flame is formed suitable for Φ measurements. The metal refluxes horizontally condensing in the cooled outer section from which it is transported back to the center section by the wick consisting of five layers of stainless steel screen (60 mesh). The reaction products tend to condense on the cooled outer sections without migrating back. During an experiment the pump line is closed. Experiments were carried out (Luria et al., 1976c) at total pressures ranging from 0.05 to 50 Torr consisting of pure alkali–metal vapor for $P < 6$ Torr. In this mode the device appears useful mainly for studies of very volatile metals. Luria et al. (1976c) have attempted to extend the usefulness of the HPOR to more refractory metals, the idea being that the metals are introduced in the polyhalide form and that the alkali atoms strip off the halide atoms, e.g., $MeCl_4 + 4\,Na \rightarrow Me + 4\,NaCl$, allowing the free refractory Me to be oxidized by the cointroduced oxidant, see Fig. 29. However, while many interesting emissions can be observed (Luria et al., 1976c,d) in these reaction systems, they tend to be dominated by those of the alkali atoms and Φ values of the refractory metal atom oxidation reactions, calculated based on complete stripping, are usually lower than in binary reactions of the same metal atom/oxidizer

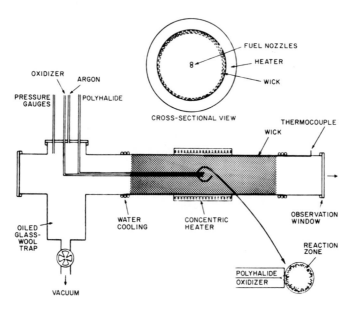

Fig. 29. A schematic diagram of the heat pipe oven reactor. [From Luria *et al.* (1976b, Fig. 1).]

combination studied in the other reactors, generally indicating a variety of competing reactions.‡ Thus, as a quantitative kinetic tool for the more refractory species, this technique does not appear too promising. Broida's work uses the oxidizer-in-excess mode which should allow somewhat more refractory species to be studied; Ba (Hesser *et al.*, 1975), Bi (Sakurai and Broida, 1976; Capelle *et al.*, 1973), and Bi_2 (Sakurai and Broida, 1976) have been reported. However, no HPOR photon yield studies have apparently been made by that group.

A representative set of reaction systems studied by the atomic diffusion flame technique is given in Table VII. The Φ values given are the maximum values under the range of conditions of the particular investigation. In comparing several pieces of work it is important to remember that Φ for many reactions is a function of pressure and sometimes also of temperature; thus disagreement between Φ values is sometimes only apparent disagreement. Φ values are again mainly mentioned when they indicate efficient production of (an) excited state(s).

‡ Molecular emissions from SnO, PO, BO, NO, CN, C_2, and probably GeO have been identified (Luria *et al.*, 1976c).

Table VII

Some Atomic Diffusion Flame Studies

Reaction	Remarks	Reference
Group IA elements		
$Na + OX \rightarrow Na(3^2P)$, $NaO + \cdots$ $OX = Cl_2, Br, I_2, HCl,$ $HBr, HI, HgCl_2$, alkyl halides	Summary of the original metal atom reaction studies. Product "NaO" in all cases is the Na halide. Recent findings on the elementary reactions occurring in some of these systems are discussed in the papers by Struve *et al.* (1975) quoted in Table VI and in Ref. *b*.	*a*
$Me + NF_3 \rightarrow Me^* + \cdots$ $Me = Li, Na, Cs$	Alkali emissions similar to those observed in these authors' ternary flame studies (see text).	*h*
$Cs + F_2 \rightarrow Cs^* + \cdots$	Evidence is given suggesting the emissions are due to formation of Me^+ followed by electron–ion recombination.	
$Na + OX + M$ $\rightarrow NaOX^* + M$ $OX = F_2, Cl_2$	Emission from NaF_2^*, $NaCl_2^*$ to stable ground state; first observation of molecules of this type.	*c*
Group IB elements		
$Cu + OX \rightarrow CuO^* + X$ $OX = Cl_2, Br_2$	The Cu monohalide products were identified both by their chemiluminescence and laser-induced fluorescence.	*d*
Group IIA elements		
$Me + OX \rightarrow MeO^* + X$ $Me = Ca, Sr;$ $OX = ClNO, BrNO$	Sr + ClNO not studied. Emissions are from the $A^2\pi$ and $B^2\Sigma$ states of the monohalides, respectively.	*e*
$Sr + OX \rightarrow SrO^* + X$ $OX = ON_2, O_3, O_2, ON,$ Br_2, Cl_2, F_2	Many emissions observed. Highest Φ (≈ 0.07) for N_2O and O_3 reactions.	*f*
$Ca + OX \rightarrow CaO^* + X$ $OX = ON_2, O_3, O_2, ON,$ ONO, Br_2, Cl_2, F_2	Many emissions observed. Highest Φ (0.06) for O_3 reaction.	*g*
$Ba + OX \rightarrow BaO^* + X$ $OX = O_2, ON_2, ONO,$ $ClNO, BrNO$	Emissions observed are from the $A^1\Sigma$ and $A'^1\pi$ states of BaO for the first three oxidizers, from the $C^2\pi$ state of the Ba halide with the latter two. For O_2, the reaction is insufficiently energetic to produce $BaO(A^1\Sigma, A'^1\pi)$ directly; production is attributed to $BaO_2^\dagger + Ba \rightarrow BaO(A^1\Sigma, A'^1\pi) + BaO(X'\Sigma)$, cf. Ref. *h*.	*e*
$Ba + N_2O \rightarrow BaO(A^1\Sigma, A'^1\pi)$ $+ N_2$	All reports agree that Φ is high; 0.08, 0.27, 0.25, according to Refs. *i, e, j*, respectively.	*i, j, k*
$Ba + O_3 \rightarrow BaO(A^1\Sigma, A'^1\pi)$ $+ O_2$	$\Phi \approx 0.36$.	*k*
$Ba + O_2 \rightarrow BaO + O$	Reaction used to determine τ_{rad} and σ_O^{He} of $BaO(A^1\Sigma)$ by laser-induced fluorescence.	*l, m*

Table VII (*continued*)

Reaction	Remarks	Reference
Me + OX → MeO* + X Me = Mg, Ca, Sr, Ba; OX = F_2 Me = Ba; OX = O_2, NOCl, ON_2		[n]
Ba + OX → BaO* + X OX = F_2, SF_6, Cl_2, Br_2, I_2	Highest $\Phi \approx 0.01$, for Ba + Cl_2; for this reaction see also Ref. *p*.	[o]
Ca + H_2CO → CaH($A^2\pi$, $B^2\pi$, $E^2\pi$) + $\cdots$	H_2CO inlet nozzle heated to ≈ 1300 K. Complicated process, Ca + HCO probably responsible for the observed emissions. Emission from CaO, CaOH, and a polyatomic species also observed.	[q]

Group IIIA elements

Al + OX → AlO* + X OX = ON_2, O_2, O_2 + O, O_3, CO_2, ONO, NO, CO	Emissions observed depend on the OX used. Observations include AlO($A^2\pi$, $B^2\Sigma$, $C^2\pi$) and a continuum probably not belonging to AlO.	[r]
Al + O_2 → AlO + O	Reaction used to determine τ_{rad} AlO($B^2\Sigma$) by laser-induced fluorescence.	[s]
Al + OX → AlO* + X OX = F_2, Cl_2, Br_2, I_2, NF_3, SF_6	Emissions observed from the $a^3\pi$ state of the monohalide and, depending upon the OX used, from several higher lying states. Al lines observed from all OX except I_2.	[t]
Al + H_2CO → AlH($A^1\Sigma$) + $\cdots$	Reaction mechanism not elucidated. H_2CO inlet nozzle heated to ≈ 1300 K.	[q]

Lanthanide elements

Me + OX → MeO* + X Me = Sm, Eu; OX = ON_2, O_3, O_2, F_2, and NF_3	Highest Φ observed in the reactions of Sm with NF_3, F_2: 0.70, 0.64, respectively. $\Phi(Sm/N_2O)$ = 0.38.	[k]

Group IVA elements

Si + N_2O → SiO($a^3\Sigma$, $b^3\pi$, $A^1\pi$) + N_2	$\Phi = 5 \times 10^{-4}$.	[u]
Ge + OX → GeO* + X OX = ON_2, O_2, ONO, ON, F_2	A number of emissions observed. All processes have a low Φ ($\leq 2 \times 10^{-3}$).	[v]
Sn + OX → SnO* + X OX = ON_2, ONO, O_2, F_2	A number of emissions observed. Highest Φ (6.7 × 10^{-2}) for Sn/N_2O.	[w]
Pb + OX → PbO* + X OX = ON_2, O, O_2, O_3	Many new PbO bands observed in the AO^+–XO^+ and al–XO^+ systems. Addition of active nitrogen (microwave discharge) greatly increased intensities.	[x]
Pb + O_3 → PbO* + O_2	Reaction with vibrationally excited O_3 to yield PbO(A) appears faster than with ground state O_3. PbO(B) formation is close to energetic threshold for Pb($^3P_0$) and its rate is increased in reaction with Pb($^3P_1$), in agreement with an observation of Ref. *z*.	[y]

(*continued*)

Table VII (*continued*)

Reaction	Remarks	Reference
Pb + OX → PbO* + X OX = Cl_2, Br_2	Weak monohalide emissions observed; also laser-induced fluorescence observations.	d
Group IVB elements Ti + OX → TiO* + X OX = O_2, ON_2, ONO, OCO	Spectroscopic observations on TiO. Addition of active nitrogen considerably increases emission intensities.	aa
Group VA elements Bi + OX → BiO + X OX = Cl_2, Br_2	Laser-induced fluorescence observations on BiCl and BiBr. For the former some chemiluminescence also observed.	d
Group VIII elements Fe + OX → FeO* + X OX = O_3, N_2O, NO_2, "active" oxygen	Infrared, orange, and blue systems of FeO.	bb
Fe + O_2 → FeO + O	FeO production established by laser-induced fluorescence; continuum of unknown origin also observed.	bb
Fe + OX → ··· OX = F_2, Cl_2, Br_2, SF_6	F_2 produces reasonably bright emission with Φ = 0.01, probably from FeF. Other emissions very weak.	bb

[a] Polanyi (1932). [b] Luria *et al.* (1976a). [c] Ham and Chang (1974).
[d] Capelle *et al.* (1973). [e] Palmer *et al.* (1975). [f] Capelle *et al.* (1975).
[g] Capelle *et al.* (1974). [h] Obenauf *et al.* (1972a). [i] Jones and Broida (1974).
[j] Field *et al.* (1974). [k] Eckstrom *et al.* (1975). [l] Sakurai *et al.* (1970).
[m] Johnson (1972). [n] Eckstrom *et al.* (1974). [o] Bradford *et al.* (1975).
[p] Menzinger and Wren (1973). [q] Sakurai *et al.* (1976). [r] Rosenwaks *et al.* (1975).
[s] Johnson *et al.* (1972). [t] Rosenwaks (1976). [u] Linton and Capelle (1977).
[v] Capelle and Brom (1975). [w] Capelle and Linton (1976). [x] Linton and Broida (1976).
[y] Kurylo *et al.* (1976). [z] Oldenborg *et al.* (1975).
[aa] Linton and Broida (1977). [bb] West and Broida (1975).

C. Traditional High Temperature Techniques

1. Self-Sustaining Flames

Up to here we have discussed the introduction of metal atoms into the reactor by direct evaporative methods from the condensed phase. An alternative approach is to thermally decompose metal "parent" compounds by introduction of metal compounds in a high temperature environment. This

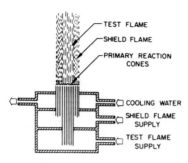

Fig. 30. Annular shield flame burner.

method has been in use, as an assaying technique, since at least the sixteenth century (Agricola, 1556). To produce free metal atoms in flames, typically, aqueous solutions of metal salts are introduced as sprays with the oxidizer flow. This manner of producing Me forms the basis of analytical flame spectroscopy (Alkemade, 1969, 1970). An alternate approach is the introduction of metal organic or carbonyl compounds. Reactions of Me atoms with species of the flame medium can lead to production of free radicals (often in excited states); flames have been widely used for spectroscopic studies of such radicals (Gaydon, 1974).

One type of flame lends itself best to kinetic studies, i.e., the fuel and oxidizer premixed flat (actually cylindrical) "test" flame with annular "shield" flame, see Fig. 30. The latter flame typically has the same fuel/oxidizer ratio and velocity as the test flame into which the Me compound is introduced. In flat flames reaction time is proportional to distance and they in effect represent wall-less flow tubes; the shield flame protects the test flame from air entrainment and sharp concentration and temperature gradients at its boundary. There are two major zones to be distinguished in premixed flames: (i) the primary reaction zone in which the main heat-release flame reactions occur and which is characterized by wide departures from equilibrium and large thermal gradients and (ii) the burned gas zone where the main flame reactions are complete, thermal gradients are small compared to those in the primary reaction zone, and equilibrium conditions of the main flame radicals are approached; in, e.g., $H_2/O_2/(N_2)$ flames these radicals are H, OH, and O. The burned gas region is the one most suited for quantitative kinetic studies. Experimentally, the premixed gases are flowed through a flow smoothing device such as a screen or a bundle of long thin tubes (cf. Fig. 30) and a laminar flow is attained in which the flow velocity of the flame gases equals the rate of propagation of the combustion reactions along the flow direction. In this way, a steady-state flame front is achieved in which the primary reaction zone (upstream of the flame front) and the burned gas zone (downstream of the flame front) are easily distinguished. With atmospheric pressure test flames, the primary reaction zone typically extends only a few millimeters in the direction of the flow, while the burned gas zone is maintained for on the order of $\gtrsim 10$ cm before mixing with the shield flame.

Because the flame medium contains many species which can interfere with the observed Me kinetics, flames are not often used for quantitative studies of neutral Me species. Nonetheless, flames are convenient media for studying reactions at elevated temperatures (≈ 1200–3000 K) and have produced considerable achievements in general kinetics measurements especially of reactions natural to the flame medium itself [see, for example, the proceedings of the biannual International Symposia on Combustion published by the Combustion Institute, Pittsburgh, PA and Fenimore (1964), Fristrom and Westenberg (1965), and Schofield and Broida (1968)].

Kinetic studies of refractory species in flames are relatively rare. An example are the studies of reactions of the type Me $+$ O$_2$ $+$ M $\rightarrow$ MeO$_2$ $+$ M, where Me is an alkali metal, in oxygen-rich H$_2$/O$_2$/N$_2$ flames by Kaskan (1965), Carabetta and Kaskan (1968), and McEwan and Phillips (1966). In those studies [Me]$_{rel}$ was determined by atomic absorption. The rate coefficients obtained by these workers appear to be somewhat too low (Kaufman, 1969), possibly due to Me oxide reactions with flame species (H, O, OH) regenerating the Me.

Rate coefficients for the excitation of Me atoms acting as third bodies in H$_2$/O$_2$/N$_2$ flame radical recombination reactions have also been obtained (Sugden, 1962; Phillips and Sugden, 1961), e.g.,

$$Tl + H + H \longrightarrow Tl^* + H_2 \tag{58}$$

A large number of such reactions has been observed; however, only when rate coefficients for quenching of Me by flame gas molecules (H$_2$O, H$_2$, O$_2$, N$_2$) at the correct temperatures are available can rate coefficients for the excitation reactions be backed out (Sugden, 1962); flames thus represent a fairly complex medium for observation of these reactions.

Flames have found wide application in studies providing equilibrium constants and reaction enthalpies, and hence heats of formation and bond energies, for high temperature reaction intermediates. The work of Sugden's group at Cambridge has led to a series of well-characterized (known concentrations of flame species and temperatures) fuel-rich H$_2$/O$_2$/N$_2$ flames suitable for such equilibrium studies (Padley and Sugden, 1958; Jensen and Padley, 1966a,b). Examples are (Jensen and Jones, 1973)

$$Fe + H_2O \; \rightleftharpoons \; FeOH + H \tag{59}$$

$$Fe + OH \; \rightleftharpoons \; FeO + H \tag{60}$$

and similar studies on Al (Jensen and Jones, 1972) and Co (Jensen and Jones, 1976). The calculational methods used are too lengthy for discussion here. Emission spectrometry is used to determine [Me] and [MeO]. Such measurements depend critically on the f numbers of the transitions observed which represent the largest potential error source. The method has been extended to ionic species, e.g. (Jensen, 1969; Jensen and Miller, 1970)

$$Sn^+ + H_2O \;\rightleftharpoons\; SnOH^+ + H \tag{61}$$

$$H_2WO_4 + e^- \;\rightleftharpoons\; HWO_4^- + H \;\rightleftharpoons\; WO_3^- + H_2O \tag{62}$$

and similar alkaline earth (Jensen, 1968; Jensen and Jones, 1975), Cr (Miller, 1972), B (Miller and Gould, 1976), and Re (Gould and Miller, 1975) studies. The tools for such studies include positive and negative ion mass spectrometers, electrostatic probes for total positive ion concentration measurements, and microwave resonant cavities for free electron concentration measurements. Discussion of these flame diagnostic techniques is outside the scope of this chapter; they are given in the original articles and have been reviewed by Jensen and Travers (1974) and by Miller (1977).

2. Shock Tubes

The temperature range covered by chemical shock tubes is comparable to that of self-sustaining flames, though higher and lower temperatures can be achieved. In dissociation studies, temperatures up to $\approx 20,000$ K have been used (Troe and Wagner, 1973). In shock tubes, isolated elementary homogeneous reactions can be studied and wall reactions (which occur on a longer time scale) need not be considered. The shock tube differs in one important experimental aspect from all methods discussed thus far, i.e., it operates in a transient mode rather than a steady-state mode. In the latter mode, reactant and product concentrations and time are invariant at any given observation point and essentially infinite observation time is available because constant gas flows are used. By contrast, the observation time over which chemical reactions at constant pressure and temperature can be observed in shock tubes is approximately limited to the 10^{-6}–10^{-3} sec range. Thus, fast response detection techniques are needed and shock tube experiments in this respect more resemble static reaction systems (see Section III,D and Chapter 4) than other flow systems. Measuring rate coefficients with reasonable accuracy in shock tubes depends on frequent repetition of the identical experiment.

The apparatus is essentially a long tube in which a shock wave is produced by the sudden bursting of a diaphragm which separates the high pressure "driver gas" from the low pressure driven (test) gas. The shock wave itself is very thin, only several mean free paths of the gas into which it propagates.

Usually, the hot gas sample to be examined is immediately behind the shock wave. The test gas is usually $\geq 90\%$ Ar, the other $\leq 10\%$ being the reagents to be investigated or their precursors. The most accurate concentration measurement technique for this environment is probably nondisperse optical absorption using filtered PMTs and intense single line light sources; other varieties of optical spectrometry as well as mass spectrometry are also in use. The output of the PMT or mass spectrometer used is fed to an oscilloscope or corresponding digital device. Initial pressure (concentrations) and temperature of the shock-heated gas are calculated from the shock front velocity. The latter is obtained by measuring the time for the shock front to travel between sets of transducers placed on the wall. Typically, these are thin film heat transfer devices or piezoelectric pressure transducers which register the change in thermal conductivity or pressure at the wall caused by passage of the shock wave. A number of texts and reviews on chemical shock tubes have appeared (e.g., Getzinger and Schott, 1973; Gaydon and Hurle, 1963; Hartunian, 1968; Bauer, 1963; Bradley, 1962; Greene and Toennies, 1964; Wray, 1970; Hague, 1969), to which we refer for details. Few studies of refractory species have been performed; the basic techniques involved in their study of this environment are essentially the same as those used for the bulk of bi- and termolecular reactions studied which are representative of those occurring in hydrocarbon, CO and H_2 combustion flames, and heated air environments.

Von Rosenberg and Wray (1972) studied the reaction

$$Fe + O_2 \longrightarrow FeO + O \qquad\qquad (63)$$

by shock-heating mixtures of $Fe(CO)_5$ and O_2 to 2400 K. Dissociation of the $Fe(CO)_5$ to Fe was determined to be complete within the time resolution of the detection optics (3×10^{-6} sec) in the absence of O_2. In the presence of O_2, FeO emission also reached a maximum within this time, allowing a limit determination of $k_{63} \geq 5 \times 10^{-12}$ mliter molecule^{-1} sec^{-1}. These experiments also allowed determination of the FeO 11.5 μ band integrated intensity. Homogeneous nucleation rates of Me atoms were determined in a series of studies by Kung and Bauer (1971), Freund and Bauer (1977), Frurip and Bauer (1977), and Bauer and Frurip (1977). Supersaturated Me vapor was obtained by shock heating of $Fe(CO)_5$, $Pb(CH_3)_4$, or $Bi(CH_3)_3$. The subsequent homogeneous condensation was followed by measuring the turbidity of the gas as a function of time. The results can be expressed in terms of parameters that specify the particle size distribution (average size 40–2000 atoms). The approximate temperature range covered in these studies is 1000–2300 K. A somewhat similar study on the formation of Pb particles and their oxidation in the presence of O_2 to PbO particles has been reported by Homer and Hurle (1972).

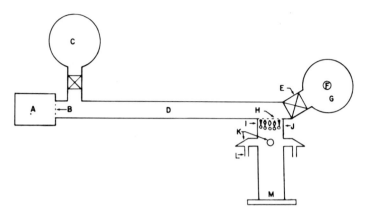

Fig. 31. Schematic diagram of powder injection shock tunnel system. A, driver; B, driver burst diaphragm; C, dummy volume tank (342 liter); D, driven section (56 liter; 12-m long); E, injection valve; F, powder chamber (43 mliter); G, injection tank (32 liter); H, dump tank burst diaphragm; I, supersonic nozzle array (37 nozzles); J, oxidant injector tubes; K, reaction zone observation window; L, sweeper purge gas; M, dump tank. [From Johnson *et al.* (1974, Fig. 1).]

An interesting device, the powder injection shock tunnel, was recently described by Johnson *et al.* (1974), see Fig. 31. In this apparatus, micron size powder particles suspended in Ar are shock heated to some 10,000 K immediately after their introduction into the driven section to produce metal atoms on a time scale of $< 10^{-5}$ sec. The Me/Ar mixture then expands through an array of supersonic nozzles and mixes with oxidizer injected at the nozzle exit. The resulting chemiluminescence is then observed in the mixing region. The high evaporation temperature allows study of virtually any metal which can be handled in fine particle form either as pure metal or as a suitable compound. As the latter MgF_2 has been used, and to produce Sr and Ba, mixtures of B and SrH_2, BaH_2 powders were employed. Reaction zone temperature is not well defined but approximates the static flow temperature (≈ 800 K) of the Me/Ar mixture at the nozzle exit, i.e., before mixing with the oxidizer. The apparatus has been used for a spectral survey of reactions of Al, B, Ba, C, Cu, Mg, Si, Sr, Ti, U, W, and Zr with one or more of the oxidants O_2, N_2O, Cl_2, and F_2 (Johnson *et al.*, 1974). Additionally, ir lasing action from AlF and BF was demonstrated (Rice *et al.*, 1976). The amount of material evaporated is adequate to produce atom densities of $\approx 1 \times 10^{15}$ mliter^{-1} in the reaction zone, although this has not been measured directly. At such densities in supersonic expansion, it appears likely that in several instances clustering of the metal atoms will occur to a significant degree. Nettleton (1977) has recently reviewed the general area of shock heating of particles and droplets.

D. Photolytic Techniques

These static system techniques represent a rich source of kinetic information on reactions of atoms of refractory species. Excited metal atoms are often used to initiate ("sensitize") photochemical reactions (Calvert and Pitts, 1966). In that work rather volatile metals are used—particularly Hg—and discussion of that major branch of photochemistry would be inappropriate here. However, flash photolysis has extensively been used for generation of refractory species. The classical example of this is the study by Erhard and Norrish (1956) on the antiknock action of PbO formed from the flash photolysis of $Pb(C_2H_5)_4$ in the presence of O_2. In its major present form based on time-resolved electro-optical measurement techniques, FPRAF [flash (or pulse) photolysis resonance absorption (or) fluorescence] represents a major source for obtaining rate coefficients of elementary reactions for both refractory and nonrefractory species (see also Lin and McDonald's discussion in Chapter 4). The accuracy of the rate coefficients obtainable is usually comparable to that of regular flow tube (≈ 10–30%) methods.

In FPRAF the parent compound (often a metal organic or carbonyl compound) and the second reactant OX are introduced with a large excess of inert buffer gas (typically in the range 1–100 Torr), which serves to thermalize the photolysis product Me and prevent wall reactions. Only a small fraction of the parent compound is dissociated with each flash and operation is in the first-order mode [OX] $\gg$ [Me]. For resonance absorption studies a long cylindrical quartz cell (e.g., 2-cm i.d., 30-cm long) is used (Husain and Littler, 1972); the long path length serves to increase sensitivity. An apparatus for such measurements is shown in Fig. 32. For fluorescence observations, short cells (Davis et al., 1972; Husain et al., 1977a,b) are of course adequate. Flash lamps are discussed by McNesby et al. (1971). In absorption studies, a single

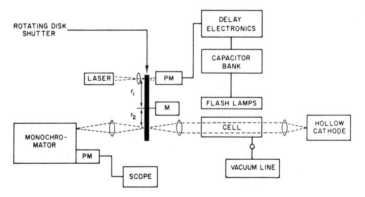

Fig. 32. Schematic diagram of flash-photolysis resonance absorption apparatus. PM, photomultiplier tube; M, motor. [From Ewing et al. (1974, Fig. 1).]

flash lamp may be used, but to obtain a uniform Me density an array of such lamps surrounding the cell is usually preferred (Ewing *et al.*, 1974). Husain and Littler (1972) have described a coaxial lamp and cell assembly, the lamp being an annular space of 0.2 cm surrounding the photolysis cell with which it shares a wall. Hollow-cathode or microwave electrical discharge lamps (Chapter 1; Davis and Braun, 1968; McNesby *et al.*, 1971) are used for the diagnostic radiation. The attenuated signal is recorded by any one of a number of fast response output devices; for low absorption or fluorescence intensity signals, averaging techniques are used.

Most FPRAF work is performed at room temperature. Studies on refractory species are assembled in Table VIII. It may be seen that temperature has been varied in only a very few studies; the widest variation ($\approx 290-570$ K) apparently has been used in the work by Trainor (1977), Trainor and Ewing (1976), and Husain and Littler (1974a). Factors limiting the temperature range over which FPRAF can be used as described are the requirements of adequate vapor pressure and good thermal stability of the parent compound. Especially for the often used metal organics, this can be a quite restrictive condition. The highest temperature at which FPRAF has been used thus far is apparently in the work of Edelstein and Davidovits (1971), Gedeon *et al.* (1971), Maya and Davidovits (1973, 1974), and Wik *et al.* (1975) on alkali and Tl atoms. In those studies the corresponding halides were heated to an adequate temperature (800–1000 K) for producing the required salt vapor density ($\approx 1 \times 10^{15}-1 \times 10^{16}$ mliter^{-1}). However, at these temperatures the halogens used as oxidizers are partially dissociated. Their concentration is then calculated using known equilibrium constants (Edelstein and Davidovits, 1971); in static reactors enough time is, of course, available for the establishment of equilibrium.

Photolytic techniques can be used to produce atoms in excited states as well as in the ground state. A considerable number of observations of excited atoms thus produced have been made. The quenching rate coefficients obtained by Me* disappearance measurements can pertain to either true physical quenching or chemical reaction and further consideration or other experimental work is needed to distinguish between them (Kolts and Setser, Chapter 3). Therefore, in a few cases, in addition to the regular FPRAF experiment, products have been identified by absorption spectrometry, as is often done in classical photochemistry. Of course, unless the time history of the products thus identified is simultaneously obtained, primary and secondary products cannot be directly distinguished by this method.

Most of the FPRAF rate coefficient measurements of refractory species are due to the work of Husain's group at Cambridge. He and Donovan have frequently reviewed this field. Further details may be found in these reviews (Donovan and Gillespie, 1975; Donovan *et al.*, 1972; Donovan and Husain,

Table VIII

Some FPRAF Reactions[a]

Me	Parent	OX	Remarks	Reference
Group IA elements				
Na, K, Rb, Cs	NaI, KI, RbI, CsI	I_2	$T = 860$ K. Ground state ($^2S_{1/2}$) observations using one of the D lines.	[b]
Na, K, Rb, Cs	NaBr, KBr, RbBr, CsBr	Br_2	$T = 973$ K. Observed as in Ref. [b].	[c]
Na, K, Rb, Cs	NaCl, KCl, RbCl, CsCl	Cl_2	$T = 1015$ K. Observed as in Ref. [b].	[d]
Group IIB elements				
Cd*($5^3P_{0,1}$)	Cd(5^1S_0)	H_2, HD, D_2, CH_4, SF_6	CdH, CdF detected in absorption.	[e]
Hg*($6^3P_{0,1}$)	Hg(6^1S_0)	H_2, HD, D_2	HgH, HgD detected in absorption.	[f, g]
Group IIIA elements				
Tl($6^2P_{1/2}$)	TlI	I_2	$T = 600$ K.	[h]
Tl*($6^2P_{3/2}$)	TlI	I_2	$T = 600$ K; this Tl state is quenched by I_2 to ground state ($6^2P_{1/2}$) which then reacts rapidly with I_2 to produce TlI.	[h]
Tl($6^2P_{1/2}$)	TlBr	Br_2	$T = 690$ K.	[i]
Tl*($6^2P_{3/2}$)	Tl(CH_3)$_3$	N_2, CO, H_2, D_2, NO, O_2, CO_2, CH_4, CF_3H, CF_4, C_2H_6, C_3H_8, C_2H_4, C_3H_6, and larger organic compounds	Data of Ref. [k, l] included for comparison.	[j]
Group IVA elements				
C(2^3P_J)	C_3O_2	CH_4, NO, O_2	Some product identification by absorption spectrography.	[m]
C(2^3P_J)	C_3O_2	H_2, O_2, NO, N_2O, CO_2, C_3O_2	$J = 0, 1, 2$ levels may be assumed in equilibrium due to small energy gaps between spin–orbit levels (also pertains to Ref. m)	[n]

Some product identification by absorption spectrography.

Species	Substrate	Reactants	Notes	Ref.
C*(2¹D₂)	C₃O₂	CH₄, N₂, NO, H₂, O₂		
C*(2¹D₂)	C₃O₂	He, Ne, Ar, Kr, Xe		o
C*(2¹D₂)	C₃O₂	NO, CO, CH₄, CO₂, N₂O, C₂H₄, O₂, H₂O, H₂, N₂		p, q
C*(2¹S₀)	C₃O₂	He, H₂, N₂, CO, CH₄, C₃O₂	Preliminary H₂ observations also in Ref. m.	r
Si(3³P_J)	SiCl₄	H₂, N₂, O₂, Cl₂, CO, NO, CO₂, N₂O, CH₄, CF₄, C₂H₂, C₂H₄, SiCl₄	Small spin–orbit splitting, J = 0, 1, 2 levels are in equilibrium during reaction.	s
Si*(3¹D₂)	SiCl₄	He, H₂, O₂, SiCl₄		t
Si*(3¹S₀)	SiCl₄	O₂, N₂O		u
Ge(4³P_{0,1,2})	GeBr₄	He, Xe, O₂, CO, NO, N₂, CO₂, N₂O, CH₄, CF₄, SF₆, C₂H₂, C₂H₄, GeBr₄		v, w
Ge*(4¹S₀)	Ge(CH₃)₄	Xe, D₂, H₂, O₂, N₂, NO, CO, N₂O, CO₂, CF₄, CH₄, SF₆, CF₃H, C₂H₂, C₂H₄, Ge(CH₃)₄		x
Sn(5³P₀)	Sn(CH₃)₄	N₂O	T varied from 341 to 377 K. Both A and E_{sl} in good agreement with HTFFR results (Ref. z) at T ≥ 500 K.	y
Sn*(5³P_{1,2})	Sn(CH₃)₄	Ar, Kr, Xe, H₂, D₂, HD, N₂, O₂, CO, NO, CO₂, CH₄, CF₃H, CF₄, C₂H₄, C₂D₄, C₂H₂, C₂D₂, Sn(CH₃)₄		aa
Sn*(5¹D₂)	Sn(CH₃)₄	He, Xe, H₂, O₂, N₂, NO, CO, CO₂, N₂O, CH₄, C₂H₄, C₂H₂, Sn(CH₃)₄		bb, cc
Sn*(5²S₀)	SnCl₄	He, Xe, H₂, O₂, SnCl₄		dd
Pb(6³P₀)	Pb(C₂H₅)₄	He, CO, CO₂, N₂O, CH₄, C₂H₄, C₂H₂		ee

(continued)

137

Table VIII (*continued*)

Me	Parent	OX	Remarks	Reference
$Pb(6^3P_0)$	$Pb(C_2H_5)_4$	$O_2 + M$, $NO + M$; $M = He$, N_2, CO, CO_2, CH_4, CF_4, SF_6, C_2H_6, C_2H_2, C_2H_4	Determination of third-order rate coefficient. See also Refs. *ee* and *gg*.	*ff*
$Pb(6^3P_0)$	$Pb(C_2H_5)_4$	$O_2 + M(= He, CO_2, C_2H_6)$; $NO + M(= CO_2, C_2H_6)$	T varied from 292 to 573 K.	*gg*
$Pb*(6^3P_{1,2})$	$Pb(C_2H_5)_4$	He, Ar, H_2, D_2, N_2, O_2, CO, NO, CO_2, N_2O, CH_4, C_2H_4, C_2H_2, CF_4, SF_6, $Pb(C_2H_5)_4$		*hh*
$Pb*(6^3P_{1,2})$	$Pb(CH_3)_4$	Ar, Xe, H_2, D_2, N_2, O_2		*ii*
$Pb*(6^3P_{1,2})$	$Pb(CH_3)_4$	Ar, Xe, N_2, D_2, H_2, Hg	T varied from 300 to 560 K. However, for Hg only 373 K was used.	*jj*
$Pb*(6^1D_2)$	$Pb(C_2H_5)_4$	He, Xe, H_2, N_2, O_2, CO, NO, CO_2, N_2O, CH_4, C_2H_4, C_2H_2, SF_6, $Pb(C_2H_5)_4$		*kk*
$Pb*(6^1S_0)$	$Pb(C_2H_5)_4$	Xe, H_2, O_2, $Pb(C_2H_5)_4$		*ll*
Group V A elements				
$Sb*(5^4S_{3,2})$	$Sb(CH_3)_3$	N_2O, NO, $Sb(CH_3)_3$; C_2H_4, C_2H_2, $O_2 + M(= He, N_2, SF_6)$	Second-order rate coefficients. Third-order rate coefficients.	*mm*
$Sb(5^4S_{3/2})$	$Sb(CH_3)_3$	NO; $O_2 + He$	This study done in resonance fluorescence to allow comparison to the absorption study of Ref. *mm*.	*nn*
$Sb*(5^2D_{3,2,5,2})$	$Sb(CH_3)_3$	He, H_2, O_2, CO, CO_2, $Sb(CH_3)_3$	Results are compared to those for the lighter nonmetallic VA elements: N, P, As.	*oo*
$Bi(6^4S_{3,2})$	$Bi(CH_3)_3$	NO. O_2, N_2O, C_2H_4, $C_2H_2 + M(= He, N_2, SF_6)$	Second-order rate coefficient. Data too complex for detailed analysis. Third-order rate coefficients. For $C_2H_4 + Bi + M$, T was varied from ≈ 290 to 500 K, showing a small, 1 kcal mole^{-1}, positive T-dependence.	*pp*

Bi($6^4S_{3/2}$)	Bi(CH$_3$)$_3$	C$_2$H$_4$, C$_2$H$_2$ + He	This study done in resonance fluorescence to allow comparison to the absorption study, see Ref. *pp*.	*qq*
Bi*($6^2D_{3/2.5/2}$)	Bi(CH$_3$)$_3$	He, Xe, H$_2$, D$_2$, O$_2$, N$_2$, CO, CO$_2$, N$_2$O, C$_2$H$_4$		*rr*
Bi*($6^2D_{3/2.5/2}$)	Bi(CH$_3$)$_3$	Ar, Xe, N$_2$, H$_2$, D$_2$, CO, O$_2$, CO$_2$, SF$_6$	T varied from 300 to 550 K.	*ss*

Group VIA elements

Te(5^3P_2)	Te(CH$_3$)$_2$	C$_2$H$_4$, C$_3$H$_6$, tetramethylethylene, Te(CH$_3$)$_2$	Measured at 298 and 353 K. Addition to C$_2$H$_4$, C$_3$H$_6$ has positive activation energy, to tetramethylethylene has negative activation energy. Results are further compared to lighter nonmetallic elements of group VIA (O, S, Se) in Ref. *uu*.	*tt*
Te*($5^3P_{0,1}$)	H$_2$Te	He, Ar, Xe, H$_2$, D$_2$, O$_2$		*uu*
Te*(5^1D_2)	D$_2$Te	Ar, D$_2$Te		*ww*

[a] Excited states of atoms are indicated by * . J, used instead of a number, indicates that spin-orbit components were not separated. Reaction temperature is approximately room temperature unless otherwise mentioned.
[b] Edelstein and Davidovits (1971). [c] Maya and Davidovits (1973). [d] Maya and Davidovits (1974).
[e] Breckenridge and Broadbent (1974). [f] Callear and McGurk (1972). [g] Callear and Wood (1972).
[h] Gedeon et al. (1971). [i] Wik et al. (1975). [j] Foo et al. (1975).
[k] Bellisio and Davidovits (1970). [l] Wiesenfeld (1973). [m] Braun et al. (1969).
[n] Husain and Young (1975). [o] Husain and Kirsch (1971a). [p] Husain and Kirsch (1971b).
[q] Husain and Kirsch (1971c). [r] Husain and Kirsch (1973–1974). [s] Husain and Norris (1978a).
[t] Husain and Norris (1978b). [u] Husain and Norris (1977a). [v] Brown and Husain (1976).
[w] Chowdhury and Husain (1977a). [x] Chowdhury and Husain (1977b). [y] Wiesenfeld and Yuen (1976).
[z] Felder and Fontijn (1978). [aa] Foo et al. (1976). [bb] Brown and Husain (1975).
[cc] Brown and Husain (1974–1975a). [dd] Brown and Husain (1974–1975b). [ee] Husain and Littler (1973–1974).
[ff] Cross and Husain (1977). [gg] Husain and Littler (1974a). [hh] Husain and Littler (1974b).
[ii] Ewing et al. (1974). [jj] Trainor and Ewing (1976). [kk] Husain and Littler (1972).
[ll] Husain and Littler (1972–1973). [mm] Husain and Slater (1977). [nn] Husain et al. (1977b).
[oo] Bevan and Husain (1975). [pp] Husain and Slater (1976–1977). [qq] Husain et al. (1977a).
[rr] Bevan and Husain (1976). [ss] Trainor (1977). [tt] Connor et al. (1971).
[uu] Donovan and Gillespie (1975). [vv] Donovan and Little (1973). [ww] Donovan et al. (1972–1973).

1970, 1971), as well as in the original literature. With the exception of the very recent studies on ground state Sb and Bi atoms (Husain *et al.*, 1977a,b), all of the work was done by the absorption method.

E. Miscellaneous Techniques

A variety of other methods has been used to produce metal atoms for kinetic studies. In the first place there are electrical discharge methods, in principle, similar to those discussed by Clyne and Nip in Chapter 1. Thus, Hager *et al.* (1974, 1975) and Swearengen *et al.* (1977) used a hollow-cathode discharge through 1 % SiH_4 and 1 % GeH_4 in He to produce Si and Ge atoms, respectively. Sridharan *et al.* (1976) used 1 % B_2H_6 in He to obtain free B atoms with a microwave discharge and Lam *et al.* (1971) and Martinoti *et al.* (1968) employed similarly diluted C_3O_2 or allene to produce free C atoms. Chou and Cool (1976) produced a number of Me species for ir laser observations by passing metal complexes through a pulsed discharge. Dilute mixtures are recommended to reduce recombination with the other reactive fragments from the discharge; nonetheless the method suffers from the presence of fragment species of the parent compounds which may include excited species (Sridharan *et al.*, 1976). Under conditions where direct evaporation cannot be used, the discharge method appears to represent an attractive alternative, provided proper attention is paid to the possible complications by other discharge-produced species.

Stripping or titration reactions could be used to produce Me atoms. This has been done successfully for C atoms as discussed in Chapter 1. Stripping of metal organics and carbonyls as well as of CH_4 and C_4H_{10} with F_2 was attempted by Black *et al.* (1974) who report it to be an inefficient method. Lee and Zare (1975) have produced intense Zn and ZnH emission from the $Zn(C_2H_5)_2/O_3$ reaction. The formation mechanism of the emitters probably involves free Zn atoms, H, and OH, similar to processes occurring in self-sustaining flames (Section III,C,1). No such emission could be observed from $Zn(CH_3)_2/O_3$,

Rice and Beattie (1973) have reported ir lasing action from MeF and MeO by exploding wires, metallic films, and graphite smears into appropriate (F_2, NF_3, O_2) oxidizing environments. Because of its irregular transient nature, this method does not seem promising for kinetic studies.

IV. CONCLUSIONS

In reviewing the literature quoted it is apparent that nearly all the quantitative kinetic information on reactions of refractory atoms is due to the techniques developed within the last ten years. Clearly this is an area of healthy growth. What further developments should be anticipated? It would be logical to extend the HTFFR technique to measurements on larger radicals by using mass spectrometry. Such an extension, which would also be useful for atoms and diatomic radicals to provide a further check on optical measurements, is being attempted at several laboratories. It would be desirable to study nonrefractory species such as O, OH, H, etc., by a single technique over as wide a temperature range as is accessible by the HTFFR. There is some doubt if a flow reactor technique would be the right approach because of rapid wall recombination of these reactants and thermal dissociation of many potential collision partners. However, we are working on an adaptation of the FPRAF technique (Section III,D) by which reactions of such nonrefractory species could be studied in a furnace resembling an HTFFR. Ultimately, this HTP (high temperature photochemistry) technique could then also be applied to refractory species; the use of two independent techniques for obtaining similar data is, of course, always a highly desirable development for increasing confidence in kinetic measurements.

Very little work has been done on reactions of free radicals of refractory species. The possibility for their reactive formation inside the same flow tube as used for the measurement of their reactions has been discussed in Section II,A,1 for the case of AlO. This approach lends itself well for further applications but has the disadvantage that, in studying reactions of an OX species other than that used in the MeO formation, the first OX species could interfere. The measurement of the absolute rate coefficient for MeO reaction with the first OX species and the relative rate coefficients of the reactions of MeO with both OX species is an obvious but sometimes time-consuming answer here. For direct evaporation of free radicals the thin film literature quoted in Section II,B,2, particularly Glang (1970, Section 6), represents a starting point. Most experience quoted there involves dissociative evaporation; the other dissociation products formed represent a potential error source for kinetic measurements. In general, information on evaporation conditions, which favor a given species from prior mass spectrometric studies of such dissociation processes, would be needed for meaningful kinetic studies. The aforementioned works by Hastie (1975) and Margrave (1967) can also be useful guides in the present context.

Finally, it is interesting to mention the existence of cocondensation techniques to produce Me-containing compounds in synthetic chemistry. Here refractory atoms are produced, usually by evaporation techniques, and are condensed together with a large excess of a molecular species, e.g., olefins, at a low (usually liquid nitrogen) temperature. Sometimes the Me atoms are condensed in an inert matrix, after which the molecular species is condensed and reaction takes place upon warming. This field originated with Skell and Westcott's work (1963) using C, C_2, C_3 produced in a carbon arc. For reviews see Skell et al. (1973), Klabunde (1975a,b), Gowenlock et al. (1976), Timms (1972, 1975), and especially Moskovits and Ozin (1976). There has been little contact between workers in this field and those in gas kinetic studies of refractory atoms, though clearly both could benefit by being more aware of what the others have developed.

ACKNOWLEDGMENTS

The preparation of this chapter has been made possible by the support of the Air Force Office of Scientific Research (AFSC), United States Air Force, under Contracts F44620-76-C-0108 (J137) and F49620-77-C-0033 (J150). The United States government is authorized to reproduce and distribute reprints for governmental purposes notwithstanding any copyright notation thereon. We also thank the Office of Naval Research (Project SQUID), the Sandia Corporation, The Defense Advanced Research Projects Agency, and The Defense Nuclear Agency for their past support of our HTFFR work.

Mr. J. J. Houghton has played a major role in translating the HTFFR concept into a practical reality. We also thank him for his comments on Section II,B and Mrs. H. Rothschild for performing the rather laborious editing involved in the preparation of this manuscript.

REFERENCES

Ackerman, M. (1967). *Bull. Cl. Sci. Acad. R. Belg.* **53**, 1311.
Agricola, G. (1556). "De Re Metallica" 1st Latin edition (translated into English by H. C. and L. H. Hoover, Book VII. Dover, New York, 1950).
Alkemade, C. Th. J. (1969). *In* "Flame Emission and Absorption Spectrometry" (J. A. Dean and T. C. Rains, eds.), Vol. I, Chapter 4. Dekker, New York.
Alkemade, C. Th. J. (1970). *In* "Analytical Flame Spectroscopy, Selected Topics" (R. Mavrodineanu, ed.), Chapter 1. Macmillan, New York.
Bauer, S. H. (1963). *Science* **141**, 867.
Bauer, S. H., and Frurip, D. J. (1977). *J. Phys. Chem.* **81**, 1015.
Bellisio, J. A., and Davidovits, P. (1970). *J. Chem. Phys.* **53**, 3474.
Bemand, P. P., and Clyne, M. A. A. (1973). *J. Chem. Soc., Faraday Trans. 2* **69**, 1643.
Benson, S. W. (1975). *Int. J. Chem. Kinet. Symp.* **1**, 359.
Bevan, M. J., and Husain, D. (1975). *J. Photochem.* **4**, 51.
Bevan, M. J., and Husain, D. (1976). *J. Phys. Chem.* **80**, 217.
Bevington, P. R. (1969). "Data Reduction and Error Analysis for the Physical Sciences." McGraw-Hill, New York.
Black, G., Luria, M., Eckström, D. J., Edelstein, S. A., and Benson, S. W. (1974). *J. Chem. Phys.* **60**, 3709.
Bolden, R. C., Hemsworth, R. S., Shaw, M. J., and Twiddy, N. D. (1970). *J. Phys. B.* **3**, 45.
Bradford, R. S., Jones, C. R., Southall, L. A., and Broida, H. P. (1975). *J. Chem. Phys.* **62**, 2060.
Bradley, J. N. (1962). "Shock Waves in Chemistry and Physics." Methuen, London.
Braun, W., Bass, A. M., Davis, D. D., and Simmons, J. D. (1969). *Proc. R. Soc. London, Ser. A* **312**, 417.
Breckenridge, W. H., and Broadbent, T. W. (1974). *Chem. Phys. Lett.* **29**, 421.
Brokaw, R. S. (1960). NASA TR-R-81. NASA Lewis Research Center, Cleveland, Ohio.
Brown, A., and Husain, D. (1974–1975a). *J. Photochem.* **3**, 37.
Brown, A., and Husain, D. (1974–1975b). *J. Photochem.* **3**, 305.

Brown, A., and Husain, D. (1975). *Int. J. Chem. Kinet.* **7**, 77.

Brown, A., and Husain, D. (1976). *Can. J. Chem.* **54**, 4.

Bruce, C. F., and Hannaford, P. (1971). *Spectrochim. Acta, Part B*, **26**, 207.

Callear, A. B., and McGurk, J. C. (1972). *J. Chem. Soc., Faraday Trans. 2* **68**, 289.

Callear, A. B., and Wood, P. M. (1972). *J. Chem. Soc., Faraday Trans. 2* **68**, 302.

Calvert, J. G., and Pitts, J. N. (1966). "Photochemistry," Chapter 2. Wiley, New York.

Campbell, I. E., and Sherwood, E. M. (1967). "High-Temperature Materials and Technology," Wiley, New York.

Capelle, G. A., and Brom, J. M. (1975). *J. Chem. Phys.* **63**, 5168.

Capelle, G. A., and Linton, C. (1976). *J. Chem. Phys.* **65**, 5361.

Capelle, G. A., Bradford, R. S., and Broida, H. P. (1973). *Chem. Phys. Lett.* **21**, 418.

Capelle, G. A., Jones, C. R., Zorskie, J., and Broida, H. P. (1974). *J. Chem. Phys.* **61**, 4777.

Capelle, G. A., Broida, H. P., and Field, R. W. (1975). *J. Chem. Phys.* **62**, 3131.

Carabetta, R., and Kaskan, W. E. (1968). *J. Phys. Chem.* **72**, 2483.

Chalek, C. L., and Gole, J. L. (1976). *J. Chem. Phys.* **65**, 2845.

Chou, M.-S., and Cool, T. A. (1976). *J. Appl. Phys.* **47**, 1055.

Chowdhury, M. A., and Husain, D. (1977a). *J. Photochem.* **7**, 41.

Chowdhury, M. A., and Husain, D. (1977b). *J. Chem. Soc., Faraday Trans. 2* **73**, 1805.

Clyne, M. A. A. and Nip, W. S. This volume, Chap. 1.

Connor, J., van Roodselaar, A., Fair, R. W., and Strausz, O. P. (1971). *J. Am. Chem. Soc.* **93**, 560.

Cross, P. J., and Husain, D. (1977). *J. Photochem.* **7**, 157.

Cvetanovic, R. J., and Singleton, D. L. (1977a). *Int. J. Chem. Kinet.* **9**, 481.

Cvetanovic, R. J., and Singleton, D. L. (1977b). *Int. J. Chem. Kinet.* **9**, 1007.

Cvetanovic, R. J., Overend, R. P., and Paraskevopoulos, G. (1975). *Int. J. Chem. Kinet. Symp.* **1**, 249.

Dagdigian, P. J., Cruse, H. W., and Zare, R. N. (1975). *J. Chem. Phys.* **62**, 1824.

Davis, C. C., and McFarlane, R. A. (1977). *J. Quant. Spectrosc. & Radiat. Transfer* **18**, 151.

Davis, D. D., and Braun, W. (1968). *Appl. Oct.* **7**, 2071.

Davis, D. D., Huie, R. E., Herron, J. T., Kurylo, M. J., and Braun, W. (1972). *J. Chem. Phys.* **56**, 4868.

deZafra, R. L., and Marshall, A. (1968). *Phys. Rev.* **170**, 28.

Dickson, C. R., and Zare, R. N. (1975). *Chem. Phys.* **7**, 361.

Dickson, C. R., George, S. M., and Zare, R. N. (1977). *J. Chem. Phys.* **67**, 1024.

Donovan, R. J., and Gillespie, H. M. (1975). *React. Kinet.* **1**, Chap. 2.

Donovan, R. J., and Husain, D. (1970). *Chem. Rev.* **70**, 489.

Donovan, R. J., and Husain, D. (1971). *Annu. Rep. Prog. Chem., Sect. A* **68**, 124.

Donovan, R. J., and Little, D. J. (1973). *J. Chem. Soc., Faraday Trans. 2* **69**, 952.

Donovan, R. J., Husain, D., and Kirsch, L. J. (1972). *Annu. Rep. Prog. Chem., Sect. A* **69**, 19.

Donovan, R. J., Little, D. J., and Konstantatos, J. (1972–1973). *J. Photochem.* **1**, 86.

Donovan, R. J., Husain, D., and Kirsch, J. J. (1975). *J. Chem. Soc., Faraday Trans. 2* **66**, 2551.

Dubois, L. H., and Gole, J. L. (1977). *J. Chem. Phys.* **66**, 779.

Dushman, S., and Lafferty, J. M. (1962). "Scientific Foundations of Vacuum Technique," 2nd ed. Wiley, New York.

Eckert, E. R. G., and Drake, R. M., Jr.(1959). "Heat and Mass Transfer," Chapter 7. McGraw-Hill, New York.

Eckstrom, D. J., Edelstein, S. A., and Benson, S. W. (1974). *J. Chem. Phys.* **60**, 2930.

Eckstrom, D. J., Edelstein, S. A., Huestis, D. C., Perry, B. E., and Benson, S. W. (1975). *J. Chem. Phys.* **63**, 3828.

Edelstein, S. A., and Davidovits, P. (1971). *J. Chem. Phys.* **55**, 5164.

Engelke, F., and Zare, R. N. (1977). *Chem. Phys.* **19**, 327.

Engelke, F., Sander, R. K., and Zare, R. N. (1976). *J. Chem. Phys.* **65**, 1146.

Erhard, K. H. L., and Norrish, R. G. W. (1956). *Proc. R. Soc. London, Ser. A* **234**, 178.

Ewing, J. J., Trainor, D. W., and Yatsiv, S. (1974). *J. Chem. Phys.* **61**, 4433.

Farragher, A. L. (1970). *Trans. Faraday Soc.* **66**, 1411.

Felder, W., and Fontijn, A. (1975). *Chem. Phys. Lett.* **34**, 398.

Felder, W., and Fontijn, A. (1976). *J. Chem. Phys.* **64**, 1977.

Felder, W., and Fontijn, A. (1978). *J. Chem. Phys.* **69**, 1112.

Felder, W., Gould, R. K., and Fontijn, A. (1976). *In* "Electronic Transition Lasers" (J. I. Steinfeld, ed.), p. 68. MIT Press, Cambridge, Massachusetts.

Felder, W., Gould, R. K., and Fontijn, A. (1977). *J. Chem. Phys.* **66**, 3256.

Fenimore, C. P. (1964). "Chemistry in Premixed Flames." Pergamon, Oxford.

Ferguson, E. E., Fehsenfeld, F. C., and Schmeltekopf, A. (1969). *Adv. At. Mol. Phys.* **5**, 1.

Field, R. W., Jones, C. R., and Broida, H. P. (1974). *J. Chem. Phys.* **60**, 4377.

Fite, W. L., and Datz, S. (1963). *Annu. Rev. Phys. Chem.* **14**, 61.

Fite, W. L., Lo, H. H., and Irving, P. (1974). *J. Chem. Phys.* **60**, 1236.

Fontijn, A. (1972). *Prog. React. Kinet.* **6**, 75.

Fontijn, A. (1974). *Pure Appl. Chem.* **39**, 287.

Fontijn, A. (1977). *Chem. Phys. Lett.* **47**, 141.

Fontijn, A., and Felder, W. (1977a). *J. Chem. Phys.* **67**, 1561.

Fontijn, A., and Felder, W. (1977b). *Chem. Phys. Lett.* **47**, 380.

Fontijn, A., and Felder, W. (1977c). *In* "Electronic Transition Lasers II" (L. E. Wilson, S. N. Suchard, and J. I. Steinfeld, eds.), p. 112. MIT Press, Cambridge, Massachusetts.

Fontijn, A., and Felder, W. (1979). *J. Phys. Chem.* **83**, 24.

Fontijn, A., and Johnson, S. E. (1973). *J. Chem. Phys.* **59**, 6193.

Fontijn, A., and Kurzius, S. C. (1972). *Chem. Phys. Lett.* **13**, 507.

Fontijn, A., Meyer, C. B., and Schiff, H. I. (1964). *J. Chem. Phys.* **40**, 64.

Fontijn, A., Kurzius, S. C., Houghton, J. J., and Emerson, J. A. (1972). *Rev. Sci. Instrum.* **43**, 726.

Fontijn, A., Kurzius, S. C., and Houghton, J. J. (1973). *Symp. (Int.) Combust. [Proc.]* **14**, 167.

Fontijn, A., Felder, W., and Houghton, J. J. (1974). *Chem. Phys. Lett.* **27**, 365.

Fontijn, A., Felder, W., and Houghton, J. J. (1975). *Symp. (Int.) Combust. [Proc.]* **15**, 775.

Fontijn, A., Felder, W., and Houghton, J. J. (1977). *Symp. (Int.) Combust. [Proc.]* **16**, 871.

Foo, P. D., Lohman, T., Podolske, J., and Wiesenfeld, J. R. (1975). *J. Phys. Chem.* **79**, 414.

Foo, P. D., Wiesenfeld, J. R., Yuen, M. J., and Husain, D. (1976). *J. Phys. Chem.* **80**, 91.

Freund, H. J., and Bauer, S. H. (1977). *J. Phys. Chem.* **81**, 1994.

Friichtenicht, J. F. (1974). *Rev. Sci. Instrum.* **45**, 51.

Friichtenicht, J. F., and Tang, S. P. (1976). *In* "Electronic Transition Lasers" (J. I. Steinfeld, ed.), p. 36. MIT Press, Cambridge, Massachusetts.

Fristrom, R. M., and Westenberg, A. A. (1965). "Flame Structure." McGraw-Hill, New York.

Froese-Fischer, C. (1976). *Can. J. Phys.* **54**, 740.

Frurip, D. J., and Bauer, S. H. (1977). *J. Phys. Chem.* **81**, 1001 and 1007.

Gardiner, W. C., Jr. (1977). *Acc. Chem. Res.* **10**, 326.

Gaydon, A. G. (1974). "The Spectroscopy of Flames," 2nd ed. Chapman & Hall, London.

Gaydon, A. G., and Hurle, I. R. (1963). "The Shock Tube in High-Temperature Chemical Physics." Van Nostrand-Reinhold, Princeton, New Jersey.

Gedeon, A., Edelstein, S. A., and Davidovits, P. (1971). *J. Chem. Phys.* **55**, 5171.

Getzinger, R. W., and Schott, G. L. (1973). *In* "Physical Chemistry of Fast Reactions," Vol. 1, "Gas Phase Reactions of Small Molecules" (B. P. Levitt, ed.), Chapter 2. Plenum Press, New York.

Glang, R. (1970). In "Handbook of Thin Film Technology" (L. I. Maissel and R. Glang, eds.) Chapter 1. McGraw-Hill, New York.

Golde, M. F., Roche, A. E., and Kaufman, F. (1973). J. Chem. Phys. 59, 3953.

Gole, J. L. (1976). Annu. Rev. Phys. Chem. 27, 525.

Gole, J. L., and Chalek, C. L. (1976). J. Chem. Phys. 65, 4384.

Gole, J. L., and Chalek, C. L. (1978). Proc. Symp. High Temp. Met. Halide Chem., 78–1, p. 278. Electrochemical Society, Princeton, New Jersey.

Gole, J. L., and Chalek, C. L. (1979). J. Chem. Phys. (submitted for publication).

Gole, J. L., and Preuss, D. R. (1977). J. Chem. Phys. 66, 3000.

Gole, J. L., and Zare, R. N. (1972). J. Chem. Phys. 57, 5331.

Gole, J. L., Preuss, D. R., and Chalek, C. L. (1977). J. Chem. Phys. 66, 548.

Gould, R. K., and Miller, W. J. (1975). J. Chem. Phys. 62, 644.

Gowenlock, B. G., Johnson, C. A. F., and Parker, J. E. (1976). In "Comprehensive Chemical Kinetics," Vol. 18, "Selected Elementary Reactions" (C. H. Bamford and C. F. H. Tipper, eds.), Chapter 4. Elsevier, Amsterdam.

Greene, E. F., and Ross, J. (1968). Science 159, 587.

Greene, E. F., and Toennies, J. P. (1964). "Chemical Reactions in Shock Waves." Academic Press, New York.

Hager, G., Wilson, L. E., and Hadley, S. G. (1974). Chem. Phys. Lett. 27, 439.

Hager, G., Harris, R., and Hadley, S. G. (1975). J. Chem. Phys. 63, 2810.

Hague, D. N. (1969). Chem. Kinet. 1, Chapter 2.

Ham, D. O., and Chang, H. W. (1974). Chem. Phys. Lett. 24, 579.

Hartunian, R. A. (1968). In "Methods of Experimental Physics," Vol. 7B, "Atomic and Electron Physics—Atomic Interactions" (B. Bederson and W. L. Fite, eds.), Chapter 7. Academic Press, New York.

Hastie, J. W. (1975). "High Temperature Vapors: Science and Technology," Academic Press, New York.

Hesser, M. M., Drulinger, R. E., and Broida, H. P. (1975). J. Appl. Phys. 46, 2317.

Hirschfelder, J. O., Curtiss, C. F., and Bird, R. B. (1963). "Molecular Theory of Gases and Liquid," 2nd printing, Chapter 8. Wiley, New York.

Homer, J. B., and Hurle, I. R. (1972). Proc. R. Soc. London, Ser. A 327, 61.

Honig, R. E. (1967). In "The Characterization of High-Temperature Vapors" (J. L. Margrave, ed.), Appendix A. Wiley, New York. Updated graphs are available from the author at the R. C. A. David Sarnoff Laboratories, Princeton, New Jersey.

Hoyermann, K. H. (1975). In "Physical Chemistry, An Advanced Treatise," (W. Jost, ed.), Vol. VIB. Academic Press, New York.

Huggins, R. W., and Cahn, J. H. (1967). J. Appl. Phys. 38, 180.

Hulett, J. R. (1964). Q. Rev., Chem. Soc. 18, 227.

Hummer, D. G. (1965). Mem. R. Astron. Soc. 70, 1.

Husain, D., and Kirsch, L. J. (1971a). Trans. Faraday Soc. 67, 2886.

Husain, D., and Kirsch, L. J. (1971b). Trans. Faraday Soc. 67, 3166.

Husain, D., and Kirsch, L. J. (1971c). Chem. Phys. Lett. 9, 412.

Husain, D., and Kirsch, L. J. (1973–1974). J. Photochem. 2, 297.

Husain, D., and Littler, J. G. F. (1972). J. Chem. Soc., Faraday Trans. 2 68, 2110.

Husain, D., and Littler, J. G. F. (1972–1973). J. Photochem. 1, 327.

Husain, D., and Littler, J. G. F. (1973–1974). J. Photochem. 2, 247.

Husain, D., and Littler, J. G. F. (1974a). Combust. Flame 22, 295.

Husain, D., and Littler, J. G. F. (1974b). Int. J. Chem. Kinet. 6, 61.

Husain, D., and Norris, P. E. (1977a). Chem. Phys. Lett. 51, 206.

Husain, D., and Norris, P. E. (1977b). J. Chem. Soc., Faraday Trans. 2 73, 415.

Husain, D., and Norris, P. E. (1978a). *J. Chem. Soc., Faraday Trans. 2* **74**, 93 and 106.
Husain, D., and Norris, P. E. (1978b). *Chem. Phys. Lett.* **53**, 474.
Husain, D., and Slater, N. K. H. (1976–1977). *J. Photochem.* **6**, 325.
Husain, D., and Slater, N. K. H. (1977). *J. Photochem.* **7**, 59.
Husain, D., and Young, A. N. (1975). *J. Chem. Soc., Faraday Trans. 2* **71**, 525.
Husain, D., Krause, L., and Slater, N. K. H. (1977a). *J. Chem. Soc., Faraday Trans. 2* **73**, 1678.
Husain, D., Krause, L., and Slater, N. K. H. (1977b). *J. Chem. Soc., Faraday Trans. 2* **73**, 1706.
Jensen, D. E. (1968). *Combust. Flame* **12**, 261.
Jensen, D. E. (1969). *J. Chem. Phys.* **51**, 4674.
Jensen, D. E., and Jones, G. A. (1972). *J. Chem. Soc., Faraday Trans. 1* **68**, 259.
Jensen, D. E., and Jones, G. A. (1973). *J. Chem. Soc., Faraday Trans. 1* **69**, 1448.
Jensen, D. E., and Jones, G. A. (1975). *J. Chem. Soc., Faraday Trans. 1* **71**, 149.
Jensen, D. E., and Jones, G. A. (1976). *J. Chem. Soc., Faraday Trans. 1* **72**, 2618.
Jensen, D. E., and Miller, W. J. (1970). *J. Chem. Phys.* **53**, 3287.
Jensen, D. E., and Padley, P. J. (1966a). *Trans. Faraday Soc.* **62**, 2133.
Jensen, D. E., and Padley, P. J. (1966b). *Trans. Faraday Soc.* **62**, 2140.
Jensen, D. E., and Travers. B. E. L. (1974). *Plasma Sci.* **2**, 34.
Johnson, S. E. (1972). *J. Chem. Phys.* **56**, 149.
Johnson, S. E., Capelle, G. A., and Broida, H. P. (1972). *J. Chem. Phys.* **56**, 663.
Johnson, S. E., Scott, P. B., and Watson, G. (1974). *J. Chem. Phys.* **61**, 2834.
Jonah, C. D., and Zare, R. N. (1971). *Chem. Phys. Lett.* **9**, 65.
Jonah. C. D., Ottinger, Ch., and Zare, R. N. (1972). *J. Chem. Phys.* **56**, 263.
Jones, C. R., and Broida, H. P. (1973). *J. Chem. Phys.* **59**, 6677.
Jones, C. R., and Broida, H. P. (1974). *J. Chem. Phys.* **60**, 4369.
Jones, R. W., and Gole, J. L. (1977). *Chem. Phys.* **20**, 311.
Kaskan, W. E. (1965). *Symp. (Int. Combust. [Proc.]* **10**, 41
Kaufman, F. (1961). *Prog. React. Kinet.* **1**, 1.
Kaufman, F. (1969). *Can. J. Chem.* **47**, 1917.
Kinsey, J. L. (1972). *In* "MPI International Review of Science" (J. C. Polanyi, ed.), Vol. 9, Chapter 6. Butterworth, London.
Kinsey, J. L. (1977). *Annu. Rev. Phys. Chem.* **28**, 349.
Klabunde, K. J. (1975a). *Acc. Chem. Res.* **8**, 393.
Klabunde, K. J. (1975b). *Angew. Chem., Int. Ed. Engl.* **14**, 287.
Kolts, J., and Setser, D. W. This volume, Chap. 3.
Kung, R. T. V., and Bauer, S. H. (1971). *Shock Tube Res., Proc. Int. Shock Tube Symp, 8th, 1970* Paper 61.
Kurylo, M. J., Braun, W., Abramowitz, S., and Krauss, M. (1976). *J. Res. Natl. Bur. Stand., Sect. A* **80**, 167.
Lam, E. Y. Y., Gaspar, P., and Wolf, A. P. (1971). *J. Phys. Chem.* **75**, 445.
Lawrence, G. M., Link, J. K., and King, R. B. (1965). *Astrophys. J.* **141**, 293.
Lee, H. U., and Zare, R. N. (1975). *Combust. Flame* **24**, 27.
Lee, H. U., and Zare, R. N. (1977). *J. Mol. Spectrosc.* **64**, 233.
Lin, C.-L., Parkes, D. A., and Kaufman, F. (1970). *J. Chem. Phys.* **53**, 3896.
Lin, M.-C. and McDonald, J. R. This volume, Chap. 4.
Lindsay, D. M., and Gole, J. L. (1977). *J. Chem. Phys.* **66**, 3886.
Linton, C., and Broida, H. P. (1976). *J. Mol. Spectrosc.* **62**, 396.
Linton, C., and Broida, H. P. (1977). *J. Mol. Spectrosc.* **64**, 382.
Linton, C., and Capelle, G. A. (1977). *J. Mol. Spectrosc.* **66**, 62.
Liu, K., and Parson, J. M. (1977). *J. Chem. Phys.* **67**, 1814.
Lovett, R. S., and Parsons, M. (1977). *Appl. Spectrosc.* **31**, 424.

Luria, M., Eckstrom, D. J., Edelstein, S. A., Perry, B. E., and Benson, S. W. (1976a). *J. Chem. Phys.* **64**, 2247.

Luria, M., Eckstrom, D. J., and Benson, S. W. (1976b). *J. Chem. Phys.* **64**, 3103.

Luria, M., Eckstrom, D. J., and Benson, S. W. (1976c). *J. Chem. Phys.* **65**, 1581.

Luria, M., Eckstrom, D. J., and Benson, S. W. (1976d). *J. Chem. Phys.* **65**, 1595.

McDanel Refractory Porcelain Co. (1970). Bull. FT70. McDanel Refract. Porcelain Co., Beaver Falls.

McEwan, M. J., and Phillips, L. F. (1966). *Trans. Faraday Soc.* **62**, 1717.

McNesby, J. R., Braun, W., and Ball, J. (1971). *In* "Creation and Detection of the Excited State" (A. A. Lomola, ed.), Vol. 1B, Chapter 2. Dekker, New York.

Manos, D. M., and Parson, J. M. (1975). *J. Chem. Phys.* **63**, 3575.

Margrave, J. L., ed. (1967). "The Characterization of High-Temperature Vapors," Wiley, New York.

Martinoti, F. F., Welch, M. J., and Wolf, A. P. (1968). *Chem. Commun.* p. 115.

Mathis, R. D. Co. (1973). "Vacuum Evaporation Source Catalog." Mathis R. D. Co., Long Beach, California.

Matthey Bishop, Inc. (1968). Technical Data Sheets MD-P3 and PGM-1. Matthey Bishop, Inc., Philadelphia, Pennsylvania.

Maya, J., and Davidovits, P. (1973). *J. Chem. Phys.* **59**, 3143.

Maya, J., and Davidovits, P. (1974). *J. Chem. Phys.* **61**, 1082.

Menzinger, M. (1974). *Chem. Phys.* **5**, 350.

Menzinger, M., and Wolfgang, R. (1969). *Angew. Chem., Int. Ed. Engl.* **8**, 438.

Menzinger, M., and Wren, D. J. (1973). *Chem. Phys. Lett.* **18**, 431.

Miller, W. J. (1972). *J. Chem. Phys.* **57**, 2354.

Miller, W. J. (1977). *Prog. Astronaut. Aeronaut.* **53**, 25.

Miller, W. J., and Gould, R. K. (1976). *Chem. Phys. Lett.* **38**, 237.

Mitchell, A. C. G., and Zemansky, M. W. (1934). "Resonance Radiation and Excited Atoms," Chapters 3 and 4. Cambridge Univ. Press, London and New York.

Moffat, R. J. (1962). *Temp. Meas. Control Sci. Ind., Proc. Symp. 4th, 1961*, Vol. 3, Part 2, Chapter 52.

Moore, C. E. (1958). *Natl. Bur. Stand. (U.S.), Circ.* **467**.

Morse, F. A., and Kaufman, F. (1965). *J. Chem. Phys.* **42**, 1785.

Moskovits, M., and Ozin, G. A. (eds.) (1976). "Cryochemistry." Wiley, New York.

Moulton, M. C., and Herschbach, D. R. (1966). *J. Chem. Phys.* **44**, 3010.

Nettleton, M. A. (1977). *Combust. Flame* **28**, 3.

Obenauf, R. H., Hsu, C. J., and Palmer, H. B. (1972a). *Chem. Phys. Lett.* **17**, 455.

Obenauf, R. H., Hsu, C. J., and Palmer, H. B. (1972b). *J. Chem. Phys.* **57**, 5607.

Ogren, P. J. (1975). *J. Phys. Chem.* **79**, 1749.

Oldenborg, R. C., Gole, J. L., and Zare, R. N. (1974). *J. Chem. Phys.* **60**, 4032.

Oldenborg, R. C., Dickson, C. R., and Zare, R. N. (1975). *J. Mol. Spectrosc.* **58**, 283.

Ottinger, Ch., and Zare, R. N. (1970). *Chem. Phys. Lett.* **5**, 243.

Padley, P. J., and Sugden, T. M. (1958). *Proc. R. Soc. London, Ser. A* **248**, 248.

Palmer, H. B., Krugh, W. D., and Hsu, C.-J. (1975). *Symp. (Int.) Combust. [Proc.]* **15**, 951.

Pasternack, L., and Dagdigian, P. J. (1977). *J. Chem. Phys.* **67**, 3854.

Penkin, N. B., and Slavenas, I. Yu. Yu. (1963). *Opt. Spectrosc. (USSR)* **15**, 83.

Perlmutter-Hayman, B. (1976). *Prog. Inorg. Chem.* **20**, 229.

Perry, R. H., Chilton, C. H., and Kirkpatrick, S. D., eds. (1963). "Chemical Engineers' Handbook," 4th ed., Sect. 5-5. McGraw-Hill, New York.

Phillips, L. F. (1976). *Chem. Phys. Lett.* **37**, 421.

Phillips, L. F., and Sugden, T. M. (1961). *Trans. Faraday Soc.* **57**, 2188.

Piepmeier, E. H., and deGalan, L. (1975). *Spectrochim. Acta, Part B* **30**, 211.

Polanyi, M. (1932). "Atomic Reactions." Williams & Norgate, London.

Preuss, D. R., and Gole, J. L. (1977a). *J. Chem. Phys.* **66**, 880.

Preuss, D. R., and Gole, J. L. (1977b). *J. Chem. Phys.* **66**, 2994.

Pruett, J. G., and Zare, R. N. (1975). *J. Chem. Phys.* **62**, 2050.

Pugh, E. M., and Winslow, G. W. (1966). "The Analysis of Physical Measurements." Addison-Wesley, Reading, Massachusetts.

Reid. R. C., and Sherwood, R. K. (1966). "The Properties of Gases and Liquids," 2nd ed. McGraw-Hill, New York.

Rice, W. W., and Beattie, W. H. (1973). *Chem. Phys. Lett.* **19**, 82.

Rice, W. W., Beattie, W. H., Oldenborg, R. C., Johnson, S. E., and Scott, P. B. (1976). *Appl. Phys. Lett.* **28**, 444.

Rosenwaks, S. (1976). *J. Chem. Phys.* **65**, 3668.

Rosenwaks, S., Steele, R. E., and Broida, H. P. (1975). *J. Chem. Phys.* **63**, 1963.

Ross, J., ed. (1966). "Molecular Beams," Adv. Chem. Phys., Vol. 10. Wiley (Interscience), New York.

Sakurai, K., and Broida, H. P. (1976). *Chem. Phys. Lett.* **38**, 234.

Sakurai, K., Johnson, S. E., and Broida, H. P. (1970). *J. Chem. Phys.* **52**, 1625.

Sakurai, K., Adams, A., and Broida, H. P. (1976). *Chem. Phys. Lett.* **39**, 442.

Sayers, M. J., and Gole, J. L. (1977). *J. Chem. Phys.* **67**, 5442.

Schofield, K., and Broida, H. P. (1968). *Methods Exp. Phys.* **B 7**, Chapter 8.

Schultz, A., and Zare, R. N. (1974). *J. Chem. Phys.* **60**, 5120.

Singleton, D. L., and Cvetanovic, R. J. (1976). *J. Am. Chem. Soc.* **98**, 6812.

Skell, P. S., and Westcott, J. D. (1963). *J. Am. Chem. Soc.* **85**, 1023.

Skell, P. S., Havel, J. J., and McGlinchy, M. J. (1973). *Acc. Chem. Res.* **6**, 97.

Slaughter, J. I., and Margrave, J. L. (1967). "High-Temperature Materials and Technology," Chapter 24. Wiley, New York.

Smith, W. H., and Liszt, H. S. (1971). *J. Opt. Soc. Am.* **61**, 938.

Sridharan, U. C., McFadden, D. L., and Davidovits, P. (1976). *J. Chem. Phys.* **65**, 5373.

Struve, W. S., Krenos, J. R., McFadden, D. L., and Herschbach, D. R. (1975). *J. Chem. Phys.* **62**, 404.

Sugden, T. M. (1962). *Annu. Rev. Phys. Chem.* **13**, 369.

Swearengen, P., Davis, S., and Niemczick, T. (1977). *In* "Electronic Transition Lasers II" (L. E. Wilson, S. N. Suchard, and J. I. Steinfeld, eds.), p. 132. MIT Press, Cambridge, Massachusetts.

Tang, S. P., Utterback, N. G., and Friichtenicht, J. F. (1976). *J. Chem. Phys.* **64**, 3833.

Tellinghuisen, J., and Clyne, M. A. A. (1976). *J. Chem. Soc., Faraday Trans. 2* **72**, 783.

Timms, P. L. (1972). *Adv. Inorg. Radiochem.* **14**, 121.

Timms, P. L. (1975). *Angew. Chem. Int. Ed. Engl.* **14**, 273.

Trainor, D. W. (1977). *J. Chem. Phys.* **67**, 1206.

Trainor, D. W. and Ewing, J. J. (1976). *J. Chem. Phys.* **64**, 222.

Troe, J., and Wagner, H. Gg. (1973). *Phys. Chem. Fast React.* **1**, Chapter 1.

Vidale, G. L. (1960). GE TIS No. R60SD330. GE Space Sci. Lab. Valley Forge, Pennsylvania.

von Rosenberg, C. W., and Wray, K. L. (1972). *J. Quant. Spectrosc. & Radiat. Transfer* **12**, 531.

Wagenaar, H. C., and deGalan, L. (1973). *Spectrochim. Acta, Part B* **28**, 157.

Walker, R. E. (1961). *Phys. Fluids* **4**, 1211.

West, C. W., and Human, H. G. C. (1976). *Spectrochim. Acta, Part B* **31**, 81.

West, J. B., and Broida, H. P. (1975). *J. Chem. Phys.* **62**, 2566.

West, J. B., Radford, R. S., Eversole, J. D., and Jones, C. R. (1975). *Rev. Sci. Instrum.* **46**, 164.

Westenberg, A. A. (1973). *Symp. (Int.) Combust. [Proc.]* **14**, 287.

Westenberg, A. A., and de Haas, N. (1967). *J. Chem. Phys.* **46**, 490.
Wicke, B. G., Tang, S. P., and Friichtenicht, J. F. (1978). *Chem. Phys. Lett.* **53**, 304.
Wiese, W. L. (1970). *Nucl. Instrum. & Methods* **90**, 25.
Wiese, W. L., Smith, M. W., and Miles, M. B. (1969), NBS-NSRDS 22. Natl. Bur. Stand., Washington, D.C.
Wiesenfeld, J. R. (1973). *Chem. Phys. Lett.* **21**, 517.
Wiesenfeld, J. R., and Yuen, M. J. (1976). *Chem. Phys. Lett.* **42**, 293.
Wik, T., Maya, J., Chao, J., and Davidovits, P. (1975). *J. Chem. Phys.* **62**, 1995.
Wilson, E. B., Jr. (1952). "An Introduction to Scientific Research." McGraw Hill, New York.
Wray, K. L. (1970) "New Experimental Techniques in Propulsion and Energetics Research" (D. Andrew and J. Surugue, eds.), AGARD-CP-38, p. 17. Technivision, Maidenhead, England.
Wren, D. J., and Menzinger, M. (1974). *Chem. Phys. Lett.* **27**, 572.
Wren, D. J., and Menzinger, M. (1975). *J. Chem. Phys.* **63**, 4557.
Yokozeki, A., and Menzinger, M. (1976). *Chem. Phys.* **14**, 427.
Yokozeki, A., and Menzinger, M. (1977). *Chem. Phys.* **20**, 9.
Zare, R. N., and Dagdigian, P. J. (1974). *Science* **185**, 739.
Zircar Products, Inc. Type ZYC. Zircar Prod., Inc., Florida, New York.

3

Electronically Excited Long-Lived States of Atoms and Diatomic Molecules in Flow Systems*

J. H. KOLTS† and D. W. SETSER

Department of Chemistry
Kansas State University
Manhattan, Kansas

* Preparation of this review was supported by the U.S. Dept. of Energy (E(11-1)-2807); J. H. Kolts also thanks Phillips Petroleum for a fellowship.

† Present address: Research and Development Department, Phillips Petroleum Company, Bartlesville, Oklahoma.

151

I. INTRODUCTION

This chapter discusses flowing afterglow techniques as applied to generation and detection of electronically excited states. The scope is limited to
electronically excited states of atoms and diatomic molecules; although,
successful studies of vibrationally excited molecules have been made using the
flow technique (Smith et al., 1977; Sung and Setser, 1978; Kwok and Cohen,
1974). Since the preceding two chapters discussed the principles of the flow
technique for kinetic measurements, only the special features related to
excited states in flow systems are presented here. The principal advantage of
the flow technique is the spatial separation of the zone used for generation
from the region where kinetic measurements are made. Thus, after generating
the excited species of interests, hopefully without any interfering species,
other reagents can be added to the reactor without affecting the generating
step. Under ideal conditions this leads to isolation of the state of interest for
definitive kinetic studies. A second advantage is that the technique frequently
allows convenient identification of product species. In order to use the flow
technique for excited states, the radiative lifetime must be sufficiently long
that the excited state species travels an appreciable distance in the flow
reactor. As a practical limit, the radiative lifetime must be ≥ 0.5 msec. In some
instances the term "metastable" is attached to states with long radiative
lifetimes. There is no special significance to the term and metastable will be
used only if this is the common usage of the literature. During recent years,
powerful spectroscopic methods have been developed for monitoring low
concentrations. Thus, the limitation to studying a specific long-lived excited
state by the flow technique frequently is a suitable generation method.

This chapter is divided into four main parts. In the first part the generation
and detection of excited states and the measurement of total quenching rate
constants are discussed. Since reactions of electronically excited states frequently produce excited state products, the assignment of product channel
rate constants by emission spectroscopy is a special point of interest and is
treated in the second part. Individual excited atomic states that have been
studied by the flow technique are reviewed in the third part and long-lived
excited diatomic states are discussed in the fourth part. The techniques used

Table I

Radiative Lifetimes and Energies of Excited States

State	Excitation energy (eV)	Radiative lifetime	Reference[a]
$e(2^1S)$	20.61	19.7 msec	Van Dyck *et al.* (1971) (*E*)
$e(2^3S)$	19.81	~ 150 min	Woodworth and Moos (1975) (*E*)
$e(^3P_0)$	16.71	>0.8 sec	Van Dyck *et al.* (1972) (*E*)
		24.8 sec	Small-Warren and Chiu (1972) (*T*)
$e(^3P_2)$	16.61	>0.8 sec	Van Dyck *et al.* (1972) (*E*)
		430 sec	Small-Warren and Chiu (1975) (*T*)
$r(^3P_0)$	11.73	1.3 sec	Van Dyck *et al.* (1972) (*E*)
		44.9 sec	Small-Warren and Chiu (1975) (*T*)
$r(^3P_2)$	11.55	1.3 sec	Van Dyck *et al.* (1972) (*E*)
		55.9 sec	Small-Warren and Chiu (1975) (*T*)
$r(^3P_0)$	10.56	1.0 sec	Van Dyck *et al.* (1972) (*E*)
		0.49 sec	Small-Warren and Chiu (1975) (*T*)
$r(^3P_2)$	9.92	1.0 sec	Van Dyck *et al.* (1972) (*E*)
		85 sec	Small-Warren and Chiu (1975) (*T*)
$e(^3P_0)$	9.45	0.078 sec	Small-Warren and Chiu (1975) (*T*)
$e(^3P_2)$	8.32	149 sec	Small-Warren and Chiu (1975) (*T*)
$1g(^3P_0)$	2.71	29 min ⎫	
$1g(^3P_1)$	2.71	2.3 msec ⎬	Wiese *et al.* (1969) (*C*)
$1g(^3P_2)$	2.72	76 min ⎭	
$(^1S_0)$	2.68	2.0 sec ⎫	
$(^1S_0-^1D_2)$		2.0 sec ⎬	Wiese *et al.* (1966) (*C*)
$(^1S_0-^3P_1)$		0.038 sec ⎭	
$(^1D_2)$	1.26	53 min ⎭	
$(^1S_0)$	1.91	1.2 sec ⎫	
$(^1S_0-^3P_1)$		28 sec ⎬	Wiese *et al.* (1969) (*C*)
$(^1S_0-^1D_2)$		1.25 sec ⎭	
$(^1D_2)$	0.78	370 sec ⎭	
$e(^1S_0)$	0.88		
$e(^1D_2)$	2.02		
$(^2P_{1/2})$	3.58	40 sec	Wiese *et al.* (1966) (*C*)
$(^2P_{3/2})$	3.58	166 sec	Wiese *et al.* (1966) (*C*)
$(^2D_{3/2})$	2.38	17 hr	Wiese *et al.* (1966) (*C*)
$(^2D_{5/2})$	2.38	40 hr	Wiese *et al.* (1966) (*C*)
$(^1S_0)$	4.19	0.76 sec	Corney and Williams (1972) (*E*)
		0.90 sec	Kernaham and Pang (1975a) (*E*)
$(^1S_0-^1D_2)$		0.94 sec	Kernaham and Pang (1975a) (*E*)
$(^1S_0-^3P_1)$		22 sec	Kernaham and Pang (1975a) (*E*)
$(^1D_2)$	1.97	110 sec	Wiese *et al.* (1966) (*C*)
		147 sec	Kernaham and Pang (1975a) (*E*)
$(^1D_2-^3P_1)$		602 sec	Kernaham and Pang (1975a) (*E*)
$(^1D_2-^3P_2)$		194 sec	Kernaham and Pang (1975a) (*E*)
$(^1S_0)$	2.75	0.51 sec	Kernaham and Pang (1975b) (*E*)
		0.56 sec	Wiese *et al.* (1969) (*C*)
$(^1S_0-^1D_2)$		0.61 sec	Kernaham and Pang (1975b) (*E*)
$(^1S_0 {}^1P_1)$		2.9 sec	Kernaham and Pang (1975b) (*E*)

(continued)

Table I (*continued*)

State	Excitation energy (eV)	Radiative lifetime	Reference[a]
$S(^1D_2)$	1.15	28 sec	Wiese et al. (1969) (C)
$I(^2P_{1/2})$	0.942	108 msec	Comes and Pionteck (1976)(E)
		127 msec	Garstang (1964) (T)
		20–45 msec	Deakin et al. (1971) (E)
			Husain et al. (1977) (E)
$Br(^2P_{1/2})$	0.456	1.12 sec	Garstang (1964) (T)
$Cl(^2P_{1/2})$	0.109	81 sec	Okabe (1978)
$F(^2P_{1/2})$	0.050	847 sec	Okabe (1978)
$Hg(^3P_2)$	5.46	6.6 sec	⎫
$Hg(^3P_0)$	4.66	5.5 sec	Garstang (1962) (T)
$Cd(^3P_2)$	3.80	144 sec	for naturally
$Cd(^3P_0)$	3.73	90 sec	occurring isotopic
$Zn(^3P_2)$	4.08	72 min	abundance[d]
$Zn(^3P_0)$	4.00	30 min	⎭
$O_2(A^3\Sigma_u^+)$	4.34	0.25–0.16 sec	⎫
$O_2(C^3\Delta_u)$	4.25	5–50(Ω = 1) sec	Slanger (1978) (E)
		10–100(Ω = 2) sec	
$O_2(c^1\Sigma_u^-)$	4.05	25–50 sec	⎭
$O_2(b^1\Sigma_g^+ - X^3\Sigma_g^-)$	1.63	12 sec	Burch and Gryvnak (1969)
$O_2(a^1\Sigma_g^- - X^3\Sigma_g^-)$	0.98	3900 sec	Badger et al. (1967) (E)
$O_2^+(a^4\Pi_u)$	4.05	not known (long)[b]	
$NF(b^1\Sigma^+)$	2.34	~0.2 sec	Clyne and White (1970) (E)
$NF(a^1\Delta)$	1.42	5.6 sec	Malins and Setser (1980) (E)
$NH(b^1\Sigma^+)$	2.63	18 msec	Gelernt et al. (1976) (E) (also see tex
$NH(a^1\Delta)$	1.56	Not known (long)[b]	
$PH(b^1\Sigma^+)$	1.78	1.3 msec	Xuan et al. (1978a,b) (E)
$N_2(a'^1\Sigma_u^+)$	8.40	~0.5 sec	Lofthus and Krupenie (1977) (C)
		0.017 sec[c]	Cassa and Golde (1979)
$N_2(A^3\Sigma_u^+)$	6.22	1.36(Σ = 0) sec	Shemansky (1969) (E)
		2.7(Σ = ±1) sec	
$CO(a^3\Pi)$	6.04	7.5 sec[c]	Lawrence (1971) (E)
$CS(a^3\Pi)$	3.42	≤16 msec[c]	Black et al. (1977) (E)
$NO(a^4\Pi)$	4.7	0.085 sec	Goodman and Brus (1978)
		in Ar matrix	
		0.1 sec	Lefebvre-Brion and Guerin (1968) (
$NO^+(A^3\Sigma^+)$	6.39	Not known (long)[b]	
$H_2(c^3\Pi_u)$	11.9	1.76 msec	Bhattacharyya and Chiu (1977) (T)
		1.02 msec	Johnson (1972) (E)
$He_2(a^3\Sigma_u^+)$	21.5	>0.1 sec	Lichten et al. (1974) (C)
$CN(^4\Sigma^+)$	5.5	Not known (long)[b]	
$Cl_2(^3\Pi_{2u})$	2.2	76 msec	Bondybey and Fletcher (1976)
		in an Ar matrix	

[a] E denotes experimental measurement, T denotes theoretical calculation, and C denotes best value fro compilation.

[b] Long denotes that the transition to the ground state of the species has never been observed, and the are no good calculated values. In some cases the lifetimes may be only moderately long, but the state m be readily quenched by the parent or by the precursors.

[c] Lifetime depends upon spin and rotational state population; value quoted is for room temperatu Boltzmann distribution.

[d] The lifetimes for these Group IIb states depend upon isotope; Bigeon (1967) has determin $\tau(Hg^{199})$ = 1.5 sec and $\tau(Hg^{201})$ = 2.2 sec.

for generating and studying excited state species resemble the flowing afterglow techniques used for ions. Therefore, an appendix was added which provides references to flowing afterglow studies of ionic species.

The excited state species that are discussed in this chapter are summarized in Table I. Okabe's (1978) recently published book contains a summary of the chemistry of several of these excited states. As already stated, this chapter is concerned with the continuous fast flow technique. For study of some long-lived excited states, pulsed excitation with time resolved monitoring techniques may be advantageous. Husain, Donovan, and co-workers have used the pulsed, time resolved method to study reactions of a large number of atomic spin–orbit excited states (Donovan and Husain, 1970, 1971; Donovan et al., 1972; Donovan and Gillespie, 1975). Although the work frequently involves long-lived states, their experimental method does not utilize flow techniques and is more closely related to the techniques discussed by Lin and McDonald in the following chapter. We will not discuss pulsed, time resolved techniques; however, if these methods have been used for the specific atomic and molecular states considered in the later sections, that work will be included in the discussion of these states. A likely development in the future is the utilization of pulsed excitation sources with the fast flow technique.

Insofar as possible, the notation of the two preceding chapters is employed. Electronically excited states are denoted by E* and the energy is given in electron volts. The units for second- and third-order rate constants are cm^3 $molecule^{-1} sec^{-1}$ and $cm^6 molecule^{-2} sec^{-1}$, respectively.

II. FLOW SYSTEMS FOR ELECTRONICALLY EXCITED STATES: TOTAL QUENCHING RATE CONSTANTS

A. Design of Apparatus

1. General Features

The flowing afterglow technique can be used to measure the total quenching rate constant k_Q of an electronically excited state E*, which may or may not be an excited state of the carrier gas, by monitoring the decay rate of E* as a function of carrier gas pressure or as a function of the concentration of some added reagent. The product formation rate from the quenching reaction normally is not used for the determination of k_Q. Before going into the details of determining k_Q, the design of the flowing afterglow apparatus utilizing helium or argon as the carrier gas will be considered.

There are many similarities, but also some differences, in flow systems used for electronically excited states relative to those discussed for atoms and radicals, Chapter 1, and refractory species, Chapter 2. The decay processes for E* fall into three broad categories. The first is deactivation upon encounter with the reactor walls; this loss rate is determined by the diffusion coefficient of the excited species in the carrier gas and is inversely proportional to the carrier gas pressure. Loss by diffusion usually establishes the low pressure limit for operation of the flow reactor and the diffusion coefficient of E^* in the carrier gas is a fundamental property. The second category includes those processes involving collisions of E^* with the carrier gas or impurities in the carrier gas, e.g., two-body collision induced emission, two-body excitation transfer, three-body quenching, and quenching by impurities. These loss processes are either first or second order with the carrier gas concentration and establish the upper limit to the pressure. A third category is radiative decay which, of course, is pressure independent. The lifetime for radiative decay must be longer than $\sim \Delta t/e$ where Δt is the required observation time in the flow reactor. As will be shown below, radiative decay does not interfere with quenching measurements since the decay rate is always adjusted to be pseudo-first-order in $[E^*]$. Generally $[E^*]$ is between 10^{10} and 10^{11} atom cm^{-3} and second-order decay rates are unimportant, relative to first-order processes, even though the rate constants for the bimolecular, $2E^* \rightarrow$ products, reactions usually are large, $\gtrsim 4 \times 10^{-10} \ cm^3 \ sec^{-1}$. Because of these general features, concentrations of excited state species are lower and flow velocities are higher than for experiments studying ground state species in flow reactors.

The loss processes mentioned above for E^* are present in all flow systems, as well as for static systems. Losses from diffusion and quenching at the walls can be reduced for ground state atoms and radicals by applying coatings of oxyacids, teflon, or other polymeric materials to the walls, as discussed by Clyne and Nip in Chapter 1. Refractory species have high rates of reaction with the reactor walls and electronically excited states frequently have unit deactivation rates (Allison et al., 1973). For refractory species the loss rates from wall reaction have been reduced by operating at higher pressure (>3 Torr) with relatively fast flow velocities and large diameter tubular reactors. This method also can be used to reduce the loss for electronically excited states from diffusion to the walls; however, at higher pressures two- and three-body processes start to severely deplete the concentration of E^*. This is especially true if the three-body rates are large. For HTFFR experiments, the metal atoms do not react with the carrier gas and the use of higher pressure is not precluded. As a general rule, the lowest pressure consistent with an acceptable loss rate from diffusion is the best operating pressure for electronically excited states in a flowing afterglow.

In addition to making allowance for the intrinsic difference in the decay processes, electronically excited states in flow systems differ from ground states in that special techniques, to be discussed in the next section, frequently are required to generate the state of interest. However, measurement of the concentration of E* does not differ significantly from methods discussed in Chapters 1 and 2 for ground states, and only a few special features are mentioned in Section II,A,4.

2. Production and Isolation of State of Interest

Two major methods have been used for producing excited states the hot- or cold-cathode discharge and microwave discharge. In a few cases the excited state atoms (usually rare gas atoms) or molecules produced by these discharge methods subsequently have been allowed to react with added reagents to produce a second long-lived excited state. Few examples can be cited that employ chemical reactions of ground state species for generation of excited states. The microwave discharge has been extensively used for the production of ground state atoms and radicals (see Chapter 1) and it also is the preferred way for generating metastable oxygen molecules. Helium metastable atoms and ions can be produced by a microwave discharge (Collins and Robertson, 1964; Fehsenfeld et al., 1966; Schmeltekopf and Broida, 1963; Cher and Hollingsworth, 1969a,b; Collins and Johnson, 1972) and was commonly used in early work. A microwave source also has been used to generate metastable argon atoms (DeJong, 1972, 1974; Suzuki and Kuchitsu, 1978; Cook et al., 1978a; Prince et al., 1964); Lin and Kaufman (1971) generated metastable nitrogen atoms in argon carrier with a microwave discharge. Suzuki et al. (1978) have reported excited metastable argon ions, $Ar^+(3s^23p^43d)$, as well as metastable atoms, in the afterglow of a microwave discharge in Ar; proof for the ions was $Cl(5p)$ emission following addition of HCl. For all studies the microwave cavity has been of the Broida type (Fehsenfeld et al., 1965), which is a cavity that fits around the quartz, alumina, or metallic tube through which the gas flows.

More recently hot- or cold-cathode discharges have been used to a somewhat larger extent than have microwave discharges for production of excited atomic states, especially the rare gas metastable atoms. The basic configurations of the cold- and hot-cathode discharges are illustrated in Fig. 1. Several descriptions of the cold-cathode type discharges have been published (Prince et al., 1964; Stedman and Setser, 1968, 1971); hot-cathode discharges mainly have been used for the production of helium metastable atoms and ions (Dunkin et al., 1968; Ferguson et al., 1969; Bolden et al., 1970). Two basic designs for cold-cathode discharges have been used by the present authors with good results. For flow reactors of relatively large diameter (> 1.5 cm), a tantalum foil cylinder rolled to a slightly smaller diameter than the inside diameter of the flow tube is used as the cathode and a tungsten wire is placed inside this for the anode; the actual length is not critical and 1.5–4 cm cylinders have been used. A second design (see Fig. 1) for smaller diameter (< 1.5 cm)

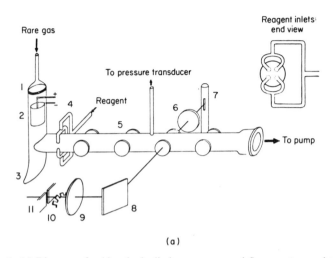

(a)

Fig. 1. (a) Diagram of cold-cathode discharge source and flow reactor used to measure quenching rate constants of heavy rare gas metastable atoms (Kolts and Setser, 1978a). The metastable atom concentrations were measured by absorption spectroscopy at the various windows. The cathode is made from rolled tantalum foil and the anode is a tungsten wire; suitable diameters for flow tubes with argon carrier gas are 3–4 cm. 1, fritted disk; 2, hollow cathode discharge; 3, light trap; 4, reagent inlets; 5, observation windows; 6, focusing lens; 7, pen-ray resonance lamp; 8, plane mirror; 9, lens; 10, mechanical light chopper; 11, monochromator. (b) Diagram of the cold-cathode discharge source and flow reactor used for observation of product emission intensities from rare gas metastable atom reactions. Both the anode and cathode are made from tantalum foil. A convenient (and satisfactory) way to make the metal-to-glass seal is to lead a sliver of tantalum through a hole in the glass, which is then sealed by epoxy. The mixing of the first reagent is done with a concentric-type reagent inlet. Additional windows, downstream from the reaction zone, are shown which would be suitable for observation of emission from long-lived product states or for study of the reactions of the product states. (Note the second reagent inlet). (c) Diagram of the hot-cathode type discharge that commonly is used with helium carrier gas for the study of $He(2^1S)$, $He(2^3S)$, He^+ or other ions. The diameter of the flow reactors is typically ~ 10 cm; the hot-cathode and the accelerating electrode are shown. The placement of these components and their mode of operation depends upon the species desired for study. The diagram was drawn to show the maximum flexibility in terms of addition of reagents. A "substrate" can be added in front of the hot-cathode discharge to generate an initial set of negative or positive ions. A reagent can be added through the fixed inlet to convert the initial positive or negative ion to a second positive or negative ion. Then a second reagent could be added through the adjustable (a slide seal) inlet. A disadvantage of the series of inlets downstream of the discharge is the introduction of flow perturbations. The pressure in the flow tube should be measured about midway in length.

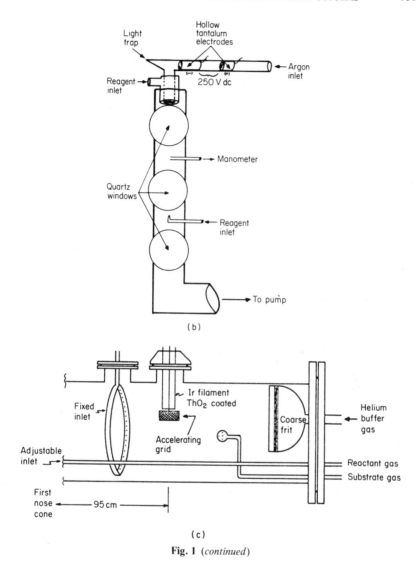

Fig. 1 (*continued*)

discharge tubes consists of two rolled tantalum foil electrodes fitted inside the flow tube. These electrodes are inserted in the mainstream of the carrier gas with spacings between the electrodes of 0.5–3.0 cm; the downstream electrode should be the cathode. These designs have been satisfactory for helium, neon, and argon metastable atoms using the parent as the carrier gas and for krypton and xenon metastable atoms with argon as the carrier gas. The cold-cathode discharge is operated at ~ 200 V and a few milliamperes for optimum production of heavy rare gas metastable atoms. The discharge is placed at right angles to the flow reactor to avoid trapping of the vacuum uv resonance lines of the carrier gas or other atoms passing through the discharge.

Hot-cathode discharges mainly have been used for producing helium metastable atoms and helium ions (Dunkin et al., 1968; Bolden et al., 1970; Ferguson et al., 1969). The hot-cathode discharge (see Fig. 1) consists of a resistively heated cathode, frequently iridium coated with thorium oxide. These cathodes have a ribbonlike geometry. For metal flow tubes the anode can be the flow tube itself. For Pyrex glass flow tubes a circular anode can be placed around the cathode or a grid may be placed perpendicular to the gas flow and 0.5–2 cm away from the heated cathode. We, and others, have used a stainless steel screen as the anode with a secondary grid placed behind the anode to monitor the current. Accelerating potentials of 60–200 V between the anode and cathode are typically seen with He as the carrier gas. The hot-cathode discharge current varies with design and with the chemical system; but, it usually is 0.1–10 mA. Since the accelerating voltage can be varied to some extent, the electron energies can be somewhat better controlled in the hot cathode than in the cold cathode and some selectivity on the species generated can be exercised. Nevertheless, for generation of metastable rare gas atoms in the absence of ions, the cold-cathode discharge is recommended based upon the experience of the authors. The vacuum uv resonance lines are no problem if the hot cathode is used to produce ions; however, if metastable atoms are being studied, a right angle bend between the discharge and flow reactor is recommended. For further comment on the hot-cathode discharge for generating positive and negative ions, the reader should see the Appendix and other general references on flowing afterglow studies of ion–molecule reactions.

Cook et al. (1978a) compared the production of $Ar(^3P_2)$ and $Ar(^3P_0)$ from a microwave and cold-cathode discharge. For optimum operating conditions the metastable atom concentrations were approximately the same. The maximum concentration for the microwave discharge occurred at ~0.8 Torr and was sharply peaked over the 0.5–1.5 Torr range. For the cold-cathode discharge the maximum concentration occurred at ~1.9 Torr and the useful pressure range was 0.5–4.0 Torr. The experience of our laboratory with the cold-cathode discharge for $Ar(^3P_2)$ is similar except that the maximum concentration normally occurs at ~1 Torr and the useful range is 0.4–8 Torr. The metastable atom concentration, as well as the pressure range, depends upon several aspects, e.g., position of electrodes relative to observation zone, discharge voltage, pumping speeds, and presence or absence of a light trap between the discharge and observation zone, etc. The high and low pressure limits of the cold-cathode discharge used for generating metastable rare gas atoms is governed mainly by the loss of the metastable atoms during the time the gas flows from the discharge to the observation zone, rather than by intrinsic limitations of the cold-cathode discharge. By increasing the pumping capacity from 1000 to 4200 liters min^{-1} (Kolts et al., 1979), experiments could be done with $Ar(^3P_2)$, $Kr(^3P_2)$, and $Xe(^3P_2)$ at 0.08 Torr. Touzeau and Pagnon (1978) report low pressure studies of metastable Ar atom reactions using a 5-cm-diameter flow tube and a hot-cathode discharge. Since loss by diffusion is even more serious in He, low pressure studies using He carrier require even larger diameter flow tubes and greater pumping capacity.

The discharge current has a large effect on the concentration of rare gas excited states which are produced by the cold cathode. For low currents the excited state population is almost entirely metastable atoms; but, at higher currents the predominant species are ions (Prince et al., 1964). In our laboratories the optimum production of heavy rare gas metastable atoms occurs at the minimum voltage (or current) which will sustain the discharge. The maximum concentration for $He(2^3S)$ occurs somewhat above the minimum sustaining voltage. For operation with low discharge currents, the thermal electron concentration in argon carrier gas is sufficiently low that the decay kinetics are not affected (Kolts and Setser, 1978a). Based upon our experience, the types of excited species produced by the cold-cathode discharge in He are very sensitive to the operating conditions. For the type of discharge shown in Fig. 1b, which is convenient and reliable for argon, considerable care must be used to avoid complications from He^+, He_2^+, and perhaps He_2^* (Piper et al., 1975a; Chang et al., 1978).

For generation of metastable rare gas atoms, cold- or hot-cathode discharges are preferred over microwave discharges for several reasons; (a) less expensive, (b) lower electron and ion densities, (c) less electrical noise, especially if high currents are needed, (d) lower photon to excited state ratios, and (e) elimination of the constriction needed for the microwave cavity, which, for large diameter flow reactors, can cause larger perturbations to the flow than does a discharge assembly positioned within the flow tube. Some of the disadvantages of the electrical discharges are the necessity of placing a foreign metal within the flow system, the presence of high voltage that can be dangerous, and immobility of the electrodes once placed in position.

Although they have been used for rare gases (Bolden et al., 1970; Schearer, 1974), pulsed discharges have not offered many advantages so far. One use of a pulsed source is to directly measure the flow velocity in the reactor (Bolden et al., 1970; Schearer, 1974). Another application may be the analysis of the shape of the concentration profile versus flow distance to obtain information on diffusion coefficients. For some species it may be necessary (or advantageous) to use a pulsed generation method (not necessarily a discharge) because a satisfactory continuous source does not exist. The use of pulsed photolytic sources with fast flow systems is likely to become more common as high power pulsed lasers become more available as excitation sources and as pulsed laser fluorescence becomes a common analytical tool.

For crossed molecular beam or beam–gas experiments, direct electron impact excitation at low pressure is used to produce the excited atoms or molecules. Representative examples using these excitation sources can be found in Foreman et al. (1977), Neynabar and Magnuson (1975), Gersh and Muschlitz (1973), Winicur and Fraites (1974, 1976), Brinkmann et al. (1967), and Lee and Martin (1976). Lin and McDonald (Chapter 4) discuss pulsed, time resolved techniques such as flash photolysis, laser excitation, and pulsed radiolysis, which have been used to generate long-lived states. Pulsed radiolysis has been especially important for studies of excited states of rare gas atoms (LeCalvé and Bourène, 1973; Yokoyama et al., 1977; Chen et al., 1976; Hurst and Klots, 1976).

No matter what excitation source is used, care must be taken to discriminate against other excited states of the parent or carrier gas and impurities that may be generated in the excitation region. The effects of resonance radiation produced in the discharge can be serious; resonance trapping can be reduced by placing a right angle bend and a Wood's horn light trap between the discharge and the reactor. A penalty for this is that the bend causes a perturbation in the flow and distance between the discharge and reaction zone is lengthened. This usually is not serious because it often is necessary to discriminate against short-lived states by delaying the observation time for a few milliseconds of flow time. This delay is equivalent to delaying the onset of the observation time in pulsed excitation systems until the short-lived species have decayed away leaving only the longer-lived states.

3. Flow Reactors Using Helium and Argon as Carrier Gases

Most studies of excited states in flow systems use helium or argon as the carrier gas because they are relatively inert toward the majority of the excited states and the cost is low. This section will discuss $He(2^3S)$ in helium and $Ar(^3P_2)$ in argon as examples; other excited states have similar properties in these carrier gases. In fact, these states place the most stringent restrictions on the design of the flow system because of the rapid diffusion of $He(2^3S)$ in He and the fast three-body decay of $Ar(^3P_0)$ and $Ar(^3P_2)$ in Ar. The special problem of atomic and molecular helium ions when using He as the carrier gas will be considered in Section IV,B. Control of impurities is important, or even critical, because many (most) excited states are readily quenched by common tank gas impurities such as O_2, H_2O, and CO_2. The most convenient purification method is passage of the carrier gas through cooled, activated molecular sieve traps; this is satisfactory for normal tank grade He and Ar. Liquid nitrogen cooled sieve traps are especially good for He because the gas can be passed through the traps at atmospheric pressure. Care must be used to prevent condensation of liquid argon into the traps and *liquid nitrogen cooled* molecular sieve traps can only be used *with Ar* at low pressure. Cooling the activated sieve traps to 195° is normally satisfactory for removing all impurities in Ar except N_2. The characteristic emission from $He(2^3S)$ and $Ar(^3P_2)$ interacting with N_2, CO_2, H_2O, and O_2 can be used to ascertain the degree of gas purity.

The specific problems of design are choosing the diameter and length of the flow tube and a pumping speed which will minimize the loss of excited states without undue capital investment in pumps or in operating costs for carrier gas. Plots of relative $He(2^3S)$ concentration versus pressure at constant time and relative $He(2^3S)$ concentration versus time at constant pressure are shown in Figs. 2 and 3. Although the details of the plots change with the diffusion coefficient and the two- and three-body loss rates of E*, the information in Figs. 2 and 3 is generally applicable to E* in He carrier gas. As is

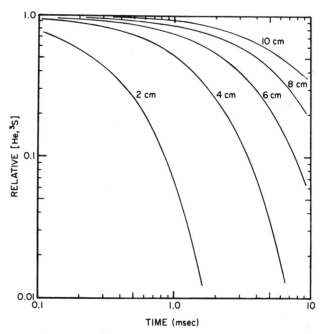

Fig. 2. The decay of $He(2^3S)$ concentration versus time in 1 Torr of He for reactors of various diameter. The rate law is $-\ln[He, 2^3S] \propto (D_0/\Lambda^2 P + k_F P^2)Z/V$; D_0 and k_F were taken from Table IV; Λ is the characteristic diffusion length and Z/V is time.

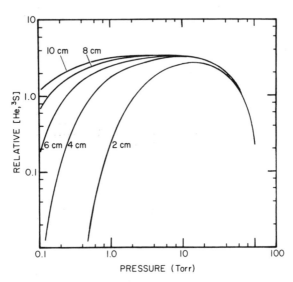

Fig. 3. The decay of He(2^3S) concentration versus pressure at a fixed time of 1 msec for reactors of various diameter. The rate law is $-\ln[\text{He, } 2^3\text{S}] \propto (D_0/\Lambda^2 P + k_E P + k_F P^2)Z/V$; D_0, k_E, and k_F were taken from Table IV; Λ is the characteristic diffusion length and Z/V is time. The concentration was normalized to unity at 70 Torr for each curve; the decay for pressure above 70 Torr is independent of tube diameter (providing the bulk flow velocity is the same in the various diameter tubes) because the three-body decay process is dominant. At lower pressure the decay becomes dependent upon tube diameter because diffusion is important. The absolute concentration decays by a factor of 0.28 at 70 Torr after 1 msec.

readily evident, for minimum loss of He(2^3S), large diameter tubes and moderate pressures are needed which dictate a large pumping capacity. There are, however, other factors which must be considered, such as the time needed to develop the flow profile (which is discussed in a Section II,B), mixing time for an added reagent and the time period over which a reaction must be observed to obtain reliable kinetic data. For quenching rate constant measurements the flow reactors using He as carrier gas are between 7 and 9 cm in diameter with flow velocities of approximately 100 m sec^{-1}. This pumping capacity is provided by a Roots booster backed by a mechanical pump.‡ For an 8-cm-diameter reactor the He(2^3S) concentration will decay by a factor of five after 10 msec (100 cm at 100 msec^{-1} flow velocity) at 1 Torr. To improve upon this, the linear flow speed must be maintained and the flow tube diameter increased, which requires increased pumping capacity. For flow tubes larger than 10 cm diameter, the cost of vacuum pumps large enough to maintain

‡ The cost for a suitable mechanical pump (300 ft^3/min) and blower (1300 ft^3/min) was ~$15,000 in 1978.

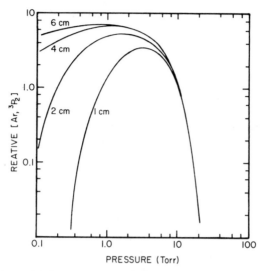

Fig. 4. The decay of $Ar(^3P_2)$ concentration versus pressure at a fixed time of 1 msec in flow tubes of various diameter. The rate law is $-\ln[Ar,\ ^3P_2] \propto (D_0/\Lambda^2 P + k_E P + k_F P^2)Z/V$; D_0, k_E and k_F were taken from Table IV, Λ is the characteristic diffusion length and Z/V is time. The concentration has been normalized to unity at 10 Torr for each curve; the decay for pressure ≥ 10 Torr is independent of tube diameter (providing the bulk flow velocity is the same in the various diameter tubes) because the three-body decay is the dominant term. At lower pressure the decay depends upon tube diameter because diffusion becomes important. The absolute $Ar(^3P_2)$ concentration will decay by a factor of 0.32 after 1 msec at 10 Torr.

high flow speeds can become prohibitive as is the operating cost of He carrier gas. Increasing the flow speed in a smaller diameter reactor also can be done, but this reduces the time over which the progress of a particular reaction can be observed. Faster flows require a longer distance to set up defined flow profiles and a longer distance to establish mixing. In practice, a flow of ~ 100 m sec^{-1} is a reasonable limit since the molecular velocities of He and Ar are 1.1×10^3 and 3.5×10^2 m sec^{-1} at 300 K. It is usually not feasible to reduce diffusion by simply increasing the pressure at constant flow velocity, because the decay from two- and three-body processes become more severe (see Figs. 2 and 3).

The consideration for E* in argon carriers are generally similar to those for helium (see Fig. 4). However, the diffusion rate is a factor of four less than for helium, which allows the use of smaller diameter flow tubes and smaller pumps.‡ For the same linear flow velocity, the loss rate of $Ar(^3P_2)$ in a 4-cm-diameter reactor with argon carrier is the same as for $He(2^3S)$ in helium

‡ We use a 1500 liter min^{-1} mechanical pump with a 4200 liter min^{-1} Roots booster for a combined cost of $\sim \$6000$ (1978).

in a 8-cm-diameter tube up to a pressure of 1 Torr (see Figs. 2 and 4). The three-body decay rate for $Ar(^3P_2)$ in argon is an order of magnitude larger than for $He(2^3S)$ in helium; the two-body decay rate for $Ar(^3P_2)$ also is significant, whereas that for $He(2^3S)$ is not. For these reasons the study of $Ar(^3P_2)$ above ~ 10 Torr in argon carrier is difficult. The two- and three-body decay rates of other E* usually are slower than that of $Ar(^3P_2)$ in argon and the concentration of E* may not decline as rapidly with increasing pressure as the plots shown in Fig. 4. The reactors used to study the decay of $Ar(^3P_2)$ versus flow distance have been 3–6 cm in diameter with pumping speeds of 90–60 m sec^{-1}.

Reagent mixing inlets are usually of the shower-head type, perforated ring inlet, or simply a tube extending into the flowing gas. The shower head and ring inlets provide good mixing; the simple point source gives moderately good mixing characteristics, but requires a somewhat longer flow distance before the reagent fills the full cross-sectional area of the flow tube. Reagents which gave strong visible emission should be used for testing the performance of mixing ports. Fast mixing is important for measuring decay rates of E* in fast flow reactors and new reagent ports *always* should be tested by visual inspection of the flow pattern. The mixing is improved by using dilute mixtures of the reagent in Ar or He and high flow rates of the dilute mixture.

Variable temperature studies of helium ions and metastable atom reactions in a flowing reactor have been done by Dunkin *et al.* (1968), Lindinger *et al.* (1974), Richardson and Setser (1973), and Clark and Setser (1975). Achieving variable temperature in a flowing afterglow was reviewed by Ferguson *et al.* (1969). Since E* frequently decays rapidly in the carrier gas, the gas must be heated or cooled to the desired temperature before entering the discharge region, because the distance required to change the temperature of the gas is too long for this to be done after E* is produced. Thus, the basic problem is to heat or cool the carrier gas to the desired temperature before entering and during the flow through the observation zone of the flow reactor. Alteration of the temperature does not apparently affect the operation of the discharges. Fontijn and Felder, Chapter 2, describe how to maintain a uniform temperature along a flow reactor. Helium is more desirable than Ar as the carrier gas because the thermal conductivity of He is approximately six times that of Ar. This lesson was painfully learned in our laboratory during some attempted studies of the reactions of CO(a) in He carrier (Clark and Setser, 1975) and $N_2(A)$ in Ar carrier (Clark and Setser, 1979). Dunkin *et al.* (1968) and Lindinger *et al.* (1974) were quite successful in using heating or cooling coils wrapped around the flow tube to achieve the desired temperature in their studies with He carrier. Richardson and Setser (1973) observed the emission spectra from $He(2^3S)$ reactions at 195 and 78 K by immersing the flow reactor and discharge (similar to Fig. 1b) into a dewar filled with coolant. The dewar had a thermally isolated window for making spectro-scopic measurements.

4. Detection of Excited States

The methods used to measure excited state concentrations either on an absolute or relative basis fall into five categories; (a) optical absorption, (b) resonance or laser fluorescence, (c) direct emission, (d) tracer techniques (addition of a reagent that gives an emission intensity proportional to [E*]), and (e) measurement of the heat released to a catalytic probe which converts the excited states to ground states. Photoionization followed by monitoring of the ions is a subcategory of (b). Laser induced magnetic resonance (see Chapter 1) is a possibility for specific cases, as is EPR spectroscopy. These techniques will not be discussed in detail because they are covered in other chapters of this book. In contrast to ground state atoms, the use of titration techniques to determine absolute concentrations of E* are difficult and somewhat unreliable if fast flow velocities are used because relatively long reaction times are required to remove E* even for reactions with rate con-stants in the 10^{-9}–10^{-10} cm^{-3} sec^{-1} range. Thus, the concentration of the reagent required to remove [E*] is not sharply defined. The methods ap-plicable to a given species will be mentioned individually in Sections IV and V.

The inherent problems in absorption spectroscopy associated with the line widths of the absorber and the source must be considered for excited states; however, the self-reversal of the source lamps is generally greatly reduced or completely absent because the concentration of the lower excited state in the lamp is very low. The lower limit of detection is determined by the oscillator strength of the relevant transition. For the strongest atomic absorption lines, concentrations of 10^{10} cm^{-3} easily can be measured. The concentration range which can be covered by absorption is about a factor of 100 but can be extended by utilizing lines of the excited states which have different oscillator strengths. Fluorescence techniques are more sensitive than absorption methods if the background light within the system is low. Laser induced fluorescence from metastable rare gas atoms has been used by Chang and Setser (1978), Bondybey and Miller (1977a), and Collins et al. (1972). This is likely to become more common in the future as more labor-atories have access to tunable dye lasers. The sensitivity allows concentra-tions of $\sim 10^7$ atoms cm^{-3} to be measured. The sensitivity of resonance fluorescence lies between that of absorption and laser induced fluorescence. The addition of reagents which react with E* to give an emission with intensity proportional to [E*] is a very useful and sensitive method for monitoring relative excited state concentrations. Specific examples of tracers will be presented with the individual discussions of each excited state. Breckenridge and Miller (1972a) used EPR techniques to measure $[Ar(^3P_2)]$; however, the sensitivity was not as good as atomic absorption and, hence, required much longer periods of time to accummulate data. This is expected to be a general difficulty in the use of EPR for monitoring most excited state concentrations.

B. Measurement of Quenching Rate Constants

Decay constants are obtained by measuring the variation of the concentration versus flow distance and then relating the distance to time. Both Clyne and Nip and Fontijn and Felder discuss the general procedure for relating flow distance to flow time and we will only mention special aspects related to excited states and the use of somewhat higher flow velocities. The effects of nonuniform velocity profiles and deactivation of E* by collision with the walls upon concentration profiles must be taken into account in relating flow distance to effective flow time. As in the previous section, the treatment will be directed to the use of helium and argon as carrier gases. Analytical solutions to the flow equations for ideal conditions have been presented by several authors. Ferguson et al. (1969) and Bolden et al. (1970) give solutions for fully developed parabolic flow of He carrier gas with unit probability for deactivation of He^+ or other ions at the wall. Bolden applied the same treatment to $He(2^3S)$ and Ar^+ in He and Huggins and Cahn (1967) did the same for $He(2^3S)$ in He. Poirer and Carr (1971) and Ogren (1975) have examined the flow equations for the case in which E* does not react at the wall with unit efficiency. If E* is deactivated with unit probability at the walls and if a fully developed parabolic profile exists, the analytical solutions are appropriate and the measured rate constants calculated for a pure plug (entrance flow) velocity profile must be increased by a factor of 1.6. This conclusion is supported by direct measurements of the flow velocity using pulsed techniques (Bolden et al., 1970; Schearer, 1974). The plug flow situation was described in Chapter 1 and holds if molecular diffusion is sufficiently rapid to eliminate the concentration gradient across the tubular reactor, because the mean velocity of the parabolic flow profile is just the plug flow velocity. In order for the concentration gradient to be negligible, the reactive species must be reflected from the wall rather than being deactivated. Some work has been done with regard to whether the method of concentration measurement (on axis, averaged over diameter of tube, etc.) affects the flow analysis. The general view is that these effects are negligible. The problem of converting the observed decay of [E*] versus distance to a decay rate lies in deciding whether or not the flow is characterized by parabolic flow, plug flow, or some intermediate case. It generally is assumed that E* is deactivated efficiently upon encounter with the walls.

The differential rate equation which describes the decay of an electronically excited state E* can be written in the general first-order form providing that the bimolecular process, $2E^* \rightarrow$ products, can be ignored

$$-d[E^*]/dt = k_{psi}[E^*].\tag{1}$$

In (1), k_{psi} is the total pseudo-first-order decay constant and may include several terms. The general form of k_{psi} will be

$$k_{psi} = (1/\eta)\{(D_0/\Lambda^2[M]) + k_E[M] + k_F[M]^2 + k_Q[Q] + \tau^{-1}\},\tag{2}$$

but other terms also may be present. The carrier gas concentration is [M], k_E and k_F are the two- and three-body deactivation rate constants, k_Q is the quenching rate constant by the reagent Q, and D_0 is the pressure independent diffusion coefficient. The characteristic diffusion length Λ is defined by $\Lambda^{-1} = (\pi/L)^2 + (4.82/l)^2$ for plug flow conditions with L and l being the length and diameter of the flow tube, respectively (Ferguson et al., 1969). The η in (2) is the correction factor that must be applied to k_{psi} if plug flow is used to relate flow distance to time. As noted previously, $\eta = 1.6$ for a fully developed parabolic flow. Estimation of the value of η for partially developed parabolic flow can be done using the treatment outlined by Fontijn and Felder in Chapter 2. The length needed to fully develop parabolic flow is given by Langaar (1942) and Dushman and Lafferty (1962):

$$L_e = 0.114 \, lR_e.\tag{3}$$

In this equation l is the flow tube diameter and R_e is the Reynolds number

$$R_e = 2MQ_m/\pi\bar{\eta}RTl = lMVP/2RT\bar{\eta},\tag{4}$$

Q_m is the carrier gas flow in mliter (STP) $\sec^{-1}$, $\bar{\eta}$ is the carrier gas viscosity in poise, and V is the carrier gas flow velocity, all other quantities have their usual meaning. For typical reactors using He as the carrier gas at 300 K, $V = 100$ m $\sec^{-1}$, $l = 8$ cm, and $\bar{\eta} = 194$ μP, the distance varies from 10 to 100 cm for 0.1–1 Torr. For Ar at 300 K with $v = 90$ m $\sec^{-1}$, $l = 4$ cm and $\bar{\eta} = 222$ μP, the distance varies from 10 to 100 cm for 0.1–1 Torr. For the pressures and reactors which are normally used for He, the assumption of a fully developed parabolic profile is good. On the other hand, for Ar the flow frequently will be in a transition between parabolic and plug flow, and the correction factor will be between 1.0 and 1.6. It normally is not feasible to wait for a longer flow time because E* decays too rapidly. Many authors (see Chapter 2, Fontijn and Felder) use $\eta = 1.3$ as an estimate of the correction. Estimation of the correction factor cannot be done by simply comparing the flow distance to L_e unless all flow perturbations are absent. Observation windows, reagent inlets, light traps, and bends in the flow tube all tend to retard the development of parabolic flow.

The correction factor for $Kr(^3P_2)$ in Ar carrier was evaluated experimentally (Kolts and Setser, 1978a) by observing how k_{psi} varied with operating conditions with $[Q] = 0$. The value of k_{psi} is constant for constant Ar pressure and any observed change in k_{psi} with variation of flow velocity arises from a change in the degree of development of parabolic flow. As can be seen from Eq. (3), the length needed to develop parabolic flow is proportional to the product of the pressure and the flow velocity. If pressure is held constant and the flow velocity is varied, the extent of development of the parabolic flow should be changed. Experiments with $[Q] = 0$ were done for $Kr(^3P_2)$ in argon carrier using the plug flow ($\eta = 1.0$) analysis to convert flow distance to flow time; $k_{psi} = (D_0/\Lambda^2[Ar]) + k_E[Ar] + k_F[Ar]^2)$ was measured using several different velocities for the 0.9–5 Torr range. The results for this particular flow tube, $l = 3.1$ cm, are presented in Table II and show that the apparent values of k_{psi}, in fact, did decrease as the flow velocity was reduced. The apparent rate constants declined nearly to the expected limiting factor of 1.6 for the slowest flow rate, i.e., the rate constants computed from plug flow should be corrected by this factor. This method can be used to estimate the value of η if k_{psi} can be measured at various flow velocities. As mentioned before, some laboratories have used direct timing techniques with pulsed discharges (Bolden *et al.*, 1970; Schearer, 1974) to measure flow velocities.

The integration of Eq. (1) under pseudo-first-order conditions, $[Q] > [E^*]$, and constant carrier gas pressure yields:

$$\ln \frac{[E^*]}{[E^*]_0} = -\frac{1}{\eta} \left\{ \tau^{-1} + \frac{D_0}{\Lambda^2[M]} + k_E[M] + k_F[M]^2 + k_Q[Q] \right\} \Delta t$$

$$= \frac{1}{\eta} \{k_1 + k_Q[Q]\}\Delta t, \tag{5}$$

The sum of all decay processes, other than those dependent upon the added reagent, is represented by k_1. Equation (5) is very useful because the absolute concentration of E^* is not needed and the zero time point can be defined at any position in the reactor. The value k_1 need not be known either, as long as it remains constant. The equation fails if deactivation at the walls is not constant with distance or is dependent upon $[Q]$; also Q must not be involved in a three-body decay process.

Two methods of obtaining rate constants from (5) may be used. The first is a fixed point analysis, which involves observation of $\ln[E^*]$ at one position as a function of added $[Q]$. In this method the effective contact time between E^* and Q must be known exactly to extract reliable rate constants. This time is not necessarily equal to Z/V (Z is the distance between reagent inlet and observation point) because in general, immediate, uniform mixing at the reagent inlet is not obtained for these fast flows. An accurate value of the "effective" Z at constant V can be determined by measuring Γ, where $\Gamma = d \ln[E^*]/d[Q]$, at several observation points (Piper *et al.*, 1973). Plotting Γ

Table II

Pseudo-First-Order Decay Constants for $Kr(^3P_2)$ *in Ar at Different Flow Velocities*

P(Torr)	$k_{psi}{}^a$ (sec^{-1})	Flow velocity[b] (cm sec^{-1})	$L_e{}^c$ (cm)
0.9[d]	240	8715	41
	239	6789	32
	201	5137	24
1.4	187	9320	68
	185	8784	64
	152	7613	56
	125	6441	47
	116	4860	35
	118	3514	26
2.0	160	9270	98
	164	8921	93
	153	8013	84
	132	6794	71
	126	5410	57
	122	4419	46
	109	3593	38
	111	3056	32
3.0	147	6343	100
	148	5697	90
	147	5105	80
	139	4486	71
	125	3922	62
	107	3482	55
	95	2821	44
5.0[d]	158	3905	102
	156	3452	90
	148	3221	84
	141	2807	74
	132	2606	68

[a] The k_{psi} values were computed using plug flow analysis to convert flow distance to flow time.

[b] The flow velocity was changed by throttling the Roots blower and adjusting the argon flow to obtain the desired pressure.

[c] The distance needed to develop full parabolic flow, Eq. (3); the $[Kr(^3P_2)]$ measurements were made in a 3.1-cm-diameter tubular reactor 35–53 cm downstream of the entrance of the gas to the flow tube (Kolts and Setser, 1978a).

[d] The limited quantity of data at low and high pressure is a consequence of the low concentration of $Kr(^3P_2)$, which limits the range of flow velocity that could be studied.

versus Z (physical distance from reagent inlet to observation points) should give a straight line with intercept on the distance axis equal to the "effective" mixing point. The distance from this "effective" mixing point to the observation zone is the value of Z used to calculate the contact time. When using this method, all conditions must be kept constant to avoid changing the "effective" Z from one experiment to the next.

An alternative method which is more time consuming, but which gives more reliable data (in the opinion of the present authors), is to measure $[E^*]$ at several positions at fixed total pressure and fixed $[Q]$. This defines k_{psi} for a given $[Q]$. A plot of k_{psi} versus $[Q]$ will be linear with slope of k_Q and intercept of k_I. There are various methods by which the time (position) can be varied. The most common is to vary the position of the reagent inlet within the flow tube. In our laboratory we have moved the detection apparatus (atomic absorption) along the flow tube. A typical set of decay curves and the resulting plot of $k_I + k_Q[Q]$ versus $[Q]$ are shown in Figs. 5 and 6. Based upon our experience (Velazco et al., 1978), the method illustrated by the data in Figs. 5 and 6 gives k_Q values with $\pm 10\%$ random error. In our laboratory the fixed point observation method gave $\sim \pm 20\%$ reproducibility. This may be improved somewhat if the concentration of $[E^*]$ can be followed over a wide dynamic range. Laboratories using the flowing afterglow technique with fixed point mass spectrometric sampling from the center of the flow tube seem to achieve $\pm 10\%$ reliability. In principle, altering the contact time by varying

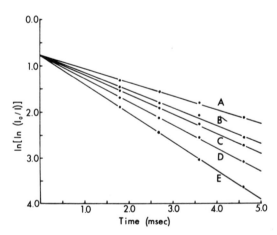

Fig. 5. Typical decay curves for heavy metastable rare gas atoms as measured in the reactor shown in Fig. 1a. The data are for $Kr(^3P_2)$ reacting with F_2 at various concentrations $k_{psi} = $ A, 0.98×10^{-11}; B, 2.2×10^{-11}; C, 2.9×10^{-11}; D, 4.2×10^{-11}; and E, 5.9×10^{-11} molecules cm^{-3}. The $[Kr, 3P_2)$ was measured by atomic absorption spectroscopy.

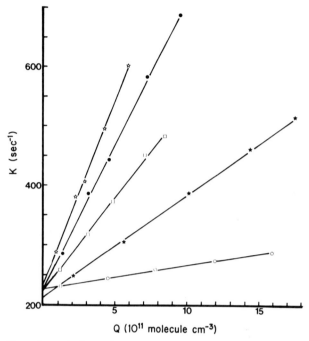

Fig. 6. Plots of pseudo-first-order rate constants, obtained from such data as shown in Fig. 5, versus reagent concentration for $Kr(^3P_2)$: open stars—F_2, solid circles—OF_2, open squares—N_2F_4, solid stars—NO and open circles—N_2. The slopes of these lines give the quenching rate constant.

the flow speed, while keeping the pressure and [Q] constant, can be used to determine k_Q, but often this method introduces complications.

Equation (5) can be used to obtain D_0, k_F, and k_F by measuring k_I values in the absence of Q for a wide pressure range; nonlinear least-squares fitting is used to assign individual rate constants (Kolts and Setser, 1978a). Using a variation of this method, Marcoux *et al.* (1977) measured the radiative lifetimes for the vibrational states of $CO(a^3\Pi)$.

III. FLOW SYSTEMS FOR ELECTRONICALLY EXCITED STATES: ASSIGNMENT OF PRODUCTS

A. General Features

In Section II the main concern was the observation of the decay rate of E* as a function of time or added reagent concentration; this section will deal with assignment of rate constants to formation of specific product states from the reaction of E* with some atom or molecule BC. There may be many exit channels, i.e., Penning ionization, associative ionization, ion-pair formation, excitation transfer, dissociation, dissociative excitation, or reactive quenching:

$$E^* + BC \xrightarrow{\quad k_{PI} \quad} E + BC^+ + e^-$$

$$\xrightarrow{\quad k_{AI} \quad} (EBC)^+ + e^-$$

$$\xrightarrow{\quad k_{IP} \quad} E + B^+ + C^-$$

$$\xrightarrow{\quad k_{ET}^* \quad} E + BC^*$$

$$\xrightarrow{\quad k_D \quad} E + B + C$$

$$\xrightarrow{\quad k_{DE} \quad} E + B + C^* \text{ (or } E + B^* + C)$$

$$\xrightarrow{\quad k_{RQ}^* \quad} EB^* \text{ (or } EC^*) + C \text{ (or B)}$$

$$\xrightarrow{\quad k_{RQ} \quad} EB \text{ (or EC)} + C \text{ (or B)}.$$

Beam experiments, whether Penning ionization electron spectroscopy (Hotop *et al.*, 1979; Yee *et al.*, 1975; Brion and Crowley, 1977; Čermák, 1976; Hotop and Niehaus, 1970) or ion collection (West *et al.*, 1975; Howard *et al.*, 1974; Čermák and Herman, 1968), provide valuable information about the first three processes. If all electrons are energy analyzed and collected, branching ratios for the ion state products are obtained. Energy analysis of the electrons also provides information about the potential curves for the entrance and exit channels. If the mass spectrometer is calibrated, branching ratios for product ions can be obtained from either beam or flowing afterglow experiments. Electronically excited state neutral and ionic products have been observed by optical methods in flowing afterglow and beam studies. The relative yields or branching ratios for such products can be obtained from the relative emission intensities. Assignment of absolute rate constants for formation of electronically excited products can be done easily too, providing that a reference reaction is available (Gundel *et al.*, 1976; Velazco *et al.*, 1976; Chang *et al.*, 1978). Determination of the neutral products that are not electronically excited is the most difficult aspect of product yield determinations. In some instances (Meyer *et al.*, 1971) it is possible to trap the stable products and subsequently make gas chromatographic measurements. Laser induced fluorescence promises to be a very important method, although absolute calibration for concentrations of the unstable species is difficult. Other techniques discussed in Chapters 1 and 2 may be applied in certain cases. In the remainder of this section the use of relative emission intensities to determine product formation rate constants will be emphasized. Although not the subject of this review, considerable insight about the reaction dynamics is provided by the quantum state distributions of the products which can be determined from the emission spectra.

B. Flow Reactors for Product State Identification

For measurement of emission intensities from products P* formed by quenching reactions, smaller diameter flow tubes with a shorter distance between the discharge and observation region and lower capacity pumps frequently are used. Figure 1b depicts a typical flow tube used in our laboratory for emission studies; this design is essentially that of Collins and Robertson (1964) and Prince et al. (1964). For this apparatus the emission appears as a Bunsen-burner-type flame in the coaxial mixing region. The length of the flame depends upon the flow velocity and the product of $[Q]k_Q$. The advantage of this technique is that most of the metastable atoms are converted to product states in front of the observation window. Observation is with a monochromator with slits aligned along the flame in the direction of the gas flow. In the visible and ultraviolet regions a quartz window is used on the apparatus. For observation in the vacuum ultraviolet the monochromator is directly attached to the flow tube and a LiF window serves as the interface; the vacuum monochromator also can be used windowless with appropriate purging or differential pumping. Although concentrations are low, *contamination of windows* with resulting changes in optical transmission can be a serious problem with some reagents. The discharge must be operated so that the vacuum uv monochromator and the ground electrode are at the same potential to prevent a secondary discharge from the gas to the monochromator chassis.

The pressure range over which the apparatus shown in Fig. 1b can be used depends upon the decay processes for E*. With a 1000 liter min^{-1} mechanical pump, this apparatus has been used from 0.5 to 100 Torr for $Xe(^3P_2)$ in Ar for which the three-body recombination rate is very slow (Kolts and Setser, 1978a). The high pressure limits for $Kr(^3P_2)$ and $Ar(^3P_2)$ in Ar and $Ne(^3P_2)$ in Ne were approximately 30, 8, and 10 Torr, respectively. Variation of pressure is the most difficult for $He(2^3S)$ in He with this apparatus; the lowest usable pressure is ~ 1 Torr and above ~ 2.5 Torr He_2^+ makes a significant contribution to the excited state concentration. To achieve pressures <0.6 Torr with Ar carrier gas, we have increased the pumping speed by adding a Roots booster pump, which tripled in the flow velocity, and shortened the distance between the discharge and mixing zone by eliminating the light trap. The discharge tube and reagent inlet tubes, which were 6- and

10-mm diameter, respectively, entered coaxially into the 20-mm-diameter main flow tube. With this apparatus, the reaction flame appeared "bushy" below 0.2 Torr and effectively filled the flow tube; at pressures above 0.2 Torr the flame was very narrow and extended along the flow tube. The high pressure limit with this apparatus was only 0.8 Torr. Touzeau and Pagnon (1978) have used a 4-cm-diameter flow tube with a hot-cathode discharge that operated down to 0.2 Torr for study of metastable argon atoms.

For reactions giving product states with long lifetimes, alternative designs to the geometry shown in Fig. 1 may be desirable. Thomas and Thrush (1977a,b) have used an integrating sphere to observe the infrared emission from E–V quenching processes with metastable molecular oxygen. This is an effective technique for increasing the signal providing that fast two- and three-body steps removing [E*] or [P*] are not important.

C. Product Formation Rate Constants by the Reference Reaction Method

Provided there is no quenching, nonradiative decay, or pumping of the excited product state P* from the observation zone, the steady-state rate of photon emission from the product excited state equals the rate of formation of the excited state from some reagent AB.

$$I_{\mathrm{P*}} = k_{\mathrm{P*}}[E^*][AB]. \tag{6}$$

If the radiative lifetime of P* $< \sim 10^{-6}$, these conditions normally are satisfied and Eq. (6) can be used. If $k_{\mathrm{P*}}$ is known for some reference molecule (AB), the formation rate constants $k_{\mathrm{P*'}}$ from other molecules Q can be obtained by comparing the integrated relative emission intensities:

$$I_{\mathrm{P*'}}/I_{\mathrm{P*}} = k_{\mathrm{P*'}}[E^*][Q]/k_{\mathrm{P*}}[E^*][AB], \qquad k_{\mathrm{P*'}} = k_{\mathrm{P*}}I_{\mathrm{P*'}}[AB]/I_{\mathrm{P*}}[Q]. \tag{7}$$

The concentration of E* cancels from the expression if [Q] and [AB] are both sufficiently small that the removal of [E*] is insignificant or if a prepared mixture of AB and Q are used. This method avoids the measurement of absolute emission intensities; but, it is necessary to record the same fraction of the total emission for each reaction and the observed intensities must be corrected for the variation of the detection efficiency with wavelength. Of course, P*' or P* must not be quenched by the reagents or the carrier gas. The best way of doing the experiments is to measure $I_{\mathrm{P*'}}$ and $I_{\mathrm{P*}}$ from prepared mixtures of AB and Q. Then [E*] and the geometry are necessarily constant. If mixtures are not used, then consecutive experiments are done with AB and Q with no changes in experimental conditions. For consecutive experiments the emission intensity must be strictly first order in the flow rate (concentration) of both AB and Q. Deviations from linearity at high flows will occur

Reference Reactions for Assignment Product Formation Rate Constants

Reaction	Rate constant[a] (10^{-11} cm³ molecule⁻¹ sec⁻¹)	Emission	References
He(2³S) + N₂ → N₂⁺(B)	2.9	N₂⁺(B-X)	Chang et al. (1978)
+ CO → CO⁺(B)	2.1	CO⁺(B-X)	Chang et al. (1978)
Ne(³P₂) + CO → CO⁺(A)	(unknown)	CO⁺(A-X)	Richardson and Setser (1980)
Ar(³P₂) + Kr → Kr(5p[3/2]₁)	0.43	Kr(5p[3/2]₁-5s[3/2]₁°)[b]	Piper et al. (1975a)
+ Cl₂ → ArCl*	21	ArCl(B-X)ᶜ	Kolts and Setser (1978b)
+ Cl₂ → Cl₂*	2.0	Cl₂*	Velazco et al.(1976)
Kr(³P₂) + Cl₂ → KrCl[b,ʌ]	73	KrCl(B-X; C-A)	Velazco et al. (1976)
Xe(³P₂) + Cl₂ → XeCl[b,c]	72	XeCl(B-X; C-A)	Velazco et al. (1976)
Hg(³P₂) + Cl₂ → HgCl(B)	d	HgCl(B-X)	Krause et al. (1975)
+ NO → NO(A)	d	NO(A-X)	Liu and Parson (1976)
N₂(A) + NO → NO(A)	11	NO(A-X)ᵉ	Meyer et al. (1972)
	4.5		Mandl and Ewing (1977)
	8.0		Callear and Wood (1971)
N₂(A) + Hg → Hg(³P₁)	29	Hg(³P₁-¹S₀)ᶠ	Callear and Wood (1971)
	8		Meyer et al. (1972)
	34		Thrush and Wild (1972)
N(²D) + N₂O → NO(B)	g	NO(B-X)	Slanger and Black (1976b)

[a] The rate constant for the emission specified in column three.

[b] The subsequent radiative cascade of Kr(5s[3/2]₁°-4p(¹S₀)) in the vacuum ultraviolet also can be used as a reference reaction.

[c] The rate constant is for only the ArCl(B-X) emission; however, both the B-X and C-A emission can be used in which case the rate constant would be 25×10^{-11} cm³ molecule⁻¹ sec⁻¹.

[d] The cross sections measured in molecular beam experiments are $\sigma_{NO} = 1.4$ Å² and $\sigma_{Cl2} = 90$ Å².

[e] According to Dreyer et al. (1974), there is a strong dependence of the rate constant on vibrational level; this may be a contributing factor to the spread in rate constants for NO.

[f] The spread in rate constants is probably related to the difficulty of measuring [Hg]. However, some interplay between Hg* spin orbit population and N₂(A) vibrational level also could be a possibility. The larger rate constants are favored.

[g] The branching fraction for NO(B) formation may be small, see Slanger and Black and associated references.

because of changes in the emission flame geometry and reduction of $[E^*]$. The concentration range for which the emission will be first order depends upon the magnitude of k_Q and upon the design of the flow apparatus. For a system similar to that shown in Fig. 1b, first-order conditions existed for $[Q] \leq 3 \times 10^{12}$ molecule cm^{-3} for reagents with $k_Q \approx 5 \times 10^{-10} cm^3$ molecule^{-1} sec^{-1}.

The reference reaction method also can be used if the readiative lifetime is longer than the residence time in the observation zone. Providing that quenching or nonradiative decay process of P^* and $P^{*\prime}$ do not exist and providing that all experimental conditions are constant, then the integrated relative intensities for very short observation times (Smith et al., 1977) are converted to rate constants in the following way:

$$\frac{I_{P^{*\prime}}/\tau_{P^{*\prime}}^{-1}}{I_P^*/\tau_{P^{*\prime}}^{-1}} = \frac{k_{P^{*\prime}}[E^*][Q]}{k_{P^*}[E^*][AB]}. \tag{8}$$

The difference between (7) and (8) is that the radiative lifetimes of P^* and $P^{*\prime}$ must be well known. This method has been used to measure relative rate constants for reactions of F, H, and Cl atoms yielding HF and HCl as products (Smith et al., 1977; Sung and Setser, 1978). For long-lived product states one must be extremely careful to ensure that neither P^* or $P^{*\prime}$ are quenched during the fraction of a millisecond that the molecules spend in the observation zone.

The key to the reference method is, of course, finding a reliable reference reaction. It is necessary that some one reaction be studied in sufficient detail that the total k_Q can be resolved into all of the product channels. The methods that can be used to accomplish this are too varied to discuss here. We have tabulated some reference reactions in Table III and original literature should be consulted for details of how these were established. In a few cases the rate constant for the particular product channel has not been measured; however, the reactions show promise for straightforward measurement (in our opinion) of the rate constants. In some cases the emission from the reference reaction also can be generated by reactions of other excited states which may be present in the flow reactor. Thus, for example, He_2^+ gives $N_2^+(B)$ and $CO^+(B)$ and experimental conditions must be adjusted to minimize $[He_2^+]$. Care also must be taken to remove $Ar(^3P_2)$ from flows of $Xe(^3P_2)$ or $Kr(^3P_2)$ using Ar carrier in order to exclude emission resulting from $Ar(^3P_{0,2})$ reactions. The rare gas resonance states also react with Cl_2 to give the excimer emission (Brashears and Setser, 1979) so the concentration of these states must be minimized when using Cl_2 as the reference for metastable atom reactions.

IV. EXCITED ATOMIC STATES IN FLOW SYSTEMS

A. Heavy Rare Gas Metastable Atoms

The lowest excited electronic configuration for Ne, Ar, Kr, and Xe is $(n)p^5(n + 1)s^1$ which gives rise to four states, two are long lived or metastable. In L–S notation the metastable states, in order of increasing energy, are denoted $^3P_2$ and $^3P_0$ and in Paschen notation as $1s_5$ and $1s_3$. The radiative lifetimes of the $^3P_2$ states are on the order of seconds or longer. Except for $Xe(^3P_0)$, which radiates to the $Xe(^3P_1)$ level with $\tau \simeq 0.1$ sec (Small-Warren and Chiu, 1975), the other $^3P_0$ states have long lifetimes. The energy separation between the $^3P_2$ and $^3P_0$ states are 0.10, 0.17, 0.65, and 1.13 eV for Ne, Ar, Kr, and Xe. The larger separation for Kr and Xe has the consequence that in flow reactors at ~ 1 Torr only $Kr(^3P_2)$ and $Xe(^3P_2)$ states are present; however, both $^3P_2$ and $^3P_0$ are present for Ar and Ne unless special precautions are taken. There is a large body of literature for Ar; however, Ne, Kr, and Xe have received less attention (Golde, 1977; King and Setser, 1976).

The decay processes which were discussed (Section II,B) for $He(2^3S)$ and $Ar(^3P_2)$ in He and Ar carrier gas are diffusion and two- and three-body deactivation. The two-body process is normally associated with collision induced emission or transfer to the nearby resonance state. The three-body decay is associated with diatomic rare gas molecule formation. The radiative lifetimes of these molecules are short, consequently the steady-state concentrations are low and reactions of rare gas molecules usually need not be considered in flow reactors. The quenching rate constants of the molecules are approximately gas kinetic for most reagents and the preceding statement may need modification for reagents with small quenching rate constants for the metastable atomic state. The values for the diffusion coefficients, two-body, and three-body rate constants are listed in Table IV. The reported two-body rate constant for $Ne(^3P_2)$ in He is surprisingly large but note the comment in the footnotes. There are several measurements for most other rate constants in the table and the agreement usually is good. Figure 7 shows a plot of K_1, the pseudo-first-order decay constant [see Eq. (5), section II], versus pressure for $Ar(^3P_2)$, $Ar(^3P_0)$, $Kr(^3P_2)$, and $Xe(^3P_0)$ in Ar and $Ne(^3P_2)$ in Ne. The K_1 values for $Ne(^3P_2)$ in He, if plotted in Fig. 7, would be off scale because the diffusion rate is four times larger than for $Ne(^3P_2)$ in Ne. The magnitude of K_1 determines the total loss rate of the metastable atoms. Somewhat surprisingly, the bimolecular decay process ($2E^* \rightarrow$ products) appears to have been studied only for He* (D. Lorents, private communication, 1978). For Ar* in Ar and Ne* in Ne, the apparent diffusion coefficients are somewhat smaller than expected because of resonance energy transfer

Table IV

Diffusion Coefficients, Two-Body, and Three-Body Rate Constants for Rare Gas Metastable Atoms and Molecules

System	D_0 (10^{18} molecule cm^{-1} sec^{-1})	k_E (10^{-15} cm^3 molecule^{-1} sec^{-1})	k_F (10^{-32} cm^6 molecule^{-2} sec^{-1})
He(2^3S) in He	15 ± 1[a]		0.025 ± 0.003[a]
	13.6 ± 0.3[b]		0.019 ± 0.004[b]
He(2^1S) in He	14[a]	6.0[a]	
He$_2$($^3\Sigma_u^+$)	9.4[a], 9.8[b]		
Ne(^{3}P$_2$) in He	20[c]	19[c,d]	
Ne(^{3}P$_2$) in Ne	5[c]	4.9[e]	0.57[e]
Ar(^{3}P$_2$) in Ar	1.8 ± 0.1[f]	2.1 ± 0.3[f]	1.1 ± 0.4[f]
Ar(^{3}P$_0$) in Ar	1.8 ± 0.1[f]	5.3 ± 0.9[f]	0.83 ± 0.3[f]
Kr(^{3}P$_2$) in Ar	2.7 ± 0.2[f]	0.69 ± 0.06[f]	0.10 ± 0.04[f]
Kr(^{3}P$_2$) in Kr	0.97 ± 0.05[g]	2.3 ± 0.1[f]	4.2 ± 0.2[g]
Xe(^{3}P$_2$) in Ar	2.5 ± 0.2[f]	0.5 ± 0.07[f]	0.03 ± 0.03[f]
Xe(^{3}P$_2$) in Xe	0.52 ± 0.08[h]	3.5 ± 0.3[h]	8.6 ± 0.4[h]

[a] Phelps (1955).
[b] Deloche et al. (1976).
[c] Phelps (1959).
[d] Leichner et al. (1975).
[d] This two-body rate constant seems surprisingly large; however, it is consistent (via detailed balance) with the rate constant measured by Leichner et al. (1975) for Ne(^{3}P$_1$) + He → Ne(^{3}P$_2$) + He.
[e] Leichner et al. (1975).
[f] Kolts and Setser (1978a).
[g] Smith and Turner (1963).
[h] Wieme (1974).

180

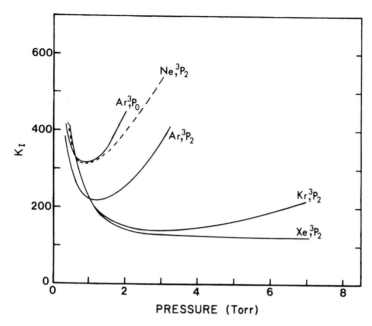

Fig. 7. Pseudo-first-order rate constants vs. pressure for $Ar(^3P_0)$, $Ar(^3P_2)$, $Kr(^3P_2)$, and $Xe(^3P_2)$ in Ar buffer gas and $Ne(^3P_2)$ in Ne buffer gas. These curves were computed from the rate constants in Table IV for a flow tube of 3.1-cm i.d. The $He(2^3S)$ in He curve would be off scale for a tube of this diameter.

(Kolts and Setser, 1978a). This, as well as the increased physical size of the metastable state, is responsible for the D_0 values of the metastable atoms being considerably *smaller* than for the ground state atoms. In ionized gases these decay processes are augmented by various time dependent formation processes (mainly diatomic-ion electron combination); such systems are discussed by Hurst and Klots (1976). These authors also present an extensive discussion of radiation trapping by the resonance states.

Metastable Ar and Ne atoms can be produced conveniently by a hollow-cathode discharge in argon or neon carrier, respectively. The ratio of the $^3P_0$ and $^3P_2$ state concentrations normally is close to the degeneracy ratio of 1 : 5. If the discharge and observation zone are separated by a long distance, the ratio will be smaller because the $^3P_0$ atoms decay more rapidly than the $^3P_2$ atoms. The $^3P_2$ concentrations in the cold-cathode discharge apparatus in our laboratory are normally $\sim 5 \times 10^9$ for $Ne(^3P_2)$ (Brom *et al.*, 1978) and $\sim 5 \times 10^{10}$ for $Ar(^3P_2)$ (Kolts and Setser, 1978a). In our apparatus the $Ne(^3P_2)$ concentration was $\sim 1 \times 10^9$ when He was used as the carrier gas. In Ne/He mixtures $Ne(^3P_{0,2})$ atoms are produced by excitation transfer from $He(2^3S)$

and $He(2^1S)$. Since these reactions are slow (6.5 and 0.38 $\times$ 10^{-11} cm^3 sec^{-1}; Schmeltekopf and Fehsenfeld, 1970), relatively high Ne flows are required to maximize $Ne(^3P_{0,2})$ formation (Bondybey and Miller, 1977a). Based upon experience in our laboratory, Ne carrier gas is preferable to Ne/He mixtures for studies of $Ne(^3P_2)$. There is no evidence for significant concentrations of Ne^+ or Ar^+ ions in these cold-cathode flowing afterglows for a wide range of pressure. This is not the case for He(see Section IV,B). Suzuki and Kuchitsu (1978) and Suzuki et al. (1978) have reported metastable Ar^+ ions in micro-wave discharge flow systems; $CH(A-X)$ from CH_4 and $Cl(5p-4s)$ emission from HCl was diagnostic of the metastable ions.

Xe and Kr metastable atoms can be produced by adding small flows of Xe and Kr to the Ar carrier gas (Stedman and Setser, 1970). Normally, the addition is done prior to the discharge, although this probably is not critical. Velazco et al. (1976) showed that xenon metastable atoms are produced by both direct excitation in the discharge and by excitation transfer from $Ar(^3P_{0,2})$. Very small flows of xenon are needed because the rate constants for quenching of $Ar(^3P_2)$ and $Ar(^3P_0)$ are large (18 and 30 $\times$ 10^{-11} cm^3 sec^{-1}; Piper et al., 1973). The excitation transfer reaction produces Xe** in highly excited levels, which subsequently cascade to either the metastable or resonance states. For moderate Xe flows only $Xe(^3P_2)$ and ground state xenon atoms are present a few millseconds downstream of the discharge, providing a light trap is placed between the discharge and observation region (J. H. Kolts, unpublished results, 1978). For high flows of xenon, $Xe(^1P_1)$ and $Xe(^3P_1)$ resonance lines can be observed which suggest radiation trapping within the flow reactor. Because $Xe(^3P_2)$ also is directly produced by the discharge, the $[Xe(^3P_2)]$ actually increases as excess xenon is added beyond that needed to quench the $Ar(^3P_{0,2})$. However, the small increase in $[Xe(^3P_2)]$ usually is not worth the added expense associated with higher Xe flows. Bondybey and Miller (1977a) report $Xe(^3P_1)/Xe(^3P_2) \leq 10^{-3}$ from laser induced fluorescence measurements in a flow reactor.

Krypton metastable atoms are produced in the same manner as xenon (Stedman and Setser, 1970); however, the $Ar(^3P_2)$ quenching rate constant (0.6 $\times$ 10^{-11} cm^3 sec^{-1}) by Kr is small (Piper et al., 1975b) and rather high flows of Kr are needed to remove $Ar(^3P_{0,2})$ and to maximize $[Kr(^3P_2)]$. If the flow time between the discharge and the observation zone is less than 5 msec, it is virtually impossible to remove all of the $Ar(^3P_{0,2})$. The produc-tion of $Kr(^3P_2)$ arises from excitation transfer with $Ar(^3P_{0,2})$ and little, if any, is directly produced by the discharge (Velazco et al., 1976). The $Kr(5p[\frac{3}{2}]_1$ and $5p[\frac{3}{2}])$ states produced by excitation transfer have been fully characterized (Piper et al., 1975b) and emission from these states forms the basis for assign-ing product formation rate constants for $Ar(^3P_{0,2})$ atom reactions. For certain operating conditions some $Kr(^3P_0)$ may be present; the $^3P_0/^3P_2$

ratio has been reported as 0.1 (Bondybey and Miller, 1976) and 0.04 (Kolts and Setser, 1978a). Krypton metastable atoms also have been produced in a helium carrier using a hot cathode discharge (Shaw and Jones, 1977); Kr^+ also was present in the flow.

One disadvantage of the heavy rare gas metastable atom flow systems is that the operating pressure range is rather limited. The exact range depends upon the physical arrangement of the observation zone relative to the discharge, the pumping capacity, and the sensitivity of the detection method. For observation of emission intensities and with moderate pumping capacity (1000 liter min^{-1}), pressures of 0.5–10 Torr can be studied with Ar and Ne. For Kr and Xe the upper end of the range may be extended to ~40 and 80 Torr respectively. With increased pumping speed, larger diameter tubes, and reduced distance of separation between the discharge and mixing zone, the pressure can be reduced to ~0.1 Torr (Kolts et al., 1979; Touzeau and Pagnon, 1978). In principle, the substitution of helium for argon as the carrier gas should permit studies of $Ar(^3P_2)$ to be extended to higher pressures because the two- and three-body decay rates are expected to be slower than in Ar. Further characterization of the decay processes for heavy metastable atoms in He carrier would be desirable; Shaw and Jones, 1977 have demonstrated the basic technique.

Atomic absorption spectroscopy has been the most commonly used method for monitoring concentrations. The method is straightforward because lamps can be made or purchased (Oriel Optics) that give non-reversed lines and the sensitivity is high because the oscillator strengths of the lines connecting the $(n)p^5(n + 1)s^1$–$(n)p^5(n + 1)p^1$ states are large. Recent work (Change and Setser, 1978, 1979, 1980) using laser induced fluorescence has provided reliable lifetimes for levels in the $(n)p^5(n + 1)p^1$ manifold. These were combined with the radiative branching ratios to yield absolute oscillator strengths for absorption lines. Table V lists these transition probabilities, which are thought to be good to $\pm 10\%$. Previous results in the literature for Ne and Ar were reasonably good; but the values in Table V for Kr and Xe are significantly better than previous results. The absorption technique is quite satisfactory for concentrations in the 10^{10}–10^{11} range. For lower concentrations, tunable dye laser induced fluorescence can be used (Bondybey and Miller, 1977a; Chang and Setser, 1978).

Tracer techniques have not been used as extensively in flow systems as for pulsed radiolysis experiments for monitoring relative concentrations of the heavy rare gas metastable atoms. However, Shaw and Jones (1977) did use the KrCl* emission intensity (from $Kr(^3P_2) + Cl_2$) to monitor the $Kr(^3P_2)$ concentration over three orders of magnitude. The XeCl emission from the $Xe(^3P_2) + Cl_2$ reaction also should be suitable for monitoring the concentrations of $Xe(^3P_2)$. However, it should be remembered that the $^3P_2$ and $^3P_1$

Table V

Transition Probabilities for the $Rg(np^5(n + 1)p-np^5(n + 1)s)$ *Array*

	Transition probability[a] $A_{ki}(10^6 \text{ sec}^{-1})$			
Transitions	Ne	Ar	Kr	Xe
$p'[\frac{1}{2}]_0 - s'[\frac{1}{2}]_1^\circ$	68.2[b]	44.1[c]	45.2[d]	(0.93)[e]
$p'[\frac{1}{2}]_0 - s[\frac{3}{2}]_1^\circ$	0.90[b]	0.09[c]	0.09[d]	(0.07)[e]
$p'[\frac{1}{2}]_1 - s'[\frac{1}{2}]_1^\circ$	21.72	16.08	14.2	(0.344)[e]
$p'[\frac{1}{2}]_1 - s'[\frac{1}{2}]_0^\circ$	13.54	12.49	21.2	(0.601)[e]
$p'[\frac{1}{2}]_1 - s[\frac{3}{2}]_1^\circ$	5.15	1.92	0.093	(<0.003)[e]
$p'[\frac{1}{2}]_1 - s[\frac{3}{2}]_2^\circ$	10.10	7.25	1.9	(<0.055)[e]
$p'[\frac{3}{2}]_2 - s'[\frac{1}{2}]_1^\circ$	23.13	22.83	35.3	32.1
$p'[\frac{3}{2}]_2 - s[\frac{3}{2}]_1^\circ$	17.07	8.80	1.6	1.07
$p'[\frac{3}{2}]_2 - s[\frac{3}{2}]_2^\circ$	10.30	4.60	0.27	0.34
$p'[\frac{3}{2}]_1 - s'[\frac{1}{2}]_1^\circ$	22.66	12.98	19.1	(0.500)[e]
$p'[\frac{3}{2}]_1 - s'[\frac{1}{2}]_0^\circ$	24.27	16.51	17.5	(0.448)[e]
$p'[\frac{3}{2}]_1 - s[\frac{3}{2}]_1^\circ$	0.677	0.02	0.07	(0.038)[e]
$p'[\frac{3}{2}]_1 - s[\frac{3}{2}]_2^\circ$	4.48	0.63	0.02	(0.008)[e]
$p[\frac{1}{2}]_0 - s'[\frac{1}{2}]_1^\circ$	0.29[b]	0.00[c]	0.04[f]	—
$p[\frac{1}{2}]_0 - s[\frac{3}{2}]_1^\circ$	60.3[b]	43.94[c]	43.8	(1.000)[e]
$p[\frac{3}{2}]_2 - s'[\frac{1}{2}]_1^\circ$	19.32	4.93	0.3[f]	
$p[\frac{3}{2}]_2 - s[\frac{3}{2}]_1^\circ$	4.21	4.19	10.4	22.4
$p[\frac{3}{2}]_2 - s[\frac{3}{2}]_2^\circ$	29.11	27.89	28.6	9.6
$p[\frac{3}{2}]_1 - s'[\frac{1}{2}]_1^\circ$	2.32	1.01	0.13[f]	e
$p[\frac{3}{2}]_1 - s'[\frac{1}{2}]_0^\circ$	11.58	2.62	0.10[f]	e
$p[\frac{3}{2}]_1 - s[\frac{3}{2}]_1^\circ$	36.05	24.93	29.0	29.1
$p[\frac{3}{2}]_1 - s[\frac{3}{2}]_2^\circ$	6.55	4.97	4.4	2.0
$p[\frac{3}{2}]_2 - s'[\frac{1}{2}]_1^\circ$	3.42	1.37	0.09[f]	e
$p[\frac{3}{2}]_2 - s[\frac{3}{2}]_1^\circ$	32.05	20.03	22.9	18.9
$p[\frac{3}{2}]_2 - s[\frac{3}{2}]_2^\circ$	17.16	9.09	8.2	6.3
$p[\frac{3}{2}]_3 - s[\frac{3}{2}]_2^\circ$	45.87	33.20	28.9	30.1
$p[\frac{1}{2}]_1 - s'[\frac{1}{2}]_1^\circ$	0.114	0.14	0.17[f]	e
$p[\frac{1}{2}]_1 - s'[\frac{1}{2}]_0^\circ$	2.43	0.85	0.15[f]	e
$p[\frac{1}{2}]_1 - s[\frac{3}{2}]_1$	9.66	5.19	4.3	(0.24)[e]
$p[\frac{1}{2}]_1 - s[\frac{3}{2}]_2$	25.82	17.60	19.8	(0.76)[e]

[a] In units of 10^6 sec^{-1}; the absorption oscillator strengths are obtained from $f_{ik} = 1.4992 \times 10^{-16} \lambda^2 (g_k/g_i)A_{ki}$ λ is in angstroms, g_k and g_i are the degeneracies of the upper and lower states, respectively, and A_{ki} is the value in the table.

[b] Taken from Inatsugu and Holmes (1975) because these Ne(3p) states could not be produced from the $(3s[\frac{3}{2}]_2)$ level by the pulsed dye laser.

states of Kr and Xe both react with Cl_2, F_2, and other simple halogen-containing compounds to give the rare gas halide excimers (Brashears and Setser, 1979). Several pulsed studies (LeCalvé and Bourène, 1973; Chen et al., 1976) have utilized the N_2(C–B) emission to ostensibly monitor the concentration of $Ar(^3P_2)$. However, a general problem with the tracer technique is that both metastable states, as well as the two resonance states, *may* give the same emission. LeCalvé et al. (1977) have clearly illustrated this point by comparing quenching of $Ar(^3P_{0,1,2})$ and $Ar(^1P_1)$ by N_2 and SF_6 when using absorption spectroscopy and the N_2 tracer method to monitor atomic concentrations.

At the present time there is not sufficient knowledge about the product state distributions from resonance and metastable atom reactions to recommend tracers that are specific for a given spin–orbit state. To our knowledge there are only two reagents for which product states have been proven to depend upon the spin–orbit state. The reaction of $Ar(^3P_{0,2})$ with H_2 gives $H_2(a^3\Sigma_g^+)$ plus H atoms (Kolts et al., 1977); whereas, $Ar(^3P_1)$ and $Ar(^1P_1)$ give a very specific vibrational–rotational distribution of $H_2(B^1\Sigma_u^+)$ and perhaps some dissociation (Fink et al., 1972; McKenney and Dubinsky, 1977). The reaction of $Kr(^3P_{0,2})$ with CO gives a broad distribution of vibrational levels in the $CO(A^1\Pi, a'^3\Sigma^+, d^3\Delta)$ states. However, the $Kr(^3P_1)$ reaction with CO shows selectivity especially for certain isotopic combinations of carbon and oxygen with specific vibrational–rotational states of $CO(A^1\Pi)$ being favored (Vikis, 1978). Based upon these examples, it seems likely that other reactions will show selectivity too, but more work is needed to identify such cases. Some workers (Chen et al., 1976) claim that the formation of $N_2(C, v = 0)$ can be used as a diagnostic test for $Ar(^3P_2)$; whereas, higher v levels of $N_2(C)$ can be used for $Ar(^1P_1)$. Sadeghi and Nguyen (1977) have shown that the low rotational levels of $N_2(C^3\Pi_u, v = 0, 1)$ were preferentially produced by transfer from $Ar(^3P_2)$ and that the $v = 2$ and 3 and the high J levels of

^{c} Taken from Lilly (1976); because these Ar(4p) states could not be produced from the $4s[\frac{3}{2}]_2$ level by the pulsed dye laser. Lilly's calculated values are in good agreement with Chang's experimental results for the other Ar(4p) levels.

^{d} Obtained from Chang's branching fractions and Lilly's (1976) calculated lifetimes, which seem to be reliable.

^{e} The numbers in parenthesis are the branching ratios measured by Chang; the transitions from the p to s′ levels were too weak to be measured. Jimenez et al. (1974) has reported lifetimes of 30.7 and 29.5 nsec for the $6p'[\frac{1}{2}]_0$ and $6p'[\frac{1}{2}]_1$ levels of Xe and Chen and Garstang (1970) report calculated A_{ki} values for some Xe(6p) levels. The values in the table for Xe* are preliminary and the published results (Chang et al., 1980) should be consulted for the best A_{ki} values.

^{f} The p–s′ transitions were very difficult to measure; the experimental values agree with calculated values (Lilly, 1976) to within a factor of 2.

$v = 0$ are produced by both $Ar(^3P_2)$ and $Ar(^3P_0)$; but, $Ar(^3P_0)$ is relatively more important. Further work is needed to establish the $N_2(C)$ vibrational distribution from excitation transfer reactions with $Ar(^1P_1)$ and $Ar(^3P_1)$. If the pressure is sufficiently low then vibrational relaxation does not occur, the presence of resonance states can be identified by utilization of thermochemical limits for reactions with N_2, HCN, or C_2N_2 (Stedman, 1970).

Pulsed tunable dye laser excitation from the metastable states is a powerful method for kinetic studies of higher atomic states of the rare gases. The only necessary modification to the apparatus shown in Fig. 1 is the addition of baffle arms holding the windows through which laser light enters and leaves the cell. The laser light passes through the flowing afterglow and fluorescence is observed perpendicular to both the laser beam and the gas flow through the normal observation window. Chang and Setser (1978) analyzed the time dependence of the laser induced fluorescence from the $Ar(4p)^{**}$ states to obtain radiative lifetimes and two-body decay constants. Similar work has been done for $Ne(3p)^{**}$ in Ne and $Xe(6p)^{**}$ and $Kr(5p)^{**}$ in argon (Chang and Setser, 1979). The first laser induced fluorescence study of rare gas atoms in a flowing afterglow seems to have been done by Collins et al. (1972) with $He(2^3S)$; they obtained a lifetime and rate constants for $He(5^3P)$.

The combination of a metastable rare gas atom flowing afterglow and a pulsed tunable dye laser is a very powerful method for making spectroscopic and kinetic measurements for a variety of reactive intermediates that can be produced by quenching reactions. If the quenching reaction is fairly specific with regard to product formation, the metastable rare gas atom concentration can be converted into a similar concentration of product species. Engleking and Smith (1975) observed laser induced fluorescence from N_2^+ and other species produced by reaction with $He(2^3S)$. Bondybey and Miller (1977b, 1978) applied the laser fluorescence technique to the $He(2^3S)$ afterglow and have observed CO^+, CO_2^+, $C_6F_3H_3^+$, and other ionic species. In a few cases the metastable rare gas atom reactions yield long-lived excited state products that become the species of interest in the afterglow. Perhaps the best example is $Ar(^3P_{0,2}) + N_2$, which provides an excellent source of the $N_2(A^3\Sigma_u^+)$ molecule. Cook et al. (1978c) have demonstrated the usefulness of laser induced fluorescence for long-lived product states in these afterglows by observation of $N_2(A)$, $CO(a^3\Pi)$, and $O_2^+(A^4\Pi)$.

A promising way of obtaining metal atoms at room temperature in a flowing afterglow is to react organometallic compounds of high vapor pressure with $Ar(^3P_{0,2})$. The reactions of $Ar(^3P_{0,2})$ with $Fe(CO)_5$ gave Fe* emission (Hartman and Winn, 1978), and Hg* emission was observed in our laboratory from reaction of $Ar(^3P_{0,2})$ with CH_3HgCH_3 (Dreiling and Setser, 1979). Lee and Zare (1976) reacted $Ar(^3P_{0,2})$ with uranocene and observed U atoms by laser induced fluorescence. Jacox (1976, 1978) has passed Ar through a microwave discharge and subsequently added reagents such as CF_2Cl_2, CF_2Br_2, $CHCl_3$, etc., and produced radicals and ionic species which were deposited in a low temperature matrix and observed by infrared spectroscopy. Jacox suggested that metastable argon atoms, as well as atoms in high energy Rydberg states, were responsible for generating the ions.

B. Helium Metastable Atoms

Helium metastable atoms, both the 2^1S and 2^3S states, can be produced with a cold-cathode, hot-cathode, or microwave discharge; however, the most commonly used method is the cold- or hot-cathode discharge. Unlike the heavy rare gas afterglows, atomic and diatomic helium ions frequently accompany the metastable atomic states. The ion concentration can be reduced by using low power in the excitation source and microwave heating between the discharge and reactant inlet region, which increases the ambipolar diffusion loss of ions and electrons. This method has been used by Bolden et al. (1970) to reduce the ion concentration by as much as an order of magnitude. Attempts to remove ions by a pair of parallel wire collector electrodes placed downstream from the discharge (Piper et al., 1975a) were only marginally effective at higher pressure. However, the use of a fine mesh screen through which the carrier gas must flow has been reported to be much more effective (Suzuki et al., 1978). One reason that the molecular ions are more important in helium flowing afterglows than for the heavy rare gases is because the electron recombination process for He_2^+ is slow ($\sim 10^{-10}$ cm^3 sec^{-1} at 1 Torr), whereas this process is fast (10^{-6} cm^3 sec^{-1}) for the heavy rare gas molecular ions. Thus, once formed, the helium molecular ions persist in the flow. Based upon experience in our laboratory with an apparatus similar to that in Fig. 1b, enhanced He_2^+ *formation processes*, relative to the heavy rare gases, also seem to be present. However, it is not so easy to identify such processes. One factor certainly is the bimolecular quenching processes, $2\,He^*$ or $2\,He_2^*$, that yield ions with rate constants of $\sim 1.5 \times 10^{-9}$ cm^3 sec^{-1} (Deloche et al., 1976). Rate constants for similar processes seem not to have been measured definitively for the heavy rare gas metastable atoms; but, the estimated values are $\sim 5 \times 10^{-10}$ cm^3 sec^{-1} (Lorents, private communication, 1978). Based upon our experience with He and Ar in the same apparatus, the bimolecular quenching processes do not explain the extreme sensitivity of $[He_2^+]$ to variation in the He pressure. For example, it is necessary to operate the apparatus shown in Fig. 1b at ≤ 3 Torr to have a flow of $He(2^3S)$ without significant concentrations of He_2^+. The presence of He_2^+ may be less serious in a large apparatus, such as Fig. 1a; at least there is less mention of the problem in the literature. In our own *limited* experience the hot cathode produces both He^+ and He^* and it is difficult (impossible) to generate either species alone. However, lowering the accelerating voltage favors the metastable atoms at the expense of the ions. A common procedure has been to add reagents which selectively react with either $He(2^1S)$, $He(2^3S)$, or He^+ (Ferguson et al., 1969). Another difference between the helium and heavy rare gases is that the lowest triplet He_2^* state also is long lived, whereas the heavy rare gas molecular states are not. Thus, for several reasons energetic species other than helium metastable atoms are more numerous in a He flowing afterglow than in the heavy rare gas afterglows.

Superelastic collisions between $He(2^1S)$ and electrons yield $He(2^3S)$ with a very high rate (Phelps, 1955; Schmeltekopf and Fehsenfeld, 1970; Deloche et al., 1976), and normally $He(2^3S)$ is the only neutral atomic state present in a He flowing afterglow. However, it is possible to study $He(2^1S)$ in a flowing afterglow (Schmeltekopf and Fehsenfeld, 1970) by careful operation of the hot-cathode discharge and by adding enough SF_6 to scavenge electrons but not so much as to quench $He(2^1S)$. Using laser induced fluorescence, Bondybey and Miller (1977a) found no $He(2^1S)$ in an apparatus similar to that in Fig. 1b, and for normal circumstances the concentration of $He(2^1S)$ in the flow is too low to affect studies of $He(2^3S)$. If, however, it becomes necessary to remove $He(2^1S)$ altogether, optical quenching techniques described by Hotop et al. (1969b) can be used.

The methods which have been most commonly used to monitor $He(2^3S)$ and $He(2^1S)$ concentrations are atomic absorption, tracer emission, and, to a lesser extent, laser induced fluorescence. Huggins and Cahn (1967) used the $O_2^+(A^2\Pi_u-X^2\Pi_g)$ emission as a measure of the relative $He(2^3S)$ concentration and also the electron density resulting from the Penning ionization of Ar. Lindinger et al. (1974) and Schmeltekopf and Fehsenfeld (1970) used neon as a tracer and monitored neon emission at 568.9 nm, which is diagnostic of $He(2^1S)$, and 703.2 nm, which is diagnostic of both $He(2^1S)$ and $He(2^3S)$; Ne has the advantage of identifying both metastable atoms while being insensitive to He^+ and He_2^+. The $N_2^+(B-X)$ emission is strong and can be used as a tracer; however, $N_2^+(B)$ also is produced by reactions of He^+ (Piper et al., 1975a; Govers et al., 1977) and He_2^+ (Piper et al., 1975a; Bearman et al., 1976; Collins and Lee, 1978). Bondybey and Miller (1977a) have demonstrated the use of laser induced fluorescence to monitor $He(2^3S)$ as well as molecular ion products from quenching. Tracer reactions also are known which identify He^+ or He_2^+. The reactions of He^+ with N_2 gives relatively strong $N_2^+(C-X)$ emission from 200 to 140 nm (Govers et al., 1975). Reactions of He_2^+ with COS gives $CO^+(A-X)$ and with N_2O gives $N_2^+(B-X)$ (Richardson and Setser, 1980).

According to the rate constants for decay of metastable helium atoms listed in Table IV (see also Figs. 2 and 3), $He(2^3S)$ could be studied up to 10–15 Torr and the three-body loss with concomitant production of $He_2(2^3\Sigma_u^+)$ should be the limiting factor affecting the decay rate. As previously emphasized, $He(2^3S)$ flowing afterglow experiments usually are done below 2 Torr to minimize the presence of atomic and molecular ions and the higher pressure regime is not fully characterized. The presence of ions is not critical for measurements of $He(2^3S)$ decay rates, if the concentration is directly monitored by atomic absorption, because the formation rate of $He(2^3S)$ due to electron recombination of He^+ or He_2^+ is slow relative to the quenching reactions (Deloche et al., 1976). However, the presence of He^+ and He_2^+ are troublesome when using product emission intensities for the purpose of identifying product channels from $He(2^3S)$ reactions.

Deloche et al. (1976) have published a valuable paper in which a model that included all known processes occurring in a helium plasma was compared to the time evolution of $He(2^3S)$, $He_2(^3\Sigma_u^+)$, He^+, He_2^+, and e^- concentrations in a pulsed static afterglow. They fitted a large amount of experimental data from 5 to 100 Torr and obtained rate constants for the important processes. A complete summary of their analysis is beyond the scope of this chapter. However, some important points are noted below.

   (i)   The rate constants for ionization by $He(2^3S) + He(2^3S)$ is $1.5 \pm 0.3 \times 10^{-9}$; the ratio of $He_2^+ : He^+$ formation is 70 : 30.

   (ii)   The rate constant for ionization by $He_2(^3\Sigma_u^+) + He_2(^3\Sigma_u^+)$ is $1.5 + 0.5 \times 10^{-9}$; more than 70% of the collisions give molecular ions.

   (iii)   Both $He(2^3S)$ and $He_2(^3\Sigma_u^+)$ are destroyed by electrons with about the same rate constant $\sim 4.2 \times 10^{-9}$ cm^3 sec^{-1}.

   (iv)   The three-body rate for He_2^+ formation from He^+ is 300 times faster than for He_2^* formation from $He(2^3S)$.

   (v)   The electron recombination rate constant for He_2^+ depends upon He density, electron concentration, and electron temperature.

Deloche et al. (1976) were able to fit the pulsed afterglow data very well. A similar effort needs to be made for a flowing afterglow and for several types of discharges as sources of the excited states and ions. The bimolecular $2\,He^*$ and $2\,He_2^*$ reactions provide a source of ions but, these are independent of pressure unless the concentration of He^* increases strongly with increasing pressure and these reactions do not explain why He_2^+ formation is so pressure dependent in the small apparatus. A possibility could be the onset of associative ionization reactions from highly excited He states produced in the discharge, e.g., $He^{**} + He \rightarrow He_2^+ + e^-$.

The three-body formation of $He_2(^3\Sigma_u^+)$ is rather slow and there is no evidence that the concentration of $He_2(^3\Sigma_u^+)$ has been sufficient to influence the studies of $He(2^1S)$ or $He(2^3S)$ with have been done at moderate pressures. Lee and Collins (1976) have reported on some termolecular quenching reactions of $He(2^3S)$ using pressures up to 2000 Torr in a pulsed experiment. Further discussion of $He_2(^3\Sigma_u^+)$ is given in Section V,F.

C. Mg(^{3}P$_J$) and Other Group IIA Metastable Atoms

Excitation of an electron to the np orbital gives ^{1}P$_1$ and ^{3}P$_{0,1,2}$ states for the ns^{1}np^1 configuration of group IIA atoms. States in the triplet manifold have sufficiently long radiative lifetimes and sufficiently slow quenching rates in Ar that they can be studied by the flow technique, and work has been done with Mg (Benard et al., 1976, 1977; Taieb and Broida, 1976). In the case of the heaviest member of the series, the Ba(6s5d ^{3}D$_{1,2,3}$) manifold is lower in energy than the states of the Ba(6s, 6p) manifold and the ^{3}D$_{1,2,3}$ levels become the long-lived species of interest. Benard et al. (1977) and Taieb and Broida (1976) generated Mg(^{3}P$_J$) using cold- and hot-cathode discharges, respectively. The flow of Mg atoms was produced by heating the metal and entraining the vapor in a flow of Ar or He. Taieb and Broida estimated the Mg(^{3}P$_2$) concentration as $\sim 1\%$ of the ground state concentration. The individual J states of Mg* have not been isolated because the energy differences are small compared to kT, and intramultiplet relaxation is fast. Benard et al. (1977) added Ca and K to the flow containing Mg(^{3}P$_J$) and observed excitation transfer to Ca and K. In a different approach to generating Mg(^{3}P$_J$), Benard et al. (1976) and Slafer et al. (1977) have discussed the formation of Mg, Ca, and Sr metastable atoms at high temperature by a two-step reaction mechanism

$$Mg + N_2O \longrightarrow MgO^\dagger + N_2$$

$$MgO^\dagger + CO \longrightarrow CO_2 + Mg(^3P_J)$$

Retention of energy from the first step is required because the equilibrium reaction of MgO with CO would be endoergic for Mg(^{3}P$_2$) formation. The additional energy most likely is provided by vibrational energy, indicated by the †, retained by MgO from the first step; a less likely possibility is the involvement of a [Mg $\cdot$ N$_2$O] complex. Blickensderfer and co-workers (1975) have directly excited Mg(^{3}P$_1$) from ground state Mg with a pulsed tunable dye laser in a static system.

The excited state magnesium atoms have been observed by direct emission from Mg(^{3}P$_1$–^{1}S$_0$) at 457.1 nm and by atomic absorption (^{3}P$_2$–^{3}S$_1$) at 518.5 nm. Laser induced fluorescence may not significantly expand the detection limits for the triplet Mg atoms because of the sensitivity of the direct emission method.

The metastable states of Ca, Sr, and Br have not received much attention in flow systems but are being studied in beam–gas experiments (Dagdigian, 1978; Dagdigian and Pasternack, 1978; Solarz et al., 1978). The excited states are produced by laser excitation or by passing the beam through a discharge. The discharge sources used for generating alkaline earth metastable atoms in beam studies show promise for use in flow systems: Ca*, Sr*, and Mg*

(Brinkmann *et al.* 1967, 1969); Ca* and Mg* (Giusfredi *et al.* (1975); and Ba* (Ishii and Ohlendorf, 1972; Dagdigian, 1978). From inspection of the energy level separations of the spin–orbit states, one expects the levels to be mixed in an Ar or He carrier gas. Except for this disadvantage, the flow technique seems to offer considerable promise for studies of these excited state alkaline earth atoms. The recent studies of the vibrational energy disposal (Solarz *et al.*, 1978) and of the electronic state branching (Dagdigian, 1978; Dagdigian and Pasternack, 1978) among the products show that the chemistry of these excited group IIA states will be intriguing.

D. C, Si, and Ge ($^1S_0$) and ($^1D_2$) States

Although the $^1S_0$ and $^1D_2$ spin–orbit states of C, Si, and Ge are long lived, these states have not been studied in flow systems. As noted by Nip and Clyne, even generating the ground states of these elements in a suitable way for quantitative studies in flow systems is difficult. Short wavelength flash photolysis has been the method used to produce the excited states. Braun *et al.* (1969) and Husain and Kirsch (1971a,b, 1974) used the flash photolysis of carbon suboxide as a source of $C(^1S_0)$ and $C(^1D_2)$. Repetitive photolysis of $SiCl_4$ (Husain and Norris, 1978) gives $Si(^1S_0)$; $Si(^1D_2)$ was also observed in low yield. Chowdhury and Husain (1977) generated $Ge(^1S_0)$ by the photolysis of $Ge(CH_3)_4$; $Ge(^1D_2)$ was not detected in their experiments. Flash photolysis of $SnCl_4$ and $Sn(CH_3)_4$ have been used as sources of $Sn(5^1S_0)$ and $Sn(5^1D_2)$, respectively, by Brown and Husain (1975). Husain and Littler (1973) produced excited state Pb atoms by photolysis of $Pb(C_2H_5)_4$. The quenching rate constants by Ar and He are small (Table VI) and do not

Table VI

Quenching Rate Constants[a] for $C(^1S_0)$, $C(^1D_2)$, $Si(^1S_0)$, $Si(^1D_2)$, and $Ge(^1S_0)$

System	Rate constant (cm^3 sec^{-1} $molecule^{-1}$)	Reference
$C(^1D_2)$ + He	$<3 \times 10^{-16}$	Husain and Kirsch (1971a)
$C(^1D_2)$ + Ar	$\leq 10^{-15}$	Husain and Kirsch (1971a)
$C(^1S_0)$ + He	$<2 \times 10^{-15}$	Husain and Kirsch (1974)
$Si(^1S_0)$ + He	$\leq 1.3 \times 10^{-15}$	Husain and Norris (1978)
$Si(^1O_2)$ + He	$\leq 10^{-15}$	Husain and Norris (1978)
$Ge(^1S_0)$ + He	4.5×10^{-15}	Chowdhury and Husain (1977)

[a] The quenching rate constants for typical tank gas impurities such as N_2 and CO_2 are rather small. For example, with $Ge(^1S_0)$, $k_{N_2} = 1.8 \times 10^{-12}$, $k_{CO_2} = 3.9 \times 10^{-13}$ and $k_{O_2} = 4.2 \times 10^{-11}$ cm^3 sec^{-1} (Chowdhury and Husain, 1977).

preclude the use of He or Ar as carrier gases for these states. It would be worthwhile to test the photolysis sources for flowing afterglow studies. Various discharge methods may prove useful; however, we found no evidence in the literature to favor a particular approach. A practical problem that is likely to be encountered with all group IV atoms is the rapid deposit of elemental material upon the walls of the reactor.

Detection of group IV atoms in the $(^1S_0)$ and $(^1D_2)$ states has been by vacuum resonance absorption spectroscopy; resonance fluorescence spectroscopy also should be feasible.

E. $N(^2D_{3/2, 5/2})$ and $N(^2P_{1/2, 3/2})$, and Analogous Group V States

Because of their importance in the upper atmosphere, the two spin–orbit, metastable, atomic states of nitrogen have received considerable attention. Experimental studies have been conducted in both pulsed static and fast flowing systems. Husain et al. (1974 and 1972) Slanger and Black (1976b), and Black et al. (1969) have used pulsed vuv photodissociation of N_2O as the source of $N(^2D)$ and $N(^2P)$. Lin and Kaufman (1971), Golde and Thrush (1972), and Kennealy et al. (1978) have used the flow technique. Lin an Kaufman produced both states by flowing N_2, diluted in He or Ar, through a microwave discharge. The reported concentrations were $<2.4 \times 10^{12}$ for $N(^2D)$ and $<4 \times 10^{11}$ for $N(^2P)$. Kennealy et al. (1978) used the same type of discharge to observe $NO(X)$ infrared chemiluminescence from the $N(^2D) + O_2$ reaction. Golde and Thrush (1972) flowed pure N_2 at 2–6 Torr through a microwave discharge and studied the quenching of $N(^2P)$ by atomic $O(^3P)$ and $N(^4S)$. Microwave discharges generate higher ground state $N(^4S)$ atom concentrations than metastable atom concentrations. Also one should remember that a microwave discharge usually enhances the $N_2(X)$ vibrational temperature (Golde and Thrush, 1973). In some types of discharges, dissociative electron combination with N_2^+ may play a role in forming the excited atomic states (McGowan et al., 1975); dissociative electron combination with NO^+ has a branching ratio of 0.76 ± 0.06 for $N(^2D)$ formation (Kley et al., 1977). The $N(^2P)$ state also is formed by the reaction of $N_2(A)$ with $N(^4S)$ in the gas phase (Meyer et al., 1970) and both states have been observed from the $N + N_2(A)$ reaction in condensed phases (Kunsch, 1978). Lin and Kaufman demonstrated that most of the excited atoms were generated in the discharge and that the production of excited atoms from $N(^4S) + N_2(A)$ was not significant in their experiments. As noted in Table VII, the decay constants of the excited nitrogen atoms in Ar and He are quite small. Since deactivation of $N(^2P)$ and $N(^2D)$ by $N(^4S)$ is moderately slow ($\sim 1 \times 10^{-11}$ cm^3 sec^{-1}, Golde and Thrush, 1972), a $N_2(A)$ flow system, see Section V,C, combined with an $N(^4S)$ flow

Table VII

Diffusion Coefficients[a] and Rate Constants[b] for $O(^1S)$, $O(^1D)$, $N(^2D)$ and $N(^2P)$

System	D_0 (10^{18} cm^{-1} sec^{-1})	k_E (cm^3 molecule^{-1} sec^{-1})	k_F (cm^6 molecule^{-2} sec^{-1})	Reference
$O(^1S)$ in He		0.7×10^{-19}		Black et al. (1975)
		$<1.2 \times 10^{-19}$		Welge and Atkinson (1976)
$O(^1S)$ in Ne	13.3	3.6×10^{-19}	$<2 \times 10^{-17}$	Corney and Williams (1972)
		3.5×10^{-19}		Welge and Atkinson (1976)
$O(^1S)$ in Ar	8.6	5.2×10^{-18}	$<4 \times 10^{-30}$	Corney and Williams (1972)
		4.7×10^{-18}		Black et al. (1975)
		4.8×10^{-18}		Welge and Atkinson (1976)
$O(^1D)$ in He		$<3.2 \times 10^{-16}$		Heidner and Husain (1974)[c]
$O(^1D)$ in Ar		3.2×10^{-13}		Heidner and Husain (1974)[c]
		5.0×10^{-13}		Davidson et al. (1978)[d]
$N(^2D)$ in He	25.5	$\leq 1.6 \times 10^{-16}$		Lin and Kaufman (1971)
		$\leq 2 \times 10^{-16}$		Black et al. (1969)
$N(^2D)$ in Ar	6.7	1×10^{-16}		Lin and Kaufman (1971)
		$\leq 2 \times 10^{-16}$		Black et al. (1969)
$N(^2P)$ in N_2		$1 \times 10^{-13}\ \exp -(1020/RT)$		Slanger and Black (1967a)
$N(^2P)$ in Ar	10.6	7×10^{-16}		Lin and Kaufman (1971)

[a] Diffusion coefficients for ground state oxygen atoms in He and Ar are 26 (Judeikis and Wun, 1978) and 9.3 (Yolles and Wise, 1968) $\times 10^{18}$ cm^1 sec^{-1}. We did not find the diffusion coefficients for $N(^4S)$ in argon and helium; however, $D_0 = 7.1 \times 10^{18}$ cm^{-1} sec^{-1} (Morgan and Schiff, 1964) for $N(^4S)$ in N_2.

[b] See Schofield (1978) for a critical evaluation of quenching rate constants for $O(^1S_0)$ and $O(^1D_2)$.

[c] The published values of Heidner and Husain have been divided by 2.2 as recommended by Phillips (1976) and by Davidson et al. (1976).

[d] The quenching rate constant for $O(^1S) + Ar$ is independent of temperature from 110 to 330 K (Davidson et al. 1978).

193

could provide a source for 2×10^{10} atoms cm^{-3} of $N(^2P)$ in the absence of $N(^2D)$. Schmatjko and Wolfrum (1978) have demonstrated that $O + CN$ gives $N(^2D)$ with a rate constant of $\sim 1-3 \times 10^{-11}$ cm^3 sec^{-1} and this could be a way to obtain $N(^2D)$ without $N(^2P)$. This approach would entail combining a flow of CN and a flow of O atoms; Schmatjko and Wolfrum used pulsed photolysis of C_2N_2 for their CN source.

The relative concentrations of $N(^2P)$ and $N(^2D)$ have been measured in several ways. With the pulsed static technique, the $NO(B^2\Pi)$ emission from the $N(^2D) + N_2O$ reaction has been popular (Husain et al., 1972; Slanger and Black, 1976b). Lin and Kaufman originally used atomic absorption; but Iannuzzi and Kaufman (1979) used resonance fluorescence spectroscopy to monitor the concentration of $N(^2D)$ and $N(^2P)$ in their study of the reactions with Cl_2. Golde and Thrush (1972) used the $N(^2P-^4S)$ emission at 346.6 as a monitor for $N(^2P)$. Summaries of quenching studies for $N(^2D)$ and $N(^2P)$ may be found in several reviews (Donovan and Gillespie, 1975; Donovan et al., 1972). The three-body decay constants of $N(^2D)$ and $N(^2P) + N(^4S)$ in Ar and He have not been reported, but are expected to be small. Lin and Kaufman (1971) found that $N(^2P)$ and $N(^2D)$ were deactivated with nearly unit efficiency upon encounter with the surface of the reactor.

The analogous ^2P and ^2D states of P, As, Sb, and Bi from the np^3 configuration have long radiative lifetimes but have not been studied by the flow technique. Pulsed photolysis of PCl_3 (Acuna and Husain, 1973), $AsCl_3$ and $As(CH_3)_3$ (Bevan and Husain, 1974), $Sb(CH_3)_3$ (Bevan and Husain, 1975) and $Bi(CH_3)_3$ (Trainor, 1977a,b) have been used as sources. Atomic absorption spectroscopy was used for monitoring concentrations in all cases. As noted in Section IV,A, the interaction of $Ar(^3P_{0,2})$ atoms with organometallic molecules may provide flow sources for some of these metal atoms.

F. $O(^1S_0)$, $O(^1D_2)$, and Analogous Group VI States

The $^1S_0$ and $^1D_2$ states of the ground $2p^4$ configuration of oxygen are long lived and have been studied by the pulsed excitation techniques for several years. We found no report which used the fast flow technique to study $O(^1S_0)$ or $(^1D_2)$ or any of the other group VI excited state atoms; however, steady-state measurements on slow flow photoexcited systems have been used (Gauthier and Snelling, 1971, 1975). In part, this is a consequence of the excellent data which have been obtained using pulsed techniques (Slanger and Black, 1978; Welge and Atkinson, 1976; Corney and Williams, 1972; Heidner and Husain, 1973; and many references cited in these papers). The production of $O(^1S_0)$ and $O(^1D_2)$ usually is by vacuum uv photolysis of precursors such as CO_2 (Slanger and Black, 1978; Young et al., 1968; Welge and Atkinson, 1976); O_2 (Slanger and Black, 1978; Filseth and Welge,

1969); N_2O (Black et al., 1975; Hampson and Okabe, 1970); and O_3 (Heidner and Husain, 1973; Davidson et al., 1976, 1978). Pulsed discharges in buffer gases containing O_2 also have been used (Cunningham and Clark, 1974). Lee and Slanger (1978) have used a H_2 laser (1 nsec pulse width at $25z$) at 160 nm to produce $O(^1D_2)$ from photodissociation of O_2. Photolytic dissociation of OCS and OCSe are favored sources of S* and Se*. Kliger et al. (1978) have generated $S(^1S)$ by a two-photon process in OCS using an ArF laser.

There are no known reactions of $He(2^3S)$ or $Ar(^3P_2)$ with precursor molecules that efficiently give $O(^1S_0)$ or $O(^1D_2)$ and discharges in O_2 give ground state atoms and excited molecular states; so, other sources for $O(^1S_0$ and $^1D_2)$ in flow systems must be sought. Continuous vacuum vu photolysis of N_2O, CO_2, or O_3 which have unit quantum yields at some wavelengths for generating $O(^1S_0)$ or $O(^1D_2)$ may be a possibility (Murray and Rhodes, 1976; Gauthier and Snelling, 1975).‡ Another method which could be useful for lower concentrations (Meyer et al., 1970) is the reaction of $N_2(A)$ with $O(^3P)$; but quenching by excess $O(^3P)$ atoms must be recognized, vide infra. Noxon (1962) observed $O(^1S_0)$ from an ozonizer discharge through nitrogen at 1 atm pressure; the excitation may have resulted from the $N_2(A)$ reaction. Another possible source for $O(^1D_2)$ is electron recombination with $O_2^+(X)$; the rate constant is large and the branching fraction is 0.9 for $O(^1D_2)$ formation (McGowan et al., 1975). Oxygen atom recombination with $O(^3P)$ acting as the third body gives $O(^1S_0)$, but ground state atoms also quench $O(^1S_0)$ with a rate constant of 1.8×10^{-11} cm^3 sec^{-1} (Slanger and Black, 1976a). The success of these or other methods depend upon small quenching rate constants of the molecular or atomic precursors and He or Ar carrier gases. The apparent two-body rate constants for $O(^1S_0)$ in He and Ar are quite small and Ar or He can serve as carrier gases, see Table VII. In contrast, $O(^1D_2)$ will decay rapidly in Ar and study of $O(^1D_2)$ in a flow system will be possible only with He carrier. Quenching of $S(^1D_2)$ and $Se(^1D_2)$ by rare gases is even faster than for $O(^1D_2)$ (Donovan and Little, 1978). The diffusion coefficient for $O(^1S_0)$ in He or Ar is similar to that of ground state atoms. The diffusion coefficients of $O(^1D_2)$ in He or Ar apparently have not been measured; but, comparison of the potential curves of $O(^3P_{0,1,2})$, $O(^1S_0)$, and $O(^1D_2)$ with argon and neon (Dunning and Hay, 1977) suggest that $O(^1D_2)$ should have larger diffusion coefficient than either $O(^3P_{0,1,2})$ or $O(^1S_0)$ because the potential curves are less repulsive. Collisional deactivation rate constants for $O(^1S)$ by such logical precursors as $O_2(3.6 \times 10^{-13})$, $CO_2(3.6 \times 10^{-13})$, and $N_2O(1.1 \times 10^{-11})$ are rather small (Murray and Rhodes, 1976). This discussion suggests that $O(^1S)$, as well as $S(^1S)$ and $Se(^1S)$, could be studied by the flow technique if a suitable sources were developed.

‡ The quantum yield for formation of $O(^1D)$ and $O_2(^1\Delta_g)$ from O_3 has been established to be unity between 250 and 300 nm (Fairchild and Lee, 1978; Amimoto et al., 1978).

The relative concentration of $O(^1S_0)$ is easily monitored using the $O(^1S_0-^1D_2)$ emission intensity at 557.7 nm. The $^1D_2$ state has been monitored using vuv atomic absorption (Heidner and Husain, 1973) and by direct emission at 630 nm (Davidson et al., 1976, 1978); although the 147 sec lifetime of $O(^1D_2)$ makes detection by direct emission quite difficult. Slanger and Black, 1974 used the infrared emission from vibrationally excited CO resulting from interaction with $O(^1D_2)$ to monitor the $O(^1D_2)$ concentration. Observing the increase in the $O(^3P)$ concentration also has been used as a measure of the concentration of the singlet states.

The apparent two-body deactivation of $O(^1S_0)$ by the rare gases, in fact, may be a three-body process (Black et al., 1975). The process corresponds to collision induced emission and the rare gas excimer $O(^1S)$, $S(^1S)$, and $Se(^1S)$ transitions have promise as a high power laser system (Powell et al., 1975; Hughes et al., 1976; Murray and Rhodes, 1976; Powell and Ewing, 1978). An important consideration for the $(^1S-^1D)$ excimer laser is the slow quenching of the upper state and more rapid quenching of the lower state. Schofield (1978) has published a critical evaluation of the rate constants for $O(^1S_0)$ and $O(^1D_2)$. Slanger and Black (1978) have investigated the product states from $O(^1S_0)$ interacting with N_2O, CO_2, H_2O, NO, and O_2. Only for H_2O, which has the largest rate constant, is chemical reaction the dominant channel. All others give physical deactivation with a mixture of $O(^1D_2)$ and $O(^3P_{0,1,2})$ states as products.

G. Halogen ($^2P_{1/2}$) Atoms

Ground state halogen ($^2P_{3/2}$) atoms have been extensively studied in flow reactors (Clyne and Nip, Chapter 1). The upper spin–orbit state ($^2P_{1/2}$) of the np^5 configuration, however, has not been studied explicitly by the flow technique, although the radiative lifetimes are quite long. Excitation of an electron to the $(n + 1)s$ orbital gives a set of ^2P_J and ^4P_J highly excited halogen states; but, even for fluorine the lifetimes of the quartet states are too short to permit study by the flow technique. Since both the $^2P_{1/2}$ and $^2P_{3/2}$ spin–orbit states frequently are present in studies of halogen atom chemistry and since the $^2P_{1/2}$ states are not rapidly quenched by carrier gases, it is worth discussing the $^2P_{1/2}$ states even though these states have not been studied by the flow technique.

Production of $^2P_{1/2}$ halogen atoms in static systems normally is by flash photolysis of the halogens or halogen-containing molecules (Husain and Donovan, 1970; Donovan et al., 1972). Production of $Br(^2P_{1/2})$ and $I(^2P_{1/2})$ by photolytic dissociation seems to be easier than production of $Cl(^2P_{1/2})$. Fletcher and Husain (1977, 1978) used pulsed photolysis of CCl_4 as a source for $Cl(^2P_{1/2})$ in their most recent studies; photolysis of CF_3Cl also gives a non-Boltzmann distribution of $Cl(^2P_{1/2})$ atoms (Donovan et al., 1969). The branching fractions for $Br(^2P_{1/2})$ formation from photolysis of alkyl bromides, CF_3Br, C_2F_5Br, and n-C_3F_7Br were determined by Ebenstein et al. (1978). Trifluoromethyl bromide yielded 66% $Br(^2P_{1/2})$ but the fractional yield of $Br(^2P_{1/2})$ decreased as the complexity of the perfluorinated molecules increased; the alkyl bromides were poor sources. Donohue and Wiesenfeld (1975) measured the branching fractions for photodissociation of some alkyl and perfluoroalkyl iodides. The fraction was greater than 0.90 for CH_3I, CD_3I, and the perfluoroiodides. Donovan et al. (1977b) studied the photolysis of GeH_3I and found that 56% of the iodine atoms were formed in the $^2P_{1/2}$ state. The photolysis of perfluoroiodides and bromides has been the preferred source of $I(^2P_{1/2})$ and $Br(^2P_{1/2})$ because of the high yield and the slow quenching; see Table VIII. Quenching by CD_3I is two orders of magnitude slower than by CH_3I, which makes CD_3I the more attractive source (Donovan et al., 1977a).

The use of tunable lasers as the photoexcitation source is a recent development. The narrow bandwidth of the laser permits study of reagents as quenching partners that would strongly absorb the light in a flash photolysis experiment. Because of the short wavelength limitation of the lasers, the photolysis of halogens and mixed halogens has received most attention. Excitation of I_2 to just above the dissociative threshold (498.9 nm) of the $^3\Pi_{0u}^+$ state gives one ground state and one excited state iodine atom (Burde et al., 1974). Lindemann and Wiesenfeld (1977) used a flash pumped dye laser to photolyze Br_2 at 476 ± 8 and 452 ± 2 nm; the ratio of $Br(^2P_{1/2})$ to $Br(^2P_{3/2})$ was 0.93 and 0.64, respectively. Petersen and Smith (1978) investigated the relative $Br(^2P_{1/2})$ yield from 568 to 444 nm using a flash pumped dye laser and found that the $Br(^2P_{1/2})$ quantum yield decreased with increasing photon energy above the threshold (510.6 nm) because of competitive absorption into Br_2^* states that dissociate to $2Br(^2P_{3/2})$. The $^2P_{1/2}$ state formation occurs at wavelengths corresponding to several kT in energy below the threshold energy for photodissociation of both I_2 and Br_2. This is attributed to collisional dissociation of the $^3\Pi_{0u}^+$ state (Petersen and Smith, 1978; Broadbent and Gallear, 1972). A N_2 pumped dye laser also has been used to generate $Br(^2P_{1/2})$ from Br_2 (Burde et al., 1975). A N_2 laser (Burde and McFarlane, 1976) and the frequency doubled output from a flash pumped laser has been used to generate $I(^2P_{1/2})$ from photolysis of n- and i-C_3F_7I (Houston, 1977; Hofmann and Leone, 1978). Laser excitation of Br* (Hariri and Witting, 1977) and I* (Grimley and Houston, 1978; Pritt and Combe, 1976) is being actively used for E–V studies.

There are several possibilities for generating halogen $^2P_{1/2}$ states in flow systems including photodissociation of precursor molecules mentioned in the two preceding paragraphs. A microwave discharge through CF_3Cl and

Table VIII

Diffusion Coefficients and Rate Constants for Cl($^2P_{1/2}$), Br($^2P_{1/2}$), and I($^2P_{1/2}$)

System	Diffusion coefficient (10^{18} cm^{-1} sec^{-1})	Rate constants (cm^3 molecule^{-1} sec^{-1})	References
Cl($^2P_{1/2}$) + He		3.8×10^{-15}	Fletcher and Husain (1978)
Cl($^2P_{1/2}$) + Ar		1.1×10^{-12}'	Fletcher and Husain (1978)
Cl($^2P_{1/2}$) + HCl	6.4^a	6×10^{-12}	Donovan et al. (1969)
			Judeikis and Wun (1978)
Br($^2P_{1/2}$) + Ar		$<2 \times 10^{-16}$	Donovan and Husain (1971)
Br($^2P_{1/2}$) + Br$_2$		8.0×10^{-13}	Wiesenfeld and Wolk (1978)
Br($^2P_{1/2}$) + HBr		1.4×10^{-12}	Leone and Wodarczyk (1974)
I($^2P_{1/2}$) + He		$<5 \times 10^{-18}$	Donovan and Gillespie (1975)
			Abrahamson et al. (1972)
I($^2P_{1/2}$) + Ar	23.2	$<2 \times 10^{-18}$	Donovan and Gillespie (1975)
	8.8		Abrahamson et al. (1972)
I($^2P_{1/2}$) + I($^2P_{3/2}$)		$<1.6 \times 10^{-14}$	Husain and Donovan (1970)
I($^2P_{1/2}$) + I$_2$		3.8×10^{-11}	Grimley and Houston (1978a, b)
I($^2P_{1/2}$) + CF$_3$I^b		$<3.5 \times 10^{-16}$	Donovan and Gillespie (1975)
I($^2P_{1/2}$) + CD$_3$I		4.2×10^{-15}	Donovan et al. (1977a)
I($^2P_{1/2}$) + CH$_3$I^b		2.6×10^{-13}	Donovan et al. (1977a)
I($^2P_{1/2}$) + HI		5.2×10^{-14c}	Donovan et al. (1976)

[a] This is the diffusion coefficient for Cl($^2P_{3/2}$).

[b] A similar trend holds for quenching efficiency of Br($^2P_{1/2}$) by CF$_3$Br and CH$_3$Br.

[c] Pritt and Coombe (1976) report $5 \pm 1 \times 10^{-14}$ for this rate constant.

CF_3Cl/Cl_2 mixtures is reported (Carrington et al., 1966) to give a non-Boltzmann population of $Cl(^2P_{1/2})$. Clyne and Nip (1976) used microwave excitation of an Ar/Cl_2 mixture and obtained 2×10^{11} cm^{-3} $Cl(^2P_{1/2})$ atoms; they attempted to obtain the rate constant for $Cl(^2P_{1/2}) + O_3$. However, as these authors pointed out, the estimated value may be just the rate constant for the $Cl(^2P_{3/2}) + O_3$ reaction because of the equilibration of $Cl(^2P_{1/2})$ and $Cl(^2P_{3/2})$ in the Ar/Cl_2 mixture of the experiment. Non-Boltzmann distributions of $Cl(^2P_{1/2})$ and $Br(^2P_{1/2})$ have been observed from the reaction of $F + HCl$ (Clyne and Nip, 1977) and $F + HBr$ (Sung and Setser, 1977), respectively. Both reactions produce $\sim 10\%$ of the $^2P_{1/2}$ state, which is also the approximate Boltzmann fraction of $F(^2P_{1/2})$ atoms in the flow. If there is a correlation between $^2P_{1/2}$ reactants and products, increasing the temperature may enhance the utility of these reactions as sources of $Cl(^2P_{1/2})$ and $Br(^2P_{1/2})$ atoms. Wiesenfeld and Wolk (1978) have measured a branching fraction of 0.8 for formation of $Br(^2P_{1/2})$ from reaction of $I(^2P_{1/2})$ with Br_2. Cadman and Polanyi (1968) used continuous photolysis of HI at 253.7 nm as a source of $I(^2P_{1/2})$ in a flow reactor and studied the $I(^2P_{1/2}) + H_2$ reaction. A three-step mechanism probably is involved in $I(^2P_{1/2})$ formation (Blauer et al., 1977):

$$HI + h\nu \longrightarrow H + I(^2P_{3/2} \text{ and } ^2P_{1/2})$$

$$H + HI \longrightarrow H_2(v) + I(^2P_{3/2} \text{ and } ^2P_{1/2})$$

$$H_2(v) + I(^2P_{3/2}) \longrightarrow H_2 + I(^2P_{1/2}).$$

The excitation transfer reaction ($k = 7.6 \times 10^{-11}$ cm^3 sec^{-1}) between $O_2(^1\Delta_g)$ and $I(^2P_{1/2})$, albeit in a rather chemically complex environment (Derwent and Thrush, 1972), is another possibility. Recent improvements in chemical generation methods of $O_2(^1\Delta_g)$ (see Section V,A) have resulted in the production of enough $I(^2P_{1/2})$ to operate a cw laser (Benard et al., 1979) and modification of this method should provide a good flow source of $I(^2P_{1/2})$. The reaction of $Ar(^3P_{0,2})$ atoms with Br_2 or I_2 (or other halogen donors) gives rare gas–halogen excimers which subsequently predissociate (Gundel et al., 1976; J. H. Kolts, unpublished results, 1978); but the $^2P_{1/2}$ concentration will be limited to $\sim 1/5$ of the initial $Ar(^3P_{0,2})$ concentration. A method, which has not been tested but could be useful, is the dissociation of molecular halogens via a microwave discharge followed by spin–orbit excitation in a cold-cathode discharge. Based upon this cursory review, numerous possibilities exist for generating the $^2P_{1/2}$ halogen states; the next question is the magnitude of the decay rates in carrier gases.

The decay constants are listed in Table VIII. As far as is known, there are no three-body processes involving two rare gas atoms with $^2P_{1/2}$ atoms, and the halogen($^2P_{3/2}$) + halogen($^2P_{1/2}$ + Ar or He process would be slow because of the low concentration of the $^2P_{3/2}$ atoms. Two-body relaxation is slow for $I(^2P_{1/2})$ and $Br(^2P_{1/2})$ in He and Ar. On the other hand, $Cl(^2P_{1/2})$ is readily quenched by Ar and could only be studied in He carrier. It is expected that $F(^2P_{1/2})$ will relax rapidly even in He and could not be studied by the flow technique. The diffusion coefficients of the $^2P_{1/2}$ states are normal and, providing that proper pumping speeds and flow tube diameters are selected, present no difficulty. High concentrations of molecular halogens should be avoided because of fast two-body quenching.

Clyne and Nip (Chapter 1) described vacuum uv atomic absorption spectroscopy and resonance fluorescence spectroscopy as detection methods for halogen atoms. These methods are directly applicable to the $^2P_{1/2}$ states and readers are referred to Clyne and Nip's discussion. Both resonance fluorescence and absorption spectroscopy have been used in the pulsed studies. Some of the discrepancies in the early quenching rate constant measurements probably arise from approximations made in relating percent absorption to concentration. For both absorption and resonance fluorescence spectroscopy, care must be exercised to account for scattered light from absorption and reemission by the $^2P_{3/2}$ atoms. The $Br(^2P_{1/2})$ and $I(^2P_{1/2})$ states can be monitored directly in the infrared using the optically forbidden $^2P_{1/2} \rightarrow {}^2P_{3/2}$ transition at 1.315 μ for $I(^2P_{1/2})$ and 2.714 μ for $Br(^2P_{1/2})$. This is quite successful for $I(^2P_{1/2})$, but the intensity from $Br(^2P_{1/2})$ is very weak because of the long lifetime, Table I. There is some uncertainty in the $Br(^2P_{1/2})$ lifetime (Wiesenfeld and Wolk, 1978) which makes it difficult to accurately relate intensity to concentration. The intensity of infrared fluorescence from vibrationally excited CO_2 resulting from E–V transfer has been used as a monitor for the $Br(^2P_{1/2})$ state (Hariri and Wittig, 1977; Petersen and Smith, 1978). Other E–V transfer reactions also may be useful under certain conditions. Intuitively, one might expect that molecular halogen emission arising from recombination of the $^2P_{1/2} + {}^2P_{3/2}$ atoms could be used to monitor the $^2P_{1/2}$ state; however, this has been shown not to be the case (Golde and Thrush, 1973). Carrington et al. (1966) used EPR to detect $F(^2P_{1/2})$ and $Cl(^2P_{1/2})$.

H. $Hg(^3P_0)$, $Hg(^3P_2)$, $Cd(^3P_J)$, and $Zn(^3P_J)$

The spin–orbit excited states of the Hg(6p, 6s) configuration have been studied extensively; see King and Setser (1976) and Krause et al. (1975) for reviews. Resonance irradiation of static or slow flow systems and pulsed resonance flash photolysis have been used to generate the excited states (Callear and McGurk, 1973; Vikis and LeRoy, 1973a,b,c; Horiguchi and Tsuchiya, 1975, 1977, 1979; and references in these papers). Molecular beam studies have used electron impact excitation (Van Itallie et al., 1972; Liu and

Parsons, 1976; Krause et al., 1975; Loh and Herm, 1976) to obtain $Hg(^3P_0)$ $Hg(^3P_2)$. Stock et al. (1977) have produced $Hg(^3P_0)$ under high pressure conditions using laser excitation at 256 nm. The $Cd(^3P_J)$ and $Zn(^3P_J)$ states have received somewhat less attention, but recent studies using pulsed resonance irradiation have been published (Breckenridge et al., 1975; Breckenridge and Renlund, 1978). The $Cd(^3P_J)$ and $Zn(^3P_J)$ states also have been produced at room temperatures by the flash photolysis of $Cd(CH_3)_2$ and $Zn(CH_3)_2$ (Young et al., 1970, 1974). Because of the small spin–orbit splitting (≈ 0.15 eV), intramultiplet relaxation of $Cd(^3P_1)$ and $Zn(^3P_J)$ is rapid in He or Ar, and quenching usually pertains to mixtures of the spin–orbit states. Since the lifetimes of $Cd(^3P_1)$ and $Zn(^3P_1)$ are $\sim 10^{-6}$ sec and since intra-multiplet relaxation is rapid, the $^3P_2$ and $^3P_0$ states of Cd and Zn cannot be studied by the flow technique.

Adaptation of the aforementioned techniques to systematically study $Hg(^3P_0)$ and $Hg(^3P_2)$ by the fast flow technique has not been done. However, resonance excitation of $Hg(^3P_1)$ and subsequent collisional relaxation to $Hg(^3P_0)$ should be a viable source. Molecular nitrogen induces the $Hg(^3P_1)$ $\rightarrow Hg(^3P_0)$ relaxation but does not quench $Hg(^3P_0)$; see Table XI. Vikis and LeRoy (1973b,c) demonstrated the suitability of the technique in their studies of the emission spectra from the interaction of $Hg(^3P_0)$ with CH, CN, and other radicals. Deactivation of $Hg(^3P_0)$ is slow in helium or argon and these gases can serve as carriers. Passing an $Ar/N_2/Hg$ mixture through a cold-cathode discharge is another source of $Hg(^3P_0)$ via the following set of reactions:

$$Ar(^3P_{0,2}) + N_2 \longrightarrow N_2(C) \text{ and } N_2(B) + Ar$$

$$N_2(C) \longrightarrow N_2(B) + h\nu \longrightarrow N_2(A) + h\nu$$

$$N_2(A) + Hg \longrightarrow Hg(^3P_1) + N_2(X)$$

$$Hg(^3P_1) + N_2(X) \longrightarrow Hg(^3P_0) + N_2(X, v).$$

A similar mechanism can be proposed using CO, but N_2 probably will be more effective because CO quenches $Hg(^3P_1)$ and $Hg(^3P_0)$ to the ground state, as well as giving intramultiplet relaxation. Pertinent cross sections for $N_2(A)$ and $CO(a)$ reacting with Hg have been measured by Callear and Wood (1971) and Lee and Martin (1976). The $Hg(^3P_0)$ concentration would be limited to $\sim 5 \times 10^{10}$ cm^{-3} by this method.

The production of $Hg(^3P_2)$ via resonance photolysis at 184.9 nm to produce $Hg(^1P_1)$ followed by collisional relaxation with N_2 is not feasible because of subsequent fast spin–orbit relaxation of $Hg(^3P_2)$. However, conversion of $Hg(^1P_1)$ to $Hg(^3P_2)$ by collisions with the Ar carrier gas (Madhavan et al., 1973) may be possible. Preliminary studies (Wren and Setser, 1980) have shown that $\approx 10^9$ atom cm^{-3} of $Hg(^3P_2)$ can be produced by

Table IX

Rate Constants or Cross Sections for Hg(6p)* *States*

Reaction	Rate constants $(cm^3\ molecule^{-1}\ sec^{-1})$ or cross sections	Reference and comments
$Hg(^1P_1) + He \rightarrow Hg(^3P_1)$	0.06	Madhavan *et al.* (1973);
$Hg(^1P_1) + Ar \rightarrow Hg(^3P_1)$	0.06	cross sections relative to that for N_2.
$Hg(^3P_2) + NO \rightarrow NO(A^2\Sigma^+\ v' = 0)$	$1.4\ Å^2$	Liu and Parsons (1976), $\langle E_T \rangle = 1.9\ kcal\ mole^{-1}$, $T_{NO} = 283$ K
$Hg(^3P_2) + N_2 \rightarrow Hg(^3P_{1,0})$	9.3×10^{-11}	Burnham and Djeu (1974)
$Hg(^3P_2) + CO \rightarrow Hg(^3P_{1,0})$	1.9×10^{-10}	Burnham and Djeu (1974)
$Hg(^3P_2) + He \rightarrow Hg(^3P_1)$	$<3 \times 10^{-14}$	Calculated from relative
$Hg(^3P_2) + Xe \rightarrow Hg(^3P_1)$	$<8 \times 10^{-15}$	cross section of Van Itallie *et al.* (1972) using absolute value of $Hg(^3P_2) + N_2$
$Hg(^3P_1) + N_2 \rightarrow Hg(^3P_0)^a$	3.9×10^{-12}	Horiguchi and Tsuchiya (1975)
$Hg(^3P_1) + CO \rightarrow Hg(^3P_0)^a$	1.1×10^{-10}	Horiguchi and Tsuchiya (1977)
$Hg(^3P_1) + Ar \rightarrow Hg(^3P_0\ or\ ^1S_0)$	$<8.3 \times 10^{-14}$	Callear and Wood (1971)
$Hg(^3P_0) + N_2 \rightarrow Hg(^1S_0)$	$<5 \times 10^{-15}$	Horiguchi and Tsuchiya (1977)
$Hg(^3P_0) + CO \rightarrow Hg(^1S_0)$	9.2×10^{-12}	Horiguchi and Tsuchiya (1977)
$Hg(^3P_0) + NO \rightarrow Hg(^1S_0)$	1.8×10^{-10}	Horiguchi and Tsuchiya (1977)
$Hg(^3P_0) + He \rightarrow Hg(^1S_0)$	$<1 \times 10^{-16}$	Estimated by the authors
$Hg(^3P_0) + Ar \rightarrow Hg(^1S_0)$	$<1 \times 10^{-16}$	Estimated by the authors

[a] Comparison of these rate constants to those for total quenching (Deech *et al.*, 1971) shows that E–V transfer is the dominant exit channel.

passing a Ar/Kr/Hg mixture through a cold-cathode discharge. The mechanism responsible for Hg* formation is a two-step excitation transfer; $Ar(^3P_{0,2})$ + Kr to form $Kr(^3P_2)$, then $Kr(^3P_2)$ + Hg to form Hg** which radiatively cascades to the $^3P_2$ and $^3P_0$ levels (Dreiling and Setser, 1979). The rate constant for quenching of $Kr(^3P_2)$ by Hg is 40×10^{-11} cm^3 sec^{-1} (Wren and Setser, 1950). The reaction $Ar(^3P_2)$ with Hg gives Penning ionization and passing Ar/Hg mixtures through a hollow-cathode discharge gives useful concentrations of both $Hg(^3P_2)$ and Hg^+. An alternative scheme is to use organomercury compounds; Dreiling and Setser (1979) have shown that Hg* is produced directly from the reaction of $Ar(^3P_{0,2})$ with $Hg(CH_3)_2$. Optical pumping methods starting from $Hg(^3P_1)$ and $Hg(^3P_0)$ is another way to produce and study $Hg(^3P_2)$ (Burnam and Djeu, 1974).

At high mercury atom concentrations, loss from binary mercury atom collisions and from three-body collisions, Hg* + 2Hg or Hg* + Hg + M, can be important. The two-body $Hg(^3P_0)$ + Hg and $Hg(^3P_1)$ + Hg processes have rate constants of 4×10^{-14} and 2.8×10^{-13} cm^3 sec^{-1} (Stock et al., 1977) at 673 K. The three-body rate constants for $2Hg(^1S_0)$ + $Hg(^3P_1)$ and $2Hg(^1S_0)$ + $Hg(^3P_0)$ were reported as 1.6×10^{-31} and 1.5×10^{-31} cm^6 sec^{-1}, respectively (Stock et al., 1977). With N_2 as the third body, the rate constants are somewhat larger. The three-body processes correspond to Hg_2^* formation and some of these molecular states are long lived (Stock et al., 1978). Complete characterization of the decay process for $Hg(^3P_0)$ and $Hg(^3P_2)$ interacting with Hg will be interesting projects for further study.

Atomic absorption is a viable technique if $Hg(^3P_0, {}^3P_2)$ concentrations are 10^{10}–10^{11} atoms cm^{-3}. Resonance fluorescence or laser induced fluorescence can be used to measure lower concentrations. Laser magnetic resonance, 1767.2 cm^{-1}, has been observed in an absorption cell placed in the cavity of a CO laser and between the poles of a 15 cm electromagnet; the $Hg(^3P_0)$ concentration was estimated as 5×10^9 cm^{-3} (Johns et al., 1977). The reaction of $Hg(^3P_2)$ with NO to give $NO(A^2\Sigma^+, v^1 = 0)$ (Liu and Parson, 1976) can be used as a specific tracer for $Hg(^3P_2)$ at thermal energies. Krause et al. (1975) used the HgCl(B–X) emission at 300–400 nm from $Hg(^3P_2)$ + Cl_2 to monitor $Hg(^3P_2)$ and emission from 400 to 590 nm to monitor the sum of $Hg(^2P_0)$ and $Hg(^3P_1)$.

V. EXCITED DIATOMIC STATES IN FLOW SYSTEMS

A. $O_2(a^1\Delta_g)$, $O_2(b^1\Sigma_g^+)$, $O_2(A^3\Sigma_u^+, C^3\Delta_u, c^1\Sigma_u^-)$, and $O_2^+(b^4\Pi_u)$

Probably no other excited states have been the object of such detailed study as the low lying singlet states of oxygen; this, of course, is because of their importance in the atmosphere and in biological systems. Unlike several previously discussed states, the $O_2(a^1\Delta_g$ and $b^1\Sigma_g^+)$ states have been studied extensively in flow reactors. Numerous reviews of $O_2(^1\Delta_g)$ and $O_2(^1\Sigma_g^+)$ are available [Wayne, 1969; a series of papers published in the Annals of the New York Academy of Science and a recently published book (Schaap, 1976)]. We will only cover the techniques pertinent to the generation and detection of singlet oxygen in flow reactors.

The most common method of producing $O_2(a^1\Delta_g)$ and $O_2(b^1\Sigma_g^+)$ has been a microwave discharge in O_2 or Ar/O_2 and He/O_2 mixtures (Wayne, 1969). The concentration of $O_2(^1\Delta_g)$ will be 5–15% of the O_2 depending on pressure, discharge power, and the residence time in the cavity (Cook and Miller, 1974; Benard and Pchelkin, 1978). The $O_2(b^1\Sigma_g^+)$ concentration is much lower than the $O_2(a^1\Delta_g)$ concentration and $O_2(b)$ mainly is formed by the energy-pooling process

$$2 O_2(a^1\Delta_g) \longrightarrow O_2(b^1\Sigma_g^-) + O_2(X^3\Sigma_g^-), \quad k = 2 \times 10^{-17}\ \mathrm{cm^3\ sec^{-1}}$$

(Derwent and Thrush, 1971). This rate is sufficiently slow that it does not significantly contribute to the decay kinetics of $O_2(a^1\Delta_g)$ (Thomas and Thrush, 1975). There also is a second bimolecular loss process, the concerted dimole emission reaction

$$2 O_2(a^1\Delta_g) \longrightarrow O_2(X^3\Sigma_g^-) + O_2(X^3\Sigma_g^-, v = 0, 1) + h\nu(634, 703\ \mathrm{nm}).$$

The major disadvantage of the electrodeless discharge method is the simultaneous generation of oxygen atoms which may interfere with the reactions to be studied. The $O + O_2 + M$ reaction also generates some ozone. The oxygen atoms can be removed by either entraining a small flow of Hg in the main gas flow or by coating the flow tube just beyond the discharge with a film of HgO (Arnold and Ogryzlo, 1967; Arnold et al., 1968). Enhancement of both the $^1\Sigma_g^+$ and $^1\Delta_g$ concentrations was observed with either the HgO or Hg methods and Wayne (1969) explains the enhancement as follows:

$$Hg + O \xrightarrow{\ \ wall\ \ } HgO$$
$$HgO + O \longrightarrow Hg + O_2^*.$$

The actual increase in concentration of the singlet states did not seem to depend on whether Hg was added through the discharge or if a HgO film was added after the discharge. Yaron and von Engel (1975) have used an AgO surface to catalyze the removal of O atoms and O_3. Oxygen atoms also can be removed by adding a small amount of NO_2 to the flow (Clyne et al., 1970; Huie and Herron, 1973). Relatively pure flows of $O_2(a)$ can be obtained by adding small amounts of H_2O which quench $O_2(b)$ but have virtually no effect upon the $O_2(a)$ concentration; see Table X. In cases where the added reagents cause interference, Wayne (1970) and Clark and Wayne (1969) diluted the flow to reduce the concentration of O or O_3 and used more sensitive techniques to monitor the $O_2(a)$ and $O_2(b)$ concentrations. Alben et al. (1978) have used the He/O_2 discharge to produce a supersonic beam of $O_2(a)$ molecules for molecular beam studies.

Chemical methods have been developed to generate $O_2(a)$ for flow reactor studies (McDermott et al., 1977, 1978). The liquid phase reaction between alkaline H_2O_2 and Cl_2 or Cl_2O gives $O_2(a)$: a cold trap is used to remove the unreacted Cl_2 and H_2O. This method has been improved to the extent that 0.35 Torr of $O_2(a)$ has been produced at a total pressure of 1 Torr (Benard et al., 1979). This chemical generator has been used to obtain laser action from $I(^2P_{1/2})$ from the transfer reaction

$$O_2(a^1\Delta_g) + I(^2P_{3/2}) \longrightarrow O_2(X^3\Sigma_g^-) + I(^2P_{1/2}).$$

The new chemical generation method appears to offer significant advantages, relative to the discharge source, for studies of $O_2(^1\Delta_g)$ chemistry by the flow technique.

Various pulsed and steady-state techniques have been used to produce $O_2(a^1\Delta_g)$ and $O_2(b^1\Sigma_g^+)$. Noxon (1970), Young and Black (1967), and Filseth et al. (1970) photolyzed O_2 in the Schumann–Runge continuum (130–175 nm) to produce $O(^1D)$; the secondary reaction, $O(^1D) + O_2(X)$, gives $O_2(b)$. This method is useful for producing $O_2(b)$ in the absence of $O_2(a)$. A closely related technique uses a H_2 laser ($\lambda \approx 160$ nm) to produce $O(^1D)$ from O_2; the subsequent reaction gives $O_2(b^1\Sigma_g^+)$ (Kohse-Hoinghaus and Stuhl, 1978). In this study the remarkably large isotope effect, ~ 60, for quenching by H_2 and D_2 was reinvestigated; differences of factors of two in quenching rate constants measured by different techniques were also pointed out. Lee and Slanger (1978) have used the H_2 laser method to measure the efficiency (0.77 ± 0.2) for conversion of the $O(^1D)$ energy into $O_2(b^1\Sigma_g^+, v = 0, 1)$. The $O_2(b^1\Sigma_g^+, v = 1)$ molecule is quenched by O_2 with a rate constant of 2.2×10^{-11} cm³ sec⁻¹, which is approximately six orders of magnitude larger than for $v = 0$, see Table X. The fast apparent quenching is a consequence of vibrational relaxation. Photolysis of ozone at 253.7 nm has been used as a source of $O_2(a^1\Delta_g)$ (Gauthier and Snelling, 1971; Wayne, 1969, and references therein). Gauthier and Snelling (1975) used the 253.7 nm photolysis of O_3 to study $O_2(b)$ by a steady-state method; the initial step produces $O(^1D)$ and $O_2(^1\Delta_g)$ and $O_2(b)$ is produced by excitation transfer of $O(^1D)$ with added O_2.‡ Findlay and Snelling (1971) produced $O_2(a)$ by irradiating oxygen–benzene mixtures at 253.7 nm; the initial photolytic step produces the triplet state of benzene which subsequently gives $O_2(a)$ by excitation transfer. An

---

‡ The quantum yield for formation of $O_2(\Delta_g) + O(^1D)$ from O_3 is unity from 250 to 300 nm (Fairchild and Lee, 1978; Amimoto et al., 1978).

Table X

Quenching Rate Constants for $O_2(b^1\Sigma_g^+)$ and $O_2(a^1\Delta_g)$

Reaction	Rate constant (cm^3 molecule^{-1} sec^{-1})	Reference
$O_2(b^1\Sigma_g^+) + He$	$\approx 10^{-17}$	Becker et al. (1971)
	$\approx 1.0 \times 10^{-16}$	Filseth et al. (1970)
$O_2(b^1\Sigma_g^+) + Ar$	1.5×10^{-17}	Becker et al. (1971)
	5.8×10^{-18}	Filseth et al. (1970)
$O_2(b^1\Sigma_g^+) + O_2$	4.6×10^{-17}	Thomas and Thrush (1975)
	3.8×10^{-17}	Lawton et al. (1977)
	3.9×10^{-17}	Lawton and Phelps (1978)
$O_2(b^1\Sigma_g^+) + O_3$	2.5×10^{-11}	Gilpin et al. (1971)
	2.3×10^{-11}	Gauthier and Snelling (1975)
$O_2(b^1\Sigma_g^+) + NO_2$	2×10^{-14}	Becker et al. (1971)
$O_2(b^1\Sigma_g^+) + H_2O$	5×10^{-12}	Becker et al. (1971)
	5.1×10^{-12}	Gauthier and Snelling (1975)
$O_2(a^1\Delta_g) + He$	$\leq 10^{-20}$	Becker et al. (1971)
	$8 \pm 3 \times 10^{-26}$	Collins et al. (1973)
$O_2(a^1\Delta_g) + Ar$	$\leq 10^{-20}$	Becker et al. (1971)
	$9 \pm 1 \times 10^{-21}$	Collins et al. (1973)
	$< 2.1 \times 30^{-19}$	Clark and Wayne (1969)
$O_2(a^1\Delta_g) + O_2$	1.6×10^{-18}	Borrell et al. (1977)
	1.7×10^{-18}	Becker et al. (1971)
$O_2(a^1\Delta_g) + O_3$	3.6×10^{-15}	Gauthier and Snelling (1971)
$O_2(a^1\Delta_g) + NO_2$	5×10^{-18}	Becker et al. (1971)
$O_2(a^1\Delta_g) + H_2O$	4×10^{-18}	Becker et al. (1971)
$2 O_2(a^1\Delta_g) \rightarrow O_2(b) + O_2(X)$	2.0×10^{-17}	Derwent and Thrush (1971)
$2 O_2(a^1\Delta_g) \rightarrow$ dimole emission	2.6×10^{-23}	Derwent and Thrush (1971)

advantage of the method is that $O(^3P)$ and O_3 are absent; however, quenching studies of $O_2(a)$ cannot be done if the added reagent deactivates the triplet benzene state. Lawton et al. (1977) and Martin et al. (1976) have used pulsed tunable dye lasers to directly produce $O_2(b)$ from $O_2(X^3\Sigma_g^-)$; this method also has the advantage that $O_2(a)$ is not present. Lawton and Phelps (1978) have used electron impact excitation in a drift tube to obtain rate coefficients for electron excitation of $O_2(b^1\Sigma_g^+)$ from $O_2(X)$.

The quenching rate constants for $O_2(a)$ and $O_2(b)$ by helium and argon are summarized in Table X, together with rate constants for impurity molecules commonly present in flow reactors. The decay rates of both singlet states are slow in helium and argon. In contrast to most of the excited states discussed previously, the probability of deactivation of $O_2(a)$ and $O_2(b)$ upon collision with the reactor walls is low. For these reasons, slow flows have been used for study of O_2(a and b) in flow reactors. The actual quenching efficiency at the wall depends upon the material of the reactor and the history of the surface. For Pyrex glass the quenching efficiency for $O_2(a)$ is $\approx 10^{-5}$, whereas it is $10^{-2}-10^{-3}$ for $O_2(b)$ (Breckenridge and Miller, 1972b; Izod and Wayne, 1968). The diffusion coefficients of $O_2(b)$ and $O_2(a)$ in oxygen are $5.0 \pm 0.3 \times 10^{18}$ (Lawton and Phelps, 1978) and $4.9 \pm 0.1 \times 10^{18}$ cm^{-1} sec^{-1} (Vidaud et al., 1976a), respectively. Within the limitations of the experimental uncertainty these are the same as for $O_2(X)$ in oxygen. Jones and Bayes (1972) showed that energy exchange between $O_2(a)$ and $O_2(X)$ occurred at least in one of every ten collisions. This rate of energy exchange may not be large enough to affect the diffusion coefficient or there may be compensating change in the intermolecular potentials (Kolts and Setser, 1978a).

Photoionization has been used by Clark and Wayne (1969) and Huie and Herron (1973) to detect $O_2(a^1\Delta_g)$. The Ar resonance line at 106.7 nm, which is ≈ 0.5 eV above the $O_2(a)$ ionization energy, is normally used as the ionization source. The $O_2(b)$ also is ionized but it makes only a small contribution to the total ion current. When using photoionization, other species that can be ionized by the 106.7 nm line cannot be present. An easy test for species with low ionization energies can be made by using Kr resonance radiation, 116.5 nm, which does not have enough energy to ionize $O_2(a)$. In principle, the Kr 116.5 line could be used to photoionize $O_2(b)$; however, this method has not been used because $[O_2(b)]$ normally is too low.

Isothermal calorimetry has been used to measure $[O_2(a^1\Sigma_g^+)]$. The calorimeter is constructed of a spiral of platinum wire, plated with silver, cobalt, nickel, manganese, or copper, then partially oxidized with atomic oxygen. Elias et al. (1959) has characterized many of these coatings. The concentration of $O_2(a)$ is proportional to the additional current required to maintain the temperature in the absence of O_2^* at the same level that was found in the presence of O_2^*. Care must be taken to remove O atoms and O_3 as these species will also contribute to the heat liberated by the detector. Derwent and Thrush (1971) found the recombination efficiency of an oxidized cobalt platinum wire calorimeter to be $\approx 35\%$. Titration techniques can be used to monitor the $O_2(a^1\Delta_g)$ concentration in flow reactors using moderate flow velocities. Huie and Herron (1973) found 2,5-dimenthylfuran (DMF) to be a suitable titrant using a mass spectrometer to follow the loss of DMF.

Direct monitoring via $O_2(b^1\Sigma_g^+-X^3\Sigma_g^-)$ or $O_2(a^1\Delta_g-X^3\Sigma_g^-)$ emission at 762 and 1270 nm, respectively, has been used by several workers. The 762 nm band is relatively easy to detect, the 1270 nm band is more difficult because of the lower transition probability and the reduced sensitivity of detectors in this wavelength region. Benard and Pchelkin (1978) have developed a method for calibrating the intensity of the 1270 nm emission in terms of the absolute $O_2(a^1\Delta_g)$ concentration. The intensity of the dimole emission at 634 nm resulting from two $O_2(a^1\Delta_g)$ molecules is more convenient to observe than the 1270 nm band: but, the intensity depends upon $[O_2(^1\Delta_g)]^2$, as Derwent and Thrush (1971) have conclusively shown. Borrell et al. (1977) showed that, if the $O_2(b^1\Sigma_g^+)$ state is formed entirely by the energy pooling reaction, the 762 nm emission intensity also is proportional to $[O_2(a)]^2$. Direct monitoring of $O_2(^1\Delta_g)$ in $v = 0$ and higher levels also has been done by vacuum uv absorption spectroscopy at 144.2 nm (Collins et al., 1973).

Gas phase EPR can be used to quantitatively measure the $O_2(a)$ concentration (Brown, 1967; Falick et al., 1965). Falick and Mahan (1967) used EPR in their study of the dimole emission and Cook and Miller (1974) used it to assign the absolute yield of $O_2(a)$ from a microwave discharge in O_2, CO_2, NO_2, and SO_2. The $O_2(b)$ state cannot be detected by EPR.

The $a^1\Delta$ and $b^1\Sigma^+$ states of S_2 and SO have not received much attention. It is known, however, that $SO(^1\Delta)$ can be produced in a flow system by the following reaction (Breckenridge and Miller, 1972b)

$$O_2(a^1\Delta_g) + SO(X^3\Sigma^-) \longrightarrow O_2(X^3\Sigma_g^-) - SO(a^1\Delta), \quad k = 2.1 \times 10^{-11} \text{ cm}^3 \text{ sec}^{-1}.$$

Ground state $SO(X^3\Sigma^-)$ was produced from reacting O atoms with COS. The quenching of $O_2(a^1\Delta_g)$ by OCS is slow and adds no complication. All four molecules in the above reaction were monitored by EPR sepctroscopy. The deactivation probability for $SO(a^1\Delta)$ by wall collisions was ~ 0.1 for Pyrex and Teflon coated Pyrex, which is four orders of magnitude larger than for $O_2(a^1\Delta_g)$. Since $O_2(a^1\Delta_g)$ can be produced so conveniently, it may be possible to use excitation transfer reactions to obtain $S_2(a^1\Delta_g)$ and $SO(a^1\Delta)$ for subsequent study in flow reactors. $SO(a^1\Delta)$ has been detected by laser magnetic spectroscopy using a CO_2 laser (Yamada et al., 1978).

The $O_2(A^3\Sigma_u^+)$, $(C^3\Delta_u^-)$, and $C^1\Sigma_u^-)$ group of states, which are between 4.05 and 4.34 eV above $O_2(X^3\Sigma_g^-)$, have long lifetimes. Little is known about these states and development of suitable flow sources would be an important advancement. Simonaities and Heicklen (1971) found indirect evidence for high lying long-lived O_2 states in mercury sensitized oxidation of CO. In studies of mercury sensitization of mixtures of O_2 with gloxyal, biacetyl, and naphthalene, Hippler et al. (1978) observed enhanced phosphorescence yields from the energy acceptors in the presence of added oxygen, which is evidence for a long-lived state of molecular oxygen. These studies suggest that mercury sensitization in a flow system might be a fruitful approach for generating these oxygen states. Golde and Thrush (1973) summarized the role of $O_2(A^3\Sigma_u^+)$ in an oxygen afterglow; catalytic recombination of $O(^3P)$ over active surfaces has been observed to enhance $O_2(A^3\Sigma_u^+)$ formation. Kenner et al. (1979) found that the distribution of $(A^3\Sigma_u^+)$ and $(c^1\Sigma_u^-)$ states depends upon the nature of the third body in O atom recombination. Slanger has observed three new band systems $(C^3\Delta_u-a^1\Delta_g, C^3\Delta_u-X^3\Sigma_g^-$, and $C^1\Sigma_g^-$–

$a^1\Delta_g$). With this new information, progress in understanding these three states of O_2 can be expected.

The $O_2^+(a^4\Pi_u)$ state is long lived and has been successfully studied in a fast flow system by Lindinger *et al.* (1975, 1979) using Penning ionization reaction of $He(2^3S)$ with O_2. Although $O_2^+(A^2\Pi_u)$, $O_2^+(b^4\Sigma_g^-)$, and $O_2^+(X^2\Pi_u)$ are formed in addition to $O_2^+(a^4\Pi_u)$, the $O_2^+(A)$ and $O_2^+(b)$ radiatively decay and the method gives a flow of $O_1^+(X)$ and $O_2^+(a)$. From observing the variation of the O_2^+ signal with various added reagents, rate constants for $O_2^+(a)$ were obtained. A more direct way to moniter $O_2^+(a^4\Pi_u)$ is laser induced fluorescence (Cook *et al.*, 1978c).

B. $NX(a^1\Delta_2)$ and $NX(b^1\Sigma^+)$; X = F, Cl, H

The NF radical is isoelectronic with O_2 and has a similar pattern of electronic states, although the energy separations between the $NF(X^3\Sigma^-)$, $NF(a^1\Delta_2)$, and $NF(B^1\Sigma^+)$ states are somewhat larger, see Table I. The NF singlet states have long radiative lifetimes and have been studied by the flow technique. Although the thermochemistry and spectroscopy of NCl and NBr are not so well characterized, their general properties resemble those of NF. The well-known NH radical also has long-lived $(a^1\Delta_2)$ and $(b^1\Sigma^+)$ states. If the analogous phosphorous and other group V atoms are included, the group V halides and hydrides represent a large number of interesting long-lived molecules. Because of the interest in visible lasers, studies of $NF(a^1\Delta)$ and $NF(b^1\Sigma^+)$ in flow reactors have been done and these molecules will be discussed first.

The reaction of NF_2 with H atoms produces $NF(a^1\Delta)$ as a major product

$$NF_2 + H \longrightarrow HF(v) + NF(a^1\Delta); \quad \Delta H° = -35 \text{ kcal mole}^{-1}.$$

The presence of NF(a) has been demonstrated by observation of the $NF(a^1\Delta-X^3\Sigma^-)$ emission (Clyne and White, 1970; Herbelin and Cohen, 1973) from EPR spectroscopy (Curran *et al.*, 1973), and from observation of the $HF(v)$ infrared chemiluminescence (Malins and Setser, 1979). The $HF(v)$ distribution (declining population with increasing v) is characteristic of unimolecular elimination reactions and supports the claim (Herbelin and Cohen, 1973) that the process is an addition followed by HF elimination rather than direct abstraction. Rate constant values of 3.3 and 1.1×10^{-11} cm^3 molecule^{-1} sec^{-1} have been reported for H + NF_2 (Herbelin, 1976; Malins and Setser, 1980). The low $v = 3$ and 4 HF population (Malins and Setser, 1979) strongly suggest that the branching fraction for $NF(X^3\Sigma^-)$ formation is minor. A flow of difluoroamino radicals can be conveniently obtained by moderate heating of a N_2F_4 reservoir (Clyne and White, 1970) and the H + NF_2 is a viable way to generate $NF(a^1\Delta)$; although the relatively slow rate should be noted. This source of $NF(a^1\Delta)$ was used to study the excitation transfer reaction to Bi atoms (Capelle $et\ al.$, 1978), which had a very large rate constant $1 \times 10^{-9}\ cm^3\ sec^{-1}$.

For some applications, a microwave discharge in Ar/NF_3 mixtures also may be an adequate NF_2 source (Herbelin and Cohen, 1973). In the $H + NF_2$ system, the $NF(b^1\Sigma^+)$ state is thought to be produced by energy pooling between vibrationally excited $HF(v)$ and $NF(a^1\Delta)$. Energy pooling by $2\ NF(a^1\Delta)$ to give $NF(b^1\Sigma^+)$ has not yet been demonstrated. The $NF(b^1\Sigma^+)$ state also can be produced by reaction of $NF(a^1\Delta)$ with $O_2(a^1\Delta_g)$ or $I(^2P_{1/2})$ (Herbelin and Kwok, 1977; Herbelin and Cohen, 1973). Interaction of NF_3 and N_2F_4 with metastable He and Ar atoms also gives $NF(b^1\Sigma^+)$ emission (Richardson and Setser, 1979); the sensitivity of the emission intensity to the operating conditions suggest the involvement of a secondary reaction. A more general source of the nitrogen halide $(b^1\Sigma^+)$ states may be reaction of halogen atoms with N_3 (Clark and Clyne, 1970). The photolysis of halogen azides also gives the $b^1\Sigma^+$ states; Piper $et\ al.$ (1978) photolyzed ClN_3 at 300 nm and observed $NCl(b^1\Sigma^+-X^3\Sigma^-)$ emission. There are no quantitative data in the published literature for the deactivation of NF* upon encounter with surfaces or of the quenching rate constants of $NF(b^1\Sigma^+)$ or $NF(^1\Delta)$ in He or Ar carrier. However, based upon the qualitative results in the literature, quenching by the carrier gas evidently is not a major problem.

Neither $NH(a^1\Delta)$ or $(b^1\Sigma^+)$ seems to have been studied by the flow technique. However, there has been many pulsed excitation studies. The discovery (Masanet $et\ al.$, 1974a,b) that vacuum ultraviolet photolysis (Ar lines at 104.8 and 106.7 and Kr lines are 116.5 and 123.6 nm) of NH_3 gives $NH(b^1\Sigma^+-X^3\Sigma^-)$ emission has led to several recent pulsed excitation studies of $NH(b^1\Sigma^+)$ and $ND(b^1\Sigma^+)$, which have been concerned with lifetimes and quenching process. Some rate constants are summarized in Table XI; the decay of $NH(b^1\Sigma^+)$ in Ar or He is slow, but quenching by tank gas impurities and the NH_3 precursor are rapid. Evidently, a large isotope effect on the quenching rate constants exists for the slower processes. The $NH(b^1\Sigma^+)$ radiative lifetime is less well established than the quenching rate constants. The most recent study (Masanet $et\ al.$, 1978) obtained 0.23 msec; however, previous measurements gave 18 and ≥ 5 msec (Gelernt $et\ al.$, 1976; Zetzsch and Stuhl, 1975, 1976b). The presence of impurities and the accounting for diffusion make this a difficult measurement; but the longer values are preferred. The interaction of metastable Ar or Kr with NH_3 mainly gives $NH(A^3\Pi)$ rather than singlet imino radicals (Stedman, 1970; Zetzsch and Stuhl, 1976a); however, NH* formation is a small fraction of the total quenching. For generation of $NH(a^1\Delta)$ in a flow apparatus, vacuum uv photolysis or possibly resonance sensitization is favored. A windowless lamp (Zetzsch and Stuhl, 1976a) apparently gives the highest concentration.

Table XI

Quenching Rate Constants for $NH(b^1\Sigma^+)$, $ND(b^1\Sigma^+)$, and $NH(a^1\Delta)$

Reaction	Rate constants 10^{-13} cm^3 molecule^{-1} sec^{-1}	Reference
$NH(b^1\Sigma^+) + Ar$	1.8×10^{-3}	Zetzsch and Stuhl (1976b)
$ND(b^1\Sigma^+) + Ar$	4.5×10^{-4}	Zetzsch and Stuhl (1976b)
$NH(b^1\Sigma^+) + Ar$	1.3×10^{-3}	Gelernt et al. (1976)
$ND(b^1\Sigma^+) + Ar$	$< 5.0 \times 10^{-5}$	Gelernt et al. (1976)
$NH(b^1\Sigma^+) + He$	4.2×10^{-4}	Zetzsch and Stuhl (1976b)
$ND(b^1\Sigma^+) + He$	1.6×10^{-4}	Zetzsch and Stuhl (1976b)
$NH(b^1\Sigma^+) + He$	7.0×10^{-4}	Gelernt et al. (1976)
$NH(b^1\Sigma^+) + NH_3$	4.1	Zetzsch and Stuhl (1977)
$ND(b^1\Sigma^+) + ND_3$	0.19	Zetzsch and Stuhl (1977)
$NH(b^1\Sigma^+) + H_2O$	4.9	Zetzsch and Stuhl (1977)
$ND(b^1\Sigma^+) + H_2O$	4.4	Zetzsch and Stuhl (1977)
$NH(b^1\Sigma^+) + H_2$	8.6	Zetzsch and Stuhl (1977)
$ND(b^1\Sigma^+) + H_2$	8.1	Zetzsch and Stuhl (1977)
$NH(b^1\Sigma^+) + N_2$	4.5×10^{-3}	Gelernt et al. (1976)
$ND(b^1\Sigma^+) + N_2$	3.8×10^{-4}	Gelernt et al. (1976)
$NH(b^1\Sigma^+) + O_2$	2.4×10^{-2}	Zetzsch and Stuhl (1976b)
$ND(b^1\Sigma^+) + O_2$	9.1×10^{-3}	Zetzsch and Stuhl (1976b)
$NH(a^1\Delta) + Ar$	~ 0.12	Piper et al. (1978)
$NH(a^1\Delta) + HN_3$[a]	1700	Piper et al. (1978)

[a] This reaction apparently gives directly the $NH_2(\tilde{A})$ state (McDonald et al., 1977; Hartford, 1978).

Vacuum uv photolysis of NH_3 also is known to give $NH(c^1\Pi)$ and $NH_2(\tilde{A}, {}^2A_1)$ (Okabe, 1978; Kawasaki et al., 1973) as well as $NH(b^1\Sigma^+)$. Since radiative cascade from $NH(c^1\Pi)$ favors $NH(a^1\Delta)$ over $NH(b^1\Sigma^-)$ by a 10/1 ratio (Smith, 1969), some $NH(a^1\Delta)$ also must be present from photolysis of NH_3. Piper et al. (1978) and McDonald et al. (1977) have photolyzed NH_3. at 265 nm to produce $NH(a^1\Delta)$, which subsequently was monitored by frequency doubled laser induced fluorescence using the $NH(c^1\Pi-a^1\Delta)$ transition. The quantum yield for $NH(a^1\Delta)$ formation from HN_3 is very high at 265 nm (McDonald et al., 1977) and photolytic excitation at these wavelengths may be a viable way to produce $NH(a^1\Delta)$ in a flow system. Multiphoton dissociation of HN_3 and DN_3 with a pulsed CO_2 laser gives $NH(^1\Delta)$ directly. Subsequent reaction with hydrozoic acid gave excited amino radicals and emission from these were used to follow the reaction (Hartford, 1978). Unfortunately, reactions of HN_3 with $N_2(A)$ or metastable rare gas atoms give NH products in the triplet manifold (Stedman, 1970). By analogy to the NF_2 reactions with H atoms, the following reaction could be a possibility for giving $NH(a^1\Delta)$

$$NH_2 + F \longrightarrow HF(v) + NH(a^1\Delta); \quad \Delta H \simeq -5 \text{ kcal mole}^{-1}.$$

The best means of detection appear to be direct emission for $NH(b^1\Sigma^+)$ and laser induced (c–a) emission for $NH(a^1\Delta)$.

The vacuum ultraviolet photolysis of PH_3 gives $PH(b^1\Sigma^+)$. Recent pulsed excitation experiments (Xuan et al., 1978a,b) give a radiative lifetime of 1.3 msec and a PH_3 quenching rate constant of 2.8×10^{-13} cm^3 sec^{-1}. Sam and Yardley (1978) observed formation of $PH(A^3\Pi)$ in a two-photon process from irradiation of PH_3 with an ArF laser. They suggest that the two-photon absorption may correspond to sequential absorption with PH_2 being involved in the second step.

C. $N_2(A^3\Sigma_u^+)$, $N_2(a'^1\Sigma_u^-)$, and $N_2(B^3\Pi_g)$

The reaction of $Ar(^3P_{0,2})$ atoms with N_2 provides a flow source of $N_2(A^3\Sigma_u^+)$ in the absence of N atoms (Stedman and Setser, 1968; Meyer et al., 1970, 1971; Clark and Setser, 1980). The $Ar(^3P_{0,2}) + N_2$ reaction produces $N_2(C^3\Pi_u)$ and $(B^3\Pi_g)$ states, which cascade to the lowest triplet $N_2(A^3\Sigma_u^+)$ state. Although there is some uncertainty regarding the primary $N_2(C^3\Pi_u)/N_2(B^3\Pi_g)$ ratio (Kolts et al., 1977; Touzeau and Pagnon, 1978; Krenos and Bel Bruno, 1977), the general features related to the production of $N_2(A)$ are understood. The only unusual aspect is the nearly equal population of the $v' = 0$ and 1 levels. This arises because $N_2(A, v)$ relaxes by V–V transfer with $N_2(X)$ with loss of two quanta; thus, the even v' levels relax to $v' = 0$ and the odd v' levels to $v' = 1$. Subsequent V–T relaxation of $v' = 1$ to 0 is slow. Two-body quenching by N_2 and Ar is very slow (Table XII) and diffusion to and quenching at the wall is the dominant decay process as shown by the P^{-1} dependence of the pseudo-first-order decay constant (Meyer et al., 1971). Thus, operation at somewhat higher pressures than the optimum for $Ar(^3P_{0,2})$ production and adding a large amount of N_2 helps maximize the $N_2(A)$ concentration, which is ~ 2–4×10^{10} cm^{-3}. This is sufficient to observe emission from the $N_2(A^3\Sigma_u^+ - X^1\Sigma_g^+)$ transition which has a lifetime of 2 sec. Although the rate constant for energy pooling,

$$2N_2(A^3\Sigma_u^+) \longrightarrow N_2(B^3\Pi_g) + N_2(X)$$
$$\longrightarrow N_2(C^3\Pi_u) + N_2(X)$$
$$\longrightarrow N_2(C'^3\Pi_u) + N_2(X),$$

Table XII

Rate Constants and Diffusion Coefficients for $N_2(A^3\Sigma_u^+)$

Reaction	D_0 (10^{18} cm^{-1} sec^{-1})	Rate constants (cm^3 molecule^{-1} sec^{-1})	Reference
$N_2(A, v' = 0)$ in Ar	5.0	$<2.0 \times 10^{-19}$	Levron and Phelps (1978)
$N_2(A, v' = 1)$ in Ar	5.5	1.3×10^{-18}[a]	Levron and Phelps (1978)
$N_2(A, v' = 0)$ in N_2	5.0	$<2.6 \times 10^{-18}$	Levron and Phelps (1978)
$N_2(A, v' = 1)$ in N_2	5.0	3.8×10^{-17}[a]	Levron and Phelps (1978)
$N_2(A, v' = 0) + H_2$		2.4×10^{-15}	Levron and Phelps (1978)
$N_2(A, v' = 1) + H_2$		4.4×10^{14}[a]	Levron and Phelps (1978)
$N_2(A, v' = 0,1) + O_2$		2.9×10^{-12}	Dunn and Young (1976)
$N_2(A, v' = 0,1) + O_2$		3.6×10^{-12}	Callear and Wood (1971)
$N_2(A, v' = 0,1) + O_2$		3.3×10^{-12}	Slanger et al. (1973)
$N_2(A, v' = 0,1) + O_2$		4.0×10^{-12}[b]	Clark and Setser (1980)
$N_2(A, v = 0.1) + H_2O$		5.5×10^{-14}	Callear and Wood (1971)
$N_2(A, v = 0,1) + CO_2$		$<1.3 \times 10^{-15}$	Callear and Wood (1971)
$N_2(A, v' = 0,1) + N(^4S)$		3.5×10^{-11}	Vidaud et al. (1976b)
$N_2(A, v' = 0,1) + N(^4S)$		3.0×10^{-11}[c]	Meyer et al. (1970)
$N_2(A, v' = 0,1) + N(^4S)$		4.8×10^{-11}	Dunn and Young (1976)
$N_2(A, v' = 0,1) + O(^3P)$		5.0×10^{-11}	Young and St. John (1968)
$N_2(A, v' = 0,1) + O(^3P)$		1.3×10^{-11}[b]	Meyer et al. (1970)
$N_2(A, v' = 0,1) + O(^3P)$		1.5×10^{-11}	Dunn and Young (1976)

[a] The rate constants for $N_2(A, c' = 1)$ with Ar, H_2, and N_2 correspond to vibrational relaxation.

[b] The quenching rate constants for $v = 0$ and $v = 1$ are nearly identical [Clark and Setser (1980) and Dunn and Young (1972)]; there was some disagreement in the early measurements for the O_2 rate constant; however, a consensus value of $3.5 \pm 0.5 \times 10^{-12}$ cm^3 molecule^{-1} sec^{-1} is now established.

[c] These rate constants were measured relative to k_{O_2}; the value in the table is for the k_{O_2} given in footnote b.

is large ($\sim 1.0 \times 10^{-9}$ cm^3 sec^{-1}; Hays and Oskam, 1973), this does not contribute significantly to the total decay because of the low $N_2(A)$ concentration. The $N_2(C^3\Pi_u)$ emission first was used to identify the energy-pooling reaction (Stedman and Setser, 1969); subsequent independent studies (Hays and Oskam, 1973; Clark and Setser, 1980) are in agreement on this rate constant (1.5×10^{-10} cm^3 sec^{-1}). The $N_2(C'^3\Pi_u)$ formation is only 10% of $N_2(C)$ and is hard to observed (Hays and Oskam, 1973). The $N_2(B)$ state emission from energy pooling is reported as being 4.4 times stronger than the $N_2(C^3\Pi_u)$ emission (Hays and Oskam, 1973) from experiments in the afterglow of a pulsed discharge. Since $N_2(B)$ can be formed in several ways, independent confirmation of $N_2(B)$ formation from energy pooling and identification of the vibrational distribution would be worthwhile. The Ar*/N_2 flow source has been used to measure a variety of quenching rate constants with small inorganic and organic molecules (Meyer et al., 1970, 1971; Clark and Setser, 1980). S. Rosenwaks and I. Nadler (personal communication, 1979) have used this source to study the reactions of $N_2(A^3\Sigma_u^+)$ with various metal atoms.

A conventional microwave discharge in N_2 gives mainly N atoms. Although these recombine to $N_2(A^3\Sigma_u^+)$ (Golde and Thrush, 1972; Gartner and Thrush, 1975), the system is not very suitable for direct studies of $N_2(A)$ chemistry. However, some rate constants have been measured using Hg as a tracer for the $N_2(A)$ concentration in active nitrogen (Meyer et al., 1972; Thrush and Wild, 1972). Viduad et al. (1976b) has reported a low power, virtually static rf discharge source of $N_2(A)$; the [$N_2(A)$] was $\sim 10^{13}$ cm^{-3} and [N]/$N_2(A)$] $\sim 10^{-2}$ at the discharge source. Such a discharge may be amenable to adaptation to flow systems.

Several pulsed discharge sources have been developed for study of $N_2(A)$. Young and St. John (1968) and Dunn and Young (1972) have used a pulsed Tesla-type discharge in a 5 liter bulb and observed decay rates over a period of 0.1 sec. Callear and Wood (1971) used a Xe or Kr flash resonance lamp to generate $N_2(A)$ from excitation transfer between N_2 and the excited states of Xe and Kr. Mandl and Ewing (1977) have developed a small, simplified version of the Xe flash resonance lamp and have used signal averaging techniques to obtain time resolved kinetics of $N_2(A)$ reactions. Slanger et al. (1973) used pulsed photolysis of N_2O at 147 nm, for which the quantum yield for $N_2(A)$ is 0.08, in their measurement of the temperature coefficient of some quenching rate constants. Pulsed radiolysis and optical absorption spectroscopy have been used (Roy et al., 1975) to study kinetics of $N_2(A, v' = 0-6, 8)$. They confirmed the $\Delta v = 2$ relaxation mechanism in N_2 and studied relaxation in all of the rare gases. Very recently Levron and Phelps (1978) have used pulsed electron impact excitation to generate $N_2(A)$; this is a very clean method that is especially suitable for studying slow reactions.

The lifetime of $N_2(A)$ is well established so absolute concentrations, as well as relative concentrations, can be obtained from measurement of absolute emission intensity of the $N_2(A^3\Sigma_u^+-X^1\Sigma_g^+)$ transition. Both Hg and NO are good monitors of $N_2(A)$, Table IV, because the excitation of $Hg(^3P_1)$ and $NO(A^2\Sigma^+)$ gives strong emission. Laser induced fluorescence from the $N_2(B^3\Pi_g \leftarrow A^3\Sigma_u^+)$ transition may be a useful monitor too (Heidner et al., 1976; Cook et al., 1978c). Viduad et al. (1976b) used a platinum coil calorimeter and photoionization to monitor $N_2(A)$; the H Lyman-α line, 121.5 nm, and Xe resonance line, 147 nm, gave comparable N_2^+ signals.

The $N_2(B^3\Pi_g)$ state has a lifetime of 8 μsec and cannot be studied in the normal sense by the flow technique. However, Heidner et al. (1976) showed that pulsed laser excitation $N_2(A^3\Sigma_u^+) \rightarrow N_2(B^3\Pi_g)$ in the $Ar/N_2(A)$ flow system provided a way of studying the kinetics of $N_2(B)$. They were primarily interested in vibrational relaxation of $N_2(B)$ and the coupling between $N_2(B^3\Pi_g)$ and the $N_2(^3\Delta_u)$ states. Cook et al. (1978c) has observed the $N_2(B-A)$ laser induced fluorescence using a standard commercial flash pumped dye laser. Since $N_2(A)$ is rather unreactive to many saturated molecules, a variety of reagents can be added to the $Ar/N_2(A)$ flow system and the subsequent decay kinetics of $N_2(B)$ can be analyzed. Such work should be of direct aid in advancing the understanding of $N_2(B)$ chemistry, which has been deduced primarily from studies in active nitrogen (Golde and Thrush, 1973; Gartner and Thrush, 1975).

The lowest excited singlet state of nitrogen, $N_2(a'^1\Sigma_u^-)$ with $\tau = 17$ msec, has been generated in a flow system by Golde (1975, 1979). A hollow-cathode discharge resembling that shown in Fig. 1b was used; however, N_2 was added to the Ar before the discharge rather than after the discharge as is the case of $N_2(A^3\Sigma_u^+)$ formation. The optimum N_2/Ar ratio was 0.005–0.05. Golde was able to directly observe the $a'^1\Sigma_u^--X^1\Sigma_g^+$ transition in the vacuum ultraviolet; the $N_2(a^1\Pi_g-X^1\Sigma_g^+)$ emission, which also was observed, probably is a consequence of excitation transfer from $N_2(a'^1\Sigma_u^-)$. Formation of $N_2(a'^1\Sigma_u^-)$ was very dependent upon operating conditions. Unless the Ar resonance states are responsible for the excitation, the $N_2(a'^1\Sigma_u^-)$ state must be formed by electron collisions within the discharge. Golde and Thrush (1974) showed that the interaction of N_2O with $(Ar^3P_{0,2})$ gives some $N_2(a')$; but, the branching fraction is less than 5×10^{-4} (Gundel et al., 1976). Vacuum uv photolytic excitation of N_2O gives no $N_2(a')$. Golde's method thus seems to offer the most promise as a source for $N_2(a')$.

D. $CO(a^3\Pi)$, $CS(a^3\Pi)$, and Higher Triplet States

The $CO(a^3\Pi)$ molecule is analogous to $N_2(B^3\Pi_g)$; however, the $CO(a^3\Pi)$ state is the lowest energy triplet state of CO; and hence, it is the long-lived (7–5 msec) triplet state. The $CO(a^3\Pi-X^1\Sigma^+)$ transition, which occurs in the ultraviolet, is relatively easy to observe and serves as a convenient monitor of the CO(a) relative concentration. Two unusual features should be noted about CO(a$^3\Pi$). The lifetimes of the different spin components vary considerably (James, 1971a,b) and, depending upon the populations in a given experiment, the mean lifetime can have different values. Second, the lifetime for the spontaneous vibrational radiative transitions, e.g., CO(a, v) $\rightarrow$ CO(a, v − 1) + $h v$, are comparable to that for the electronic transition (Marcoux et $al.$, 1977). Thus, the vibrational distributions can change with time even in the absence of collisional relaxation and early claims that these population variations were due to vibrational relaxation were incorrect.

The CO(a$^3\Pi$) molecule can be generated in a flow apparatus using the reaction of Ar($^3P_{0,2}$) with various precursors (Taylor and Setser, 1971) and He(2^3S) with CO_2 (Wauchop and Broida, 1972a,b). The He(2^3S) + CO_2 reaction yields CO(a) via dissociative electron recombination with the CO_2^+, which is formed initially by He(2^3S) Penning ionization. The best precursor with Ar($^3P_{0,2}$) also is CO_2 (Taylor and Setser, 1973). The He(2^3S)/ CO_2 source is generally superior because the concentrations of CO(a) are higher, it is easier to remove the impurities (N_2 is especially bothersome, vide infra) and the distribution of v levels facilitates the study of the $v' = 0$–4 levels. The He(2^3S) + CO_2 source also works well at reduced temperature (Clark and Setser, 1975). The only disadvantage of He(2^3S) + CO_2, relative to Ar($^3P_{0,2}$) + CO_2, is the greater loss by the more rapid diffusion of CO(a$^3\Pi$). No quenching or vibrational relaxation of CO(a$^3\Pi$, v) by He has been observed in the experiments done to data. The dominant loss of CO(a) is by diffusion and quenching at the walls, $D_0 = 21 \times 10^{18}$ cm^{-1} sec^{-1}, which is $\sim 20\%$ larger than for CO(X) in He (Clark and Setser, 1975) and the flow system is usually operated at ~ 5 Torr to reduce the diffusion rate. Fortunately, the He(2^3S) quenching rate constant by CO_2 is 30 times larger than that for CO(a) by CO_2 (2×10^{-11} cm^3 sec^{-1}) and the He(2^3S) is quenched without causing loss of CO(a) by reaction with CO_2.

Except for perfluorinated molecules, nearly all other reagents have quenching rate constants in the 10^{-10} cm^3 sec^{-1} range, even at 77 K, for CO(a$^3\Pi$) (Taylor and Setser, 1973; Clark and Setser, 1975). Bimolecular energy-pooling reactions have not yet been identified for CO(a). Molecular nitrogen is an especially troublesome impurity because of the reversible exchange reaction

$$CO(a, v) + N_2(X) \longrightarrow N_2(A) + CO(X)$$

and the presence of N_2 in the He carrier gas must be avoided. Under normal operating conditions of 1–5 Torr, the reaction of CO(a) with NO gives both NO(A) and NO(B), although these excitation transfer channels correspond to only $\sim 20\%$ of the quenching (Taylor and Setser, 1973). Recent molecular beam experiments (Ottinger *et al.*, 1978) show, however, that the CO(a) + NO excitation transfer reaction under collision free conditions gives only NO(A), just like the N_2(A) excitation transfer reaction. This raises questions about the reaction mechanism in the flow reactor. Perhaps the CO(a) + NO reaction gives some excitation into the elusive NO quartet manifold and secondary collisions with He or NO give NO(B)? Slanger *et al.* (1975) have investigated the product channels from the fast $(1.1 \times 10^{-10} \text{ cm}^3 \text{ sec}^{-1})$ quenching of CO(a) by CO. They utilized direct photolysis from an I_2 lamp to produce CO(A) and measured the conversion rate of electronic to vibrational energy of CO(X). They find a very high (0.89 ± 0.24) branching fraction for E–V transfer. Electronic to vibrational conversion also may be an important mechanism for quenching of CO(a) by other molecules.

Laser induced fluorescence from $CO(a^3\Pi-d^3\Delta)$ has been observed in a flow apparatus (Cook *et al.*, 1978c) and at high resolution in a molecular beam experiment (Hemminger *et al.*, 1977). Several other CO triplet states are accessible by laser induced fluorescence and such studies should add to our understanding of the chemistry of the CO(a) triplet states.

Dissociative electron combination with CS_2^+ also is a source of $CS(a^3\Pi)$ in a He(2^3S) metastable flow apparatus (Coxon *et al.*, 1976; Taylor *et al.*, 1972), probably superior to precursors $Ar(^3P_{0,2}) + CS_2$ such as (Piper *et al.*, 1972). The $CS(a^3\Pi)$ lifetime has been set as ≤ 16 msec (Black *et al.*, 1977). Although more direct measurements would still be desirable, the lifetime is sufficiently long for study of CS(a) by the flow technique. Black used the pulsed photolysis of CS_2, which has a high quantum yield for $CS(a^3\Pi)$ formation at 125–140 nm, to measure several rate constants. Quenching by CS_2 and other likely precursors (Piper *et al.*, 1972) are likely to be fast; however, the quenching rate constants for He and Ar are $< 10^{-15} \text{ cm}^3 \text{ sec}^{-1}$ and these could serve as satisfactory carrier gases. Laser induced excitation from $CS(a^3\Pi)$ would provide a direct way to study the higher CS triplet states. Since $CS^+(X)$ also is produced by the He(2^3S) + CS_2 reaction (Coxon *et al.*, 1976), laser fluorescence studies of $CS^+(X)$ could be done.

E. NO($a^4\Pi$) and NO$^+(a^3\Sigma^+)$

Nitric oxide is isoelectronic with O_2^+ and has a similar energy level pattern with the $a^4\Pi$ state being the lowest quartet level in both cases. Both have known $b^4\Sigma^- - a^4\Pi$ transitions; however, that for NO is less well characterized because it occurs in the far red region (Miescher and Huber, 1976; Möhlmann and DeHeer, 1977). A similar analogy holds between the NO$^+$ and N_2 molecules. The $a^3\Sigma^+$ states are 6.2 and ~ 7 eV above the ground states for N_2 and NO, respectively. In contrast to N_2, relatively little spectroscopic information is available concerning the higher NO$^+$ triplet states and there is no known emission that is analogous to the well-characterized $N_2(B^3\Pi_g-A^3\Sigma_u^+)$ transition. The reactive nature of the NO^+ states with NO and the difficulty of working in the infrared region of the spectrum (Chang et al., 1978) probably are responsible for this. The $NO^+(a^3\Sigma^+)$ state has been studied successfully in a flow apparatus and it will be discussed first.

The reaction of $He(2^3S)$ with NO gives both triplet and singlet states of NO^+ (Coxon et al., 1975a; Yee et al., 1975). The triplet states radiatively cascade to $NO^+(a^3\Sigma^+)$. Albritton et al. (1979) have utilized this $He(2^3S)$ Penning ionization source and monitored the NO^+ ion signal to measure rate constants for $NO^+(a^3\Sigma^+)$ reactions. Coxon et al. (1975a) observed strong neutral NO(B–X) emission in a $He(2^3S)$ Penning ionization optical spectroscopy study of NO. The origin of this relatively low energy neutral emission was suggested to be excitation transfer between long-lived triplet NO^+ states (probably $NO^+(a^3\Sigma^+)$) and NO. If this suggestion is correct, the NO(B–X) emission could serve as a convenient tracer for $NO^+(A^2\Sigma^+)$.

The $NO(a^4\Pi)$ state presents a challenge for both kinetic and spectroscopic study. Although the NO(a–X) intercombination emission has never been observed in the gas phase, the transition is relatively well known in rare gas matrices and has a lifetime of 0.085 sec (Goodman and Brus, 1978) in an argon matrix. Although the gas phase value is expected to be somewhat longer, emission can be readily observed from states with lifetimes that are approximately two orders of magnitude longer, and the failure to record the spectrum implies rapid quenching by collisions with NO. Support for this expectation is provided by the large $(1.0 \times 10^{-10} \text{ cm}^3 \text{ sec}^{-1})$ quenching rate constant for $NO(b^4\Sigma^-)$ by NO(X) measured by Möhlmann and DeHeer (1977). Thus, a sensitive detection method is needed; laser induced fluorescence utilizing the $NO(a^4\Pi-b^4\Sigma^-)$ transition (Miescher and Huber, 1976; Möhlmann and DeHeer, 1977) may be one possibility.

Heicklen and Cohen (1968) summarized the chemical evidence in favor of $NO(a^4\Pi)$ formation in mercury photosensitization of NO. However, Horiguchi and Tsuchiya (1979) have demonstrated the direct E–V transfer from Hg* to NO(X) without an intermediate state. Thus, at least part of the quenching does not involve $NO(a^4\Pi)$. Since the total quenching rate constant for $Hg^*(^3P_0)$ by NO, Table IX, is much larger than for CO. Horiguchi and Tsuchiya's results do not eliminate concurrent excitation transfer to the $NO(a^4\Pi)$ state and additional attempts to directly observe $NO(a^4)$ from mercury photosensitization is worthwhile. Various other excitation transfer schemes yielding $NO(a^4)$ can be envisioned, e.g., see Section V,D on CO(a). Ground state N and O atoms correlate to $NO(a^4\Pi)$ and the afterglow emissions from N + O combination are thought to involve crossings from high v levels of $a^4\Pi$ to $b^4\Sigma^-$ and to the doublet manifold (Golde and Thrush, 1973). In matrix isolation work (Goodman and Brus, 1978), the $a^4\Pi$ state is formed by intersystem crossing from $NO(B^2\Pi)$. If a sufficiently sensitive detection method was available, gas phase collisional transfer from the doublet manifold to the quartet manifold might (?) be a possible method for studying $NO(a^4\Pi)$.

F. $H_2(c^3\Pi_u)$, $He_2(a^3\Sigma_u^+)$, $CN(^4\Sigma^+)$, and Other Miscellaneous States

The optically metastable $H(n = 2, {}^2S)$ state is rapidly collisionally transferred to the $H(^2P)$ state, which radiatively decays, and studies of metastable H atoms by the flow technique are not feasible. The situation may be more favorable for metastable molecular hydrogen, the $H_2(c^3\Pi_u)$ state. Since only one-half of the $H_2(c^3\Pi_u)$ levels are predissociated (Meierjohann and Vogler, 1978) and since the propensity for conservation of nuclear spin prohibits the collisional transfer of populations between the predissociating and nonpredissociating levels, predissociation does not preclude the study of the even rotational levels of parahydrogen and odd rotational levels of orthoydrogen of $H_2(c^3\Pi_u)$ by the flow technique. The $H_2(c)$ state can be made by electron impact on $H_2(X)$ and studies of spectroscopy (Lichten and Wik, 1978) and Penning ionization and competing reactions (Hotop et al., 1969a) have been made by the molecular beam technique. Bethune et al. (1978) observed a concentration of 10^{12} cm^{-3} of $H_2(c^3\Pi_u)$ in a pulsed discharge in hydrogen; the concentration was monitored by absorption spectra from a broadband dye laser. The $(a^3\Sigma_g^+)$ state, which radiatively decays to $H_2(b^3\Sigma_u^+)$, is only 291 cm^{-1} above the $c^3\Pi_u$ state, and collisional transfer from $H_2(c)$ to $H_2(b)$ by the carrier gas may constitute a serious two-body loss mechanism. A measurement of this rate constant in He is necessary before a decision can be made about the possibility of studying $H_2(c^3\Pi_u)$. It is unlikely that argon could be used as a carrier gas because formation of $Ar(^3P_{0,1,2})$, as well as $H_2(a)$, is possible:

$$H_2(c^3\Pi_u) + Ar \longrightarrow H_2(a^3\Sigma_g^+) + Ar - 0.04 \text{ eV}$$

$$\longrightarrow H_2(X) + Ar(^3P_{0,1,2}) + 0.03, 0.13, 0.20 \text{ eV}$$

In fact, the reverse of the last reaction *may* be involved in the quenching of the $Ar(^3P_{0,2})$ states (Kolts et al., 1977; Huo and Dillon, 1977).

The lowest energy triplet molecular state of helium, $He_2(a^3\Sigma_u^+)$, is long lived and correlates to $He + He(2^3S)$. The ground state of He_2 is totally repulsive and electron impact is not an appropriate method for generating $He_2^+(a^3\Sigma_u^+)$. However, three-body recombination of $He(2^3S)$ leads naturally to $He_2(^3\Sigma_u^+)$ formation and has been used in high pressure pulsed static afterglows to form $He_2(^3\Sigma_u^+)$ (Lee and Collins, 1977; Myers and Cunningham, 1977; Pitchford and Deloche, 1978; Lee et al., 1978). At lower pressure, a recombination process involving two electrons and He_2^+, that is not well understood, seems to be the more important formation process for $He_2(^3\Sigma_u^+)$. These authors used absorption spectroscopy, $He_2(a^3\Sigma_u^+-e^3\Pi_g)$, to monitor the time dependence of the $He_2(a)$ concentration. These pulsed static afterglows also have been used for laser excitation of the $He_2(e^3\Pi_g)$ state; electronic quenching and rotational relaxation (Collins and Johnson, 1976; Gauthier et al., 1976; Delpech et al., 1977) were studied. Rotational relaxation rates of $He_2(e^3\Pi_u)$ are normal for a diatomic molecule. However, Callear and Hedges (1970) found that the relaxation of $He(a^3\Sigma_u^+)$ to be abnormally slow, which they attributed to the virtually spherical potential of the $\sigma_g(2s)$ Rydberg orbital of $He_2(a)$. Cellear and Hedges used a single pulse microwave discharge to produce $He_2(a^3\Sigma_u^+)$. Nonequilibrium rotational distributions of $He_2(a^3\Sigma_u^+)$ are observed even at late times in the pulsed afterglows; this seems to be a consequence of both a formation mechanism giving molecules in high J states and the slow rotational relaxation. Without doubt, conditions can be adjusted so that $He_2(a^3\Sigma_u^+)$ is present in He flowing afterglows. However, measurement of rate constants and identification of products is difficult because of the simultaneous formation steps [the $He(2^3S) + 2He$ reaction is relatively slow] and because of the presence of the atomic and molecular ions and $He(2^3S)$; see Section IV,B. One improvement may be to adapt the high pressure hollow-cathode source used by Lichten et al. (1974) in their molecular beam study of the spectroscopic properties of $He_2(a^3\Sigma_u^+)$. The only significant loss of $He_2(a)$ in pure He is from diffusion to wall (and quenching) and bimolecular ionization (Deloche et al., 1976; Phelps, 1955).

The lowest quartet state of O_2^+ and NO were discussed in preceding sections. Since there are a large number of radicals with doublet ground states, the lowest energy, long-lived state of such radicals frequently will be a quartet level. Few of these states have been characterized; however, recent work has led to a source and detection method for the $CN(^4\Sigma^+)$ state. Coxon et al. (1973) discovered an unusual intensity anomaly in the $CN(B, v' = 11, N' = 20 \rightarrow X)$ emission spectra excited by collisions of $Ar(^3P_{0,2})$ with BrCN. The origin of the intensity anomaly was suggested to be the perturbations provided by high population in a nearby quartet level. Further support was provided by high resolution spectra (Coxon et al., 1975b). The anticrossing experiments of Miller et al. (1976) identified perturbations for several additional rotational levels of $CN(B, v' = 11)$. which was persuasive evidence for the quartet perturbation. Cook et al. (1978b) have used the anticrossing method to search for the $CN(^4\Sigma^+)$ state in active nitrogen + chloromethanes as sources of CN. Some quartet state populations was found in nearly every case; although, the $Ar(^3P_{0,2})$ + BrCN has the highest concentration. If the approximate location of the second quartet state could be found, the $Ar(^3P_{0,2})$ + BrCN source appears suitable for laser induced fluorescence studies of the CN quartet levels.

The lowest triplet states of the group IIA and IIB metal oxides and sulfides and tin and lead oxides have been mentioned as long lived reservoir states in oxidation studies of these metals. The quartet states of the group IIA and IIB metal halides may play a similar role. This work is discussed by Fontijin and Felder in Chapter 2 and will not be pursued further here. As knowledge develops about the upper states of more refractory species, other classes of long-lived states may be discovered.

In concluding this section, we will note that metastable molecular halogen states are mentioned in the literature (Coombe et al., 1978) as possibly being involved in some kinetic processes. The $^3\Pi_{2u}$ and $^3\Pi_{0_u^-}$ states, which correlate with halogen (ns^2np^5) atoms, are expected to be very long lived, have never been spectroscopically observed in the gas phase.‡ However, Bondybey and Fletcher (1976) observed pulsed laser emission from $Cl_2(^3\Pi_{2u})$ following irradiation of Cl_2 in an argon matrix. The lifetime was 76 msec.

VI. SUMMARY

After completing this review, one feature of overriding significance emerges. With only a few exceptions, all of the long-lived excited states studied by the flow technique are generated by electrical discharges. Yet there seems to be little, if any, fundamental understanding of the microscopic processes occurring in these operating discharges. The barrier to the development of new electrical discharge sources is the lack of understanding of interactions of electrons with precursor molecules, as well as the subsequent ion–electron interaction processes. Up to now trial and error has been the approach to the development of practical electrical discharge sources. At the present time, the analytical methods for monitoring species are more advanced than the methods of preparing states of interest.

We hope to have conveyed to the reader that for some species the fast flow technique is an excellent way to systematically study long-lived excited states. However, we do not wish to leave the impression that the fast flow technique is intrinsically either inferior or superior to pulsed methods for study of long lived states. Each technique has its own advantages depending upon the type of measurements which are to be made. Frequently the choice of method will depend upon the best way of generating the species to be studied.

APPENDIX: FLOWING AFTERGLOW STUDIES OF ION–MOLECULE REACTIONS

As was noted in the text, detailed kinetic studies of positive and negative ion–molecule reactions by the helium flowing afterglow technique is a fully

‡ Even the $^3\Pi_1$ states have relatively long radiative lifetimes; the $ICl(^3\Pi_1)$ state has been measured as 0.2 msec (Havey and Wright, 1978).

developed field. For some recent reviews the reader may consult Ferguson (1975), Bohme and Schiff (1975), Fehsenfeld (1975), Tanaka et al. (1976), and Bone (Chapter 5). Total rate constants, branching fractions, equilibrium constants, negative–positive ion recombination, and neutral molecule clustering to negative and positive ions all can be studied. A review of this field is far beyond the scope of this chapter, even though the flowing afterglow technique basically employs a fast flow apparatus nearly identical to that shown in Fig. 1c. A hot-cathode discharge generates the ions which are detected at the end of the flow tube with a mass spectrometer. In some instances, optical spectroscopic measurements have been made of products of the reactions, especially for charge transfer processes. Fehsenfeld (1975) has provided recent overview of applications and recent developments of the technique. An important extension has been the addition of a drift tube section to the flow reactor which allows rate constants to be measured as a function of collision energy (0.05–5 eV). Another development (Adams and Smith, 1976) is the capability of injecting mass selected low energy positive ions from a discharge source into the flowing gas. This, of course, provides isolated positive ion for subsequent kinetic studies. Important advantages are the absence of the light emission associated with the helium flowing afterglow generated by the hot-cathode discharge and the absence of electrons.

Bone, Chapter 5, mentions the flowing afterglow as a way of generating positive ions. Numerous possibilities exist for generating positive ions from added reagents: (i) charge transfer from He^+, (ii) Penning ionization by $He(2^3S)$, (iii) direct electron impact in the hot-cathode discharge source, and (iv) successive ion–molecule reactions from reagents added downstream from the discharge region. Selective formation of negative ions is somewhat more difficult since one must rely upon electron attachment or reactions of negative ions with added molecules to generate the desired negative ions. The possibilities for producing a given positive or negative ion are rather large and selection of the " best " way is not necessarily obvious without some experience. Fortunately, Albritton (1978) has published a compendium of ion–neutral rate constants and product distributions for both negative and positive ions that should be of enormous assistance in selecting the best reagent and operating conditions for preparing a given ion.

REFERENCES

Abrahamson, E. W., Andrews, L. J., Husain, D., Wiesenfeld, J. R. (1972). J. Chem. Soc., Faraday Trans. 2, **68**, 48.
Acuna, A. U., and Husain, D. (1973). J. Chem. Soc., Faraday Trans. 2, **69**, 585.
Adams, H. G., and Smith D. (1976). Int. J. Mass Spectrom. Ion Phys., **21**, 349.
Alben, K. T., Auerbach, A., Ollison, W. M., Weiner, J., and Cross, R. J., Jr. (1978). J. Am. Chem. Soc., **100**, 3274.

Albritton, D. L. (1978). *At. Data. Nucl. Data Tables* **22**, 1.

Albritton, D. L. *et al.* (1979). *J. Chem. Phys.* (in press).

Allison, W., Dunning, F. B., and Smith, A. E. H. (1973). *J. Phys. B* **5**, 1175.

Amimoto, S. T., Force, A. P., and Wiesenfeld, J. R. (1978). *Chem. Phys. Lett.* **60**, 40.

Arnold, S. J., and Ogryzlo, E. A. (1967). *Can. J. Chem.* **45**, 2053.

Arnold, S. J., Jubo, M., and Ogryzlo, E. A. (1968). *Adv. Chem. Ser.* **77**, 133.

Badger, R. M., Wright, A. D., and Whitlock, R. F. (1967). *J. Chem. Phys.* **43**, 3341 (1967).

Bearman, G. H., Earl, J. D., Pieper, R. J., Harris, H. H., and Leventhal, J. J. (1976). *Phys. Rev. A* **13**, 1734.

Becker, K. H., Groth, W., and Schurath, U. (1971). *Chem. Phys. Lett.* **8**, 259.

Benard, D. J., and Pchelkin, N. R. (1978). *Rev. Sci. Instrum.* **49**, 794.

Benard, D. J., Slafer, W. D., and Lee, P. H. (1976). *Chem. Phys. Lett.* **43**, 69.

Benard, D. J., Love, P. J., and Slafer, W. D. (1977). *Chem. Phys. Lett.* **48**, 321.

Benard, D. J. McDermott, W. E., Pchelkin, N. R., and Bousek, R. R. (1979). *Appl. Phys. Lett.* **34**, 40.

Bethune, D. S., Lankard, J. R., and Sorokin, P. P. (1978). *J. Chem. Phys.* **69**, 2076.

Bevan, M. J., and Husain, D. (1974). *J. Photochem.* **3**, 1.

Bevan, M. J., and Husain, D. (1975). *J. Photochem.* **4**, 51.

Bhattacharyya, D. K., and Chiu, L. Y. C. (1977). *J. Chem. Phys.* **67**, 5727.

Bigeon, M. C. (1967). *J. Phys.* **28**, 51.

Black, G., Slanger, T. G., St. John, G. A., and Young, R. A. (1969). *J. Chem. Phys.* **51**, 116.

Black, G., Sharpless, R. L., and Slanger, T. G. (1975). *J. Chem. Phys.* **63**, 4546.

Black, G., Sharpless, R. L., and Slanger, T. G. (1977). *J. Chem. Phys.* **66**, 2113.

Blauer, J. A., Janiak, D., and Solomon, W. C. (1977). *In* "Electronic Transitions Lasers II" (L. E. Wilson, S. N. Suchard, and J. I. Steinfeld, eds.), p. 283. MIT Press, Cambridge, Massachusetts.

Blickensderfer, R. P., Breckenridge, W. H., and Moore, D. S. (1975). *J. Chem. Phys.* **63**, 3681.

Bohme, D. K., and Schiff, H. I. (1975). *Int. J. Mass Spectrom. Ion Phys.* **16**, 167.

Bolden, R. C., Hemsworth, R. S., Shaw, M. I., and Twiddy, N. D. (1970). *J. Phys. B* **3**, 145.

Bondybey, V. E., and Fletcher, C. (1976). *J. Chem. Phys.* **64**, 3615.

Bondybey, V. E., and Miller, T. A. (1977a). *J. Chem. Phys.* **66**, 3337.

Bondybey, V. E., and Miller, T. A. (1977b). *J. Chem. Phys.* **67**, 1790.

Bondybey, V. E., and Miller, T. A. (1978). *J. Chem. Phys.* **69**, 3597.

Borrell, P., Borrell, P. M., and Pedley, M. D. (1977). *Chem. Phys. Lett.* **51**, 300.

Brashears, H. C., Jr., and Setser, D. W. (1979). *J. Phys. Chem.* (submitted for publication).

Braun, W., Bass, A. M., Davis, D. D., and Simmons, J. D. (1969). *Proc. R. Soc. London, Ser. A* **312**, 417.

Breckenridge, W. H., and Miller, T. A. (1972a). *Chem. Phys. Lett.* **12**, 437.

Breckenridge, W. H., and Miller, T. A. (1972b). *J. Chem. Phys.* **56**, 465.

Breckenridge, W. H., and Renlund, A. M. (1978). *J. Phys. Chem.* **82**, 1474, 1484.

Breckenridge, W. H., Broadbent, T. W., and Moore, D. S. (1975). *J. Phys. Chem.* **79**, 1233.

Brinkmann, U., Steudel, A., and Walther, H. (1967). *Z. Agnew. Phys.* **22**, 223.

Brinkmann, U., Gaschler, J., Steudel, A., and Walther, H. (1969). *Z. Phys.* **228**, 427.

Brion, C. E., and Crowley, P. (1977). *J. Electron. Spectrosc. Relat. Phenom.* **11**, 399.

Broadbent, T. W., and Callear, A. B. (1972). *J. Chem. Soc., Faraday Trans. 2* **68**, 1367.

Brom, J. M., Jr., Kolts, J. H., and Setser, D. W. (1978). *Chem. Phys. Lett.* **55**, 44.

Brown, A., and Husain, D. (1975). *J. Chem. Soc., Faraday Trans. 2* **71**, 699.

Brown, R. L. (1967). *J. Phys. Chem.* **71**, 2492.

Burch, D. E., and Gryvnak, D. A. (1969). *Appl. Opt.* **8**, 1493.

Burde, D. H., and McFarlane, R. A. (1976). *J. Chem. Phys.* **64**, 1850.
Burde, D. H., McFarlane, R. A., and Wiesenfeld, J. R. (1974).
Burde, D. H., McFarlane, R. A., and Wiesenfeld, J. R. (1975). *Chem. Phys. Lett.* **32**, 296.
Burnham, R., and Djeu, N. (1974). *J. Chem. Phys.* **61**, 5158.
Busch, Y. A., McFarland, M., Albritton, D. L., and Schmeltekopf, A. L. (1973). *J. Chem. Phys.* **59**, 4020.
Cadman, P., and Polanyi, J. C. (1968). *J. Phys. Chem.* **72**, 3715.
Callear, A. B., and Hedges, R. E. M. (1970). *Trans. Faraday Soc.* **66**, 92.
Callear, A. B., and McGurk, J. C. (1973). *J. Chem. Soc., Faraday Trans. 2* **69**, 97.
Callear, A. B., and Wood, P. M. (1871). *Trans. Faraday Soc.* **67**, 272 and 598.
Capelle, G. A., Sutton, P. G., and Steinfeld, J. J. (1978). *J. Chem. Phys.* **69**, 5140.
Carrington, A., Levy, D. H., and Miller, T. A. (1966). *J. Chem. Phys.* **45**, 4093.
Cassa, M. P., and Golde, M. F. (1979). *Chem. Phys. Lett.* **60**, 281.
Čermák, V. (1976). *J. Electron. Spectrosc. Relat. Phenom.* **9**, 419.
Čermák, V., and Herman, Z. (1968). *Collect. Czech. Chem. Commun.* **33**, 468.
Chang, R. S. F., and Setser, D. W. (1978). *J. Chem. Phys.* **69**, 3885.
Chang, R. S. F., and Setser, D. W. (1979). *J. Chem. Phys.* (in press).
Chang, R. S. F., Taylor, W. G., and Setser, D. W. (1978). *Chem. Phys.* **35**, 201.
Chang, R. S. F., Horiguchi, H., and Setser, D. W. (1980), *J. Chem. Phys.* (to be published).
Chen, C. H., Payne, M. G., Hurst, G. S., and Judish, J. P. (1976). *J. Chem. Phys.* **65**, 4028.
Chen, C. J., and Garstang, R. H. (1970). *J. Quant. Spectrosc.& Radiat. Transfer* **10**, 1347.
Cher, M., and Hollingsworth, C. S. (1969a) *Adv. Chem. Ser.* **80**, 118.
Cher, M., and Hollingsworth, C. S. (1969b). *J. Chem. Phys.* **50**, 4942.
Chowdhury, M. A. and Husain, D. (1977). *J. Chem. Soc., Faraday Trans. 2* **73**, 1805.
Clark, I. D., and Wayne, R. P. (1969). *Proc. Soc. London. Ser. A* **314**, 111.
Clark, I. D., Masson, A. J., and Wayne, R. P. (1972). *Mol. Phys.* **23**, 995.
Clark, T. C., and Clyne, M. A. A. (1970). *Trans. Faraday Soc.* **66**, 877.
Clark, W. G., and Setser, D. W. (1975). *Chem. Phys. Lett.* **33**, 71.
Clark, W. G., and Setser, D. W. (1980). *J. Phys. Chem.* (to be published).
Clyne, M. A. A., and Nip, W. S. (1976). *J. Chem. Soc., Faraday Trans. 2*, **72**, 838.
Clyne, M. A. A., and Nip, W. S. (1977). *Int. J. Chem. Kinet.* **10**, 365.
Clyne, M. A. A., and White, I. F. (1970). *Chem. Phys. Lett.* **6**, 465.
Clyne, M. A. A., Coxon, J. A., and Cruse, H. W. (1970). *Chem. Phys. Lett.* **6**, 57.
Collins, C. B., and Johnson, B. W. (1972). *J. Chem. Phys.* **57**, 5317.
Collins, C. B., and Johnson, B. W. (1976). *J. Chem. Phys.* **64**, 2605.
Collins, C. B., and Lee, F. W. (1978). *J. Chem. Phys.* **68**, 1391.
Collins, C. B., and Robertson, W. W. (1964). *J. Chem. Phys.* **40**, 701.
Collins, C. B., Johnson, B. W., and Shaw, M. J. (1972). *J. Chem. Phys.* **57**, 5310.
Collins, R. J., Husain, D., and Donovan, R. J. (1973). *J. Chem. Soc., Faraday Trans. 2*, **69**, 145.
Comes, F. J., and Pionteck, S. (1976). *Chem. Phys. Lett.* **42**, 558.
Cook, J. M., Miller, T. A., and Bondybey, V. E. (1978a). *J. Chem. Phys.* **68**, 2001.
Cook, J. M., Zegarski, B. R., and Miller, T. A. (1978b). *J. Chem. Phys.* **68**, 4763.
Cook, J. M., Miller, T. A., and Bondybey, V. E. (1978c). *J. Chem. Phys.* **69**, 2562.
Cook, T. J., and Miller, T. A. (1974). *Chem. Phys. Lett.* **25**, 396.
Coombe, R. D., Pilipovich, D., and Horne, R. K. (1978). *J. Phys. Chem.* **82**, 2484.
Corney, A., and Williams, O. M. (1972). *J. Phys. B* **5**, 686.
Coxon, J. A., Setser, D. W., and Duewer, W. H. (1973). *J. Chem. Phys.* **58**, 2224.
Coxon, J. A., Clyne, M. A. A., and Setser, D. W. (1975a). *Chem. Phys.* **7**, 255.
Coxon, J. A., Ramsay, D. A., and Setser, D. W. (1975b). *Can. J. Phys.* **53**, 1587.

Coxon, J. A., Marcoux, P. J., and Setser, D. W. (1976). *Chem. Phys.* **17**, 403.

Cunningham, D. L., and Clark, K. C. (1974). *J. Chem. Phys.* **61**, 1118.

Curran, A. H., MacDonald, R. B., Stone, A. I., and Thrush, B. A. (1973). *Proc. Soc. London, Ser. A* **332**, 355.

Dagdigian, P. J. (1978). *Chem. Phys. Lett.* **55**, 239.

Dagdigian, P. J., and Pasternack, L. (1978). *Chem. Phys.* **33**, 1.

Davidson, J. A., Sadowski, C. M. Schiff, H. I., Streit, G. E., Howard, C. J., Jennings, D. A., and Schmeltekopf, A. L. (1976). *J. Chem. Phys.* **64**, 57.

Davidson, J. A., Schiff, H. I., Brown, T. J., Streit, G. E., and Howard, C. J. (1978). *J. Chem. Phys.* **69**, 1213.

Deakin, J. J. and Husain, D., and Wiesenfeld, J. R. (1971). *Chem. Phys. Lett.* **10**, 146.

Deech, J. S., Pitre, J., and Krause, L. (1971). *Can. J. Phys.* **49**, 1976.

DeJong, H. J. (1972). *Chem. Phys. Lett.* **15**, 414.

DeJong, H. J. (1974). *Chem. Phys. Lett.* **25**, 129.

Deloche, R., Monchicourt, P., Cheret, M., and Lambert, F. (1976). *Phys. Rev. A* **13**, 1140.

Delpech, J. F., Gauthier, J. C., and Devos, F. (1977)..*J. Chem. Phys.* **67**, 5934.

Derwent, R. G., and Thrush, B. A. (1971). *Trans. Faraday Soc.* **67**, 2036.

Derwent, R. G., and Thrush, B. A. (1972). *Faraday Discuss. Chem. Soc.* **53**, 162.

Donohue, T., and Wiesenfeld, J. R. (1975). *J. Chem. Phys.* **63**, 3130.

Donovan, R. J., and Gillespie, H. M. (1975). *React. Kinet.* **1**, 14.

Donovan, R. J., and Husain, D. (1966). *Trans. Faraday Soc.*, **62**, 11, 1050.

Donovan, R. J., and Husain, D. (1970). *Chem. Rev.* **70**, 489.

Donovan, R. J., and Husain, D. (1971). *Ann. Rep. Chem. Soc.* **68A**, 123.

Donovan, R. J., and Little, D. J. (1978). *Chem. Phys. Lett.* **53**, 394.

Donovan, R. J., Husain, D., Bass, A. M., Braun, W., and Davis, D. D. (1969). *J. Chem. Phys.* **50**, 4115.

Donovan, R. J., Husain, D., and Kirsch, L. J. (1972). *Ann. Rep. Chem. Soc.* **69A**, 19.

Donovan, R. J., Fotakis, C., and Golde, M. F. (1976). *J. Chem. Soc., Faraday Trans. 2* **72**, 2055.

Donovan, R. J., Gillespie, H. M., and Strain, R. H. (1977a). *J. Chem. Soc., Faraday Trans 2* **73**, 1553.

Donovan, R. J., Fotakis, C., and Gillespie, H. M. (1977b). *J. Photochem.* **6**, 193.

Dreiling, T., and Setser, D. W. (1980). *J. Chem. Phys.* (to be published).

Dreyer, J. W., Perner, D., and Roy, C. R. (1974). *J. Chem. Phys.* **61**, 3164.

Dunkin, D. B., Fehsenfeld, F. C., Schmeltekopf, A. L., and Ferguson, E. E. (1968). *J. Chem. Phys.* **49**, 1365.

Dunn, O. J., and Young, R. A. (1972). *Int. J. Chem. Kinet.* **8**, 161.

Dunning, T. H., and Hay, P. J. (1977). *J. Chem. Phys.* **66**, 3767.

Dushman, S., and Lafferty, J. M. (1962). "Scientific Foundations of Vacuum Techniques," 2nd ed, Wiley, New York.

Ebenstein, W. L., Wiesenfeld, J. R., and Wolk, G. L. (1978). *Chem. Phys. Lett.* **53**, 185.

Elias, L., Ogryzlo, E. A., and Schiff, H. I. (1959). *Can. J. Chem.* **37**, 1680.

Engelking, P. C., and Smith, A. L. (1975). *Chem. Phys. Lett.* **36**, 21.

Fairchild, P. W., and Lee, E. K. C. (1978). *Chem. Phys. Lett.* **60**, 37.

Falick, A. M., and Mahan, B. H. (1967). *J. Chem. Phys.* **47**, 4778.

Falick, A. M., Mahan, B. H., and Myers, R. J. (1965). *J. Chem. Phys.* **42**, 1837.

Fehsenfeld, F. C. (1975). *Int. J. Mass Spectrom. Ion Phys.* **16**, 151.

Fehsenfeld, F. C., Evenson, K. M., and Broida, H. P. (1965). *Rev. Sci. Instrum.* **36**, 294.

Fehsenfeld, F. C., Schmeltekopf, A. L., Goldan, P. D., Schiff, H. I., and Ferguson, E. E. (1966). *J. Chem. Phys.* **44**, 4087.

Ferguson, E. E. (1975). *Annu. Rev. Phys. Chem.* **26**, 17.

Ferguson, E. E., Fehsenfeld, F. C., and Schmeltekopf, A. L. (1969). *Adv. At. Mol. Phys.* **5**, 1.
Filseth, S. V., and Welge, K. H. (1969). *J. Chem. Phys.* **51**, 839.
Filseth, S. V., Zia, A., and Welge, K. H. (1970). *J. Chem. Phys.* **52**, 5502.
Findlay, F. D., and Snelling, D. R. (1971). *J. Chem. Phys.* **55**, 545.
Fink, E. H., Wallach, D., and Moore, C. B. (1972). *J. Chem. Phys.* **56**, 3608.
Fletcher, I. S., and Husain, D. (1977). *Chem. Phys. Lett.* **49**, 516.
Fletcher, I. S., and Husain, D. (1978). *J. Chem. Soc., Faraday Trans. 2* **74**, 203.
Foreman, P. B., Parr, T. P., and Martin, R. M. (1977). *J. Chem. Phys.* **67**, 5591.
Garstang, R. H. (1962). *J. Opt. Soc. Am.* **52**, 845.
Garstang, R. H. (1964). *J. Res. Natl. Bur. Stand. Sect. A* **8**, 61.
Gartner, E. M., and Thrush, B. A. (1975). *Proc. Soc. London, Ser. A* **346**, 103 and 121.
Gauthier, J. C., Geindre, J. P., Moy, J. P., and Delpech, J. F. (1976). *Phys. Rev. A* **13**, 1781.
Gauthier, M., and Snelling, D. R. (1971). *J. Chem. Phys.* **54**, 2750 and 4317.
Gauthier, M., and Snelling, D. R. (1975). *J. Photochem.* **4**, 27.
Gelernt, B., Filseth, S. V., and Carrington, T. (1976). *J. Chem. Phys.* **65**, 4940.
Gersh, M. E., and Muschlitz, E. E., Jr., (1973). *J. Chem. Phys.* **59**, 3538.
Gilpin, R., Schiff, H. I., and Welge, K. H. (1971). *J. Chem. Phys.* **55**, 1087.
Giusfredi, G., Minguzzi, P., Strumia, F., and Tonelli, M. (1975). *Z. Phys. A* **274**, 279.
Golde, M. F. (1975). *Chem. Phys. Lett.* **31**, 349.
Golde, M. F. (1977). *Spec. Period. Rep.: Gas Kinet. Energy Transfer* **2**, 123.
Golde, M. F., and Thrush, B. A. (1972). *Faraday Discuss. Chem. Soc.* **53**, 233.
Golde, M. F., and Thrush, B. A. (1973). *Rep. Prog. Phys.* **36**, 1285.
Golde, M. F., and Thrush, B. A. (1974). *Chem. Phys. Lett.* **29**, 486.
Goodman, J., and Brus, L. E. (1978). *J. Chem. Phys.* **69**, 1853.
Govers, T. R., Van de Runstraat, C. A., and de Heer, F. J. (1975). *Chem. Phys.* **9**, 285.
Govers, T. R., Gerard, M., Mauclaire, G., and Marx, R. (1977). *Chem. Phys.* **23**, 411.
Grimley, A. J., and Houston, P. L. (1978a). *J. Chem. Phys.* **68**, 3366.
Grimley, A. J., and Houston, P. L. (1978b). *J. Chem. Phys.* **69**, 2339.
Gundel, L. A., Setser, D. W., Clyne, M. A. A., Coxon, J. A., and Nip, W. (1976). *J. Chem. Phys.* **64**, 4390.
Hampson, R. F., Jr., and Okabe, H. (1970). *J. Chem. Phys.* **52**, 1930.
Hariri, A., and Wittig, C. (1977). *J. Chem. Phys.* **67**, 4454.
Hartford, A., Jr., (1978). *Chem. Phys. Lett.* **57**, 352.
Hartman, D. C., and Winn, J. S. (1978). *J. Chem. Phys.* **68**, 2990.
Havey, M. D., and Wright, J. J. (1978). *J. Chem. Phys.* **68**, 4754.
Hays, G. H., and Oskam, H. J. (1973). *J. Chem. Phys.* **59**, 1507 and 6088.
Heicklen, J. and Cohen, N. (1968). *Ad. Photochem.* **5**, 157.
Heidner, R. F., III, and Husain, D. (1973). *Int. J. Chem. Kinet.* **5**, 819.
Heidner, R. F., III, and Husain, D. (1974). *Int. J. Chem. Kinet.* **6**, 77.
Heidner, R. F., III, Sutton, D. G., and Suchard, S. N. (1976). *Chem. Phys. Lett.* **37**, 243.
Hemminger, J. C., Cavanagh, R., Lisy, J. M., and Klemperer, W. (1977). *J. Chem. Phys.* **67**, 4952.
Herbelin, J. M. (1976). *Chem. Phys. Lett.* **42**, 367.
Herbelin, J. M., and Cohen, N. (1973). *Chem. Phys. Lett.* **20**, 605.
Herbelin, J. M., and Kwok, M. A. (1977). *In* " Electronic Transitions Lasers II " (L. E. Wilson, S. N. Suchard and J. I. Steinfeld, eds.), MIT Press, Cambridge, Massachusetts.
Hippler, H., Wendt, H. R., and Hunziker, H. E. (1978). *J. Chem. Phys.* **68**, 5103.
Hoffman, H., and Leone, S. R. (1978). *J. Chem. Phys.* **69**, 641.
Horiguchi, H., and Tsuchiya, S. (1975). *J. Chem. Soc., Faraday Trans. 2* **71**, 1164.
Horiguchi, H., and Tsuchiya, S. (1977). *Bull. Chem. Soc. Jpn.* **50**, 1657 and 1661.

Horiguchi, H., and Tsuchiya, S. (1979). *J. Chem. Phys.* **70**, 762.
Hotop, H., and Niehaus, A. (1970). *Int. J. Mass Spectrom. Ion Phys.* **5**, 415.
Hotop, H., Lampe, F. W., and Niehaus, A. (1969a). *J. Chem. Phys.* **51**, 593.
Hotop, H., Niehaus, H., and Schmeltekopf, A. L. (1969b). *Z. Phys.* **229**, 1.
Hotop, H., Kolb, E., and Lorensen, J. (1979). *J. Electron Spectrosc. Relat. Phenom.* **16**, 213.
Houston, P. L. (1977). *Chem. Phys. Lett.* **47**, 137.
Howard, J. S., Riola, J. P., Rundel, R. D., and Stebbings, R. F. (1974). *J. Phys. B* **7**, 376.
Huggins, R. W., and Cahn, J. H. (1967). *J. Appl. Phys.* **38**, 180.
Hughes, M. M., Olson, N. T., and Hunter, R. (1976). *Appl. Phys. Lett.* **28**, 81.
Huie, R. E., and Herron, J. T. (1973). *Int. J. Chem. Kinet.* **5**, 197.
Huo, W. M., and Dillon, M. A. (1977). *J. Chem. Phys.* **66**, 3588.
Hurst, G. S., and Klots, C. E. (1976). *Adv. Radiat. Chem.* **5**, 1.
Husain, D., and Donovan, R. J. (1970). *Adv. Photochem.* **8**, 1.
Husain, D., and Kirsch, L. J. (1971a). *Trans. Faraday Soc.* **67**, 2886.
Husain, D., and Kirsch, L. J. (1971b). *Trans. Faraday Soc.*, **67**, 3166.
Husain, D., and Kirsch, L. J. (1974). *J. Photochem.* **2**, 297.
Husain, D., and Littler, J. G. F. (1973). *J. Chem. Soc. Faraday Trans.* 2 **69**, 842.
Husain, D., and Norris, P. E. (1978). *J. Chem. Soc., Faraday Trans.* 2 **74**, 106 and 1483.
Husain, D., Kirsch, L. J., and Wiesenfeld, J. R. (1972). *Discuss. Faraday Soc.* **53**, 201.
Husain, D., Shantanu, K., Yoring, M. and Young, A. N. (1974). *J. Chem. Soc., Faraday Trans.* 2 **70**, 1721.
Husain, D., Slater, N. K. H., and Wiesenfeld, J. R. (1977). *Chem. Phys. Lett.* **51**, 201.
Iannuzzi, M. P., and Kaufman, F. (1979). *J. Am. Chem. Soc.* **101**, 4003.
Inatsugu, S., and Holmes, J. R. (1975). *Phys. Rev. A* **11**, 26.
Ishii, S., and Ohlendorf, W. (1972). *Rev. Sci. Instrum.* **43**, 1632.
Izod, T. P. J., and Wayne, R. P. (1968). *Proc. R. Soc. London, A* **308**, 81.
Jacox, M. E. (1976). *Chem. Phys.* **12**, 51.
Jacox, M. E. (1978). *Chem. Phys. Lett.* **54**, 176.
James, T. C. (1971a). *J. Chem. Phys.* **55**, 4118.
James, T. C. (1971b). *J. Mol. Spectrosc.* **40**, 545.
Jiminez, E., Campos, J., and Sanchez, del Rio, C. (1974). *J. Opt. Soc. Am.* **64**, 1009.
Johns, J. W. C., McKellar, A. R. W., and Riggin, M. (1977). *J. Chem. Phys.* **66**, 3962.
Johnson, C. E. (1972). *Phys. Rev. A* **5**, 1026.
Jones, I. T. N., and Bayes, K. D. (1972). *J. Chem. Phys.* **57**, 1003.
Judeikis, H. S., and Wun, M. (1978). *J. Chem. Phys.* **68**, 4123.
Kawasaki, M., Hirata, Y., and Tanaka, I. (1973). *J. Chem. Phys.* **59**, 648.
Kennealy, J. P., Del Greco, I. P., Caledonia, G. E., and Green, B. D. (1978). *J. Chem. Phys.* **69**, 1574.
Kenner, R. D., Ogryzlo, E. A., and Turley, S. (1979). *J. Photochem.* **10**, 199.
Kernaham, J. A., and Pang, P. H-L. (1975a). *Can. J. Phys.* **53**, 455.
Kernaham, J. A., and Pang, P. H.-L. (1975b). *Can. J. Phys.* **53**, 1114.
King, D. L., and Setser, D. W. (1976). *Annu. Rev. Phys. Chem.* **27**, 407.
Kley, D., Lawrence, G. M., and Stone, E. I. (1977). *J. Chem. Phys.* **66**, 4157.
Kligler, D. J., Pummer, H., Bischel, W. K., and Rhodes, C. K. (1978). *J. Chem. Phys.* **69**, 4652.
Kohse-Hoinghaus, K. and Stuhl, F. (1978). *Ber. Bunsenges. Phys. Chem.* **82**, 828.
Kolts, J. H., and Setser, D. W. (1978a). *J. Chem. Phys.* **68**, 4848.
Kolts, J. H., and Setser, D. W. (1978b). *J. Phys. Chem.* **82**, 1766.
Kolts, J. H., Brashears, H. C., and Setser, D. W. (1977). *J. Chem. Phys.* **67**, 2931.
Kolts, J. H., Velazco, J. E., and Setser, D. W. (1978). *J. Chem. Phys.* **69**, 4357.
Kolts, J. H., Velazco, I. E., and Setser, D. W. (1979). *J. Chem. Phys.* **71**, 1247.

Krause, L. (1975). *Adv. Chem. Phys.* **28**, 267.
Krause, H. F., Johnson, S. G., Datz, S., and Schmidt-Bleek, F. K. (1975). *Chem. Phys. Lett.* **31**, 577.
Krenos, J., and Bel Bruno, J. (1977). *Chem. Phys. Lett.* **49**, 447.
Kunsch, P. L. (1978). *J. Chem. Phys.* **68**, 4564.
Kwok, M. A., and Cohen, N. (1974). *J. Chem. Phys.* **61**, 5221.
Langaar, H. L. (1942). *J. Appl. Mech.* **9**, A-55.
Lawrence, G. M. (1971). *Chem. Phys. Lett.* **9**, 575.
Lawton, S. A., and Phelps, A. V. (1978). *J. Chem. Phys.* **69**, 1055.
Lawton, S. A., Novick, S. E., Broida, H. P., and Phelps, A. V. (1977), *J. Chem. Phys.* **66**, 1381.
LeCalvé, J., and Bourène, M. (1973). *J. Chem. Phys.* **58**, 1446 and 1452.
LeCalvé, J., Gutcheck, R. A., and Dutuit, O. (1977). *Chem. Phys. Lett.* **47**, 470.
Lee, F. W., and Collins, C. B. (1976). *J. Chem. Phys.* **65**, 5189.
Lee, F. W., and Collins, C. B. (1977). *J. Chem. Phys.* **67**, 2799.
Lee, F. W., Collins, C. B., Pitchford, L. C., and Deloche, R. (1978). *J. Chem. Phys.* **68**, 3025.
Lee, L. C., and Slanger, T. G. (1978). *J. Chem. Phys.* **69**, 4053.
Lee, H. U., and Zare, R. N. (1976). *J. Chem. Phys.* **64**, 431.
Lee, W., and Martin, R. M. (1976). *J. Chem. Phys.* **64**, 678.
Lefebvre-Brion, H., and Guérin, F. (1968). *J. Chem. Phys.* **49**, 1446.
Leichner, P. K., Cook, J. D., and Luerman, S. J. (1975). *Phys. Rev. A* **12**, 2501.
Leone, S. R., and Wodarczyk, F. J. (1974). *J. Chem. Phys.* **60**, 314.
Levron, D., and Phelps, A. V. (1978). *J. Chem. Phys.* **69**, 2260.
Lichten, W., and Wik, T. (1978). *J. Chem. Phys.* **69**, 5428.
Lichten, W., McCusker, M. V., and Vierima, T. L. (1974). *J. Chem. Phys.* **61**, 2200.
Lilly, R. A. (1976). *J. Opt. Soc. Am.* **66**, 245.
Lin, C. L., and Kaufman, F. (1971). *J. Chem. Phys.* **55**, 3760.
Lindemann, T. G., and Wiesenfeld, J. R. (1977). *Chem. Phys. Lett.* **50**, 364.
Lindinger, W., Schmeltekopf, A. L., and Fehsenfeld, F. C. (1974). *J. Chem. Phys.* **61**, 2890.
Lindinger, W., Albritton, D. L., McFarland, M., Fehsenfeld, F. C., Schmeltekopf, A. L., and Ferguson, E. E. (1975). *J. Chem. Phys.* **62**, 4101.
Lindinger, W., Albritton, D. L., and Fehsenfeld, F. C. (1979). *J. Chem. Phys.* **70**, 2038.
Liu, K., and Parsons, J. M. (1976). *J. Chem. Phys.* **65**, 815.
Lofthus, A., and Krupenie, P. H. (1977). *J. Phys. Chem. Ref. Data* **6**, 113.
Loh, L. C. H., and Herm, R. R. (1976). *Chem. Phys. Lett.* **38**, 263.
McDermott, W. E., Lotz, K. E., DeLong, M. L., and Thomas, D. M. (1977). *In* "Electronic Transition Lasers II" (L. E. Wilson, S. N. Suchard, and J. I. Steinfeld, eds.), p. 302. MIT Press, Cambridge, Massachusetts.
McDermott, W. E., Pchelkin, N. R., Benard, D. J., and Bousek, R. R. (1978). *Appl. Phys. Lett.* **32**, 469.
McDonald, J. R., Miller, R. G., and Baronavski, A. P. (1977). *Chem. Phys. Lett.* **51**, 57.
McDonald, J. R., Miller, R. G., and Baronavski, A. P. (1978). *Chem. Phys.* **30**, 133.
McGowan, J. W., Kummler, R. H., and Gilmore, F. R. (1975). *In* "The Excited State in Chemical Physics" (J. W. McGowan, ed.), p. 379, Wiley, New York.
McKenney, D. J., and Dubinsky, R. N. (1977). *Chem. Phys.* **26**, 141.
Madhavan, V., Lichtin, N. N., and Hoffman, M. Z. (1973). *J. Phys. Chem.* **77**, 875.
Malins, R. J., and Setser, D. W. *J. Chem. Phys.* (to be published).
Mandl, A., and Ewing, J. J. (1977). *J. Chem. Phys.* **67**, 3490.
Marcoux, P. J., Piper, L. G., and Setser, D. W. (1977). *J. Chem. Phys.* **66**, 351.
Martin, L. R., Cohen, R. B., and Schatz, J. (1976). *Chem. Phys. Lett.* **41**, 394.
Masanet, J., Gillen, A., and Vermeil, C. (1974a). *Chem. Phys. Lett.* **25**, 346.

Masanet, J., Gillen, A., and Vermeil, C. (1974b). *Int. J. Photochem.* **3**, 417.

Masanet, F., Lalo, C., Durand, G., and Vermeil, C. (1978). *Chem. Phys.* **33**, 123.

Meierjohann, B., and Vogler, M. (1978). *Phys. Rev. A* **17**, 47.

Meyer, J. A., Setser, D. W., and Stedman, D. H. (1970). *J. Phys. Chem.* **74**, 2238.

Meyer, J. A., Klosterboer, D. H., and Setser, D. W. (1971). *J. Chem. Phys.* **55**, 2084.

Meyer, J. A., Setser, D. W., and Clark, W. G. (1972). *J. Phys. Chem.* **76**, 1.

Miescher, E., *Int. Rev. Sci.* **3**, and Huber, K. P. (1976). *Phys. Chem. Ser. Two.*

Miller, T. A., Freund, R. S., and Field, R. W. (1976). *J. Chem. Phys.* **65**, 3790.

Möhlmann, G. R., and DeHeer, J. J. (1977). *Chem. Phys. Lett.* **49**, 588.

Morgan, J. E., and Schiff, H. I. (1964). *Can. J. Chem.* **42**, 2300.

Murray, J. R., and Rhodes, C. K. (1976). *J. Appl. Phys.* **47**, 5041.

Myers, G., and Cunningham, A. J. (1977). *J. Chem. Phys.* **67**, 1942, 3353.

Neynaber, R. H., and Magnuson, G. D. (1975). *Phys. Rev. A* **11**, 865.

Noxon, J. F. (1962). *J. Chem. Phys.* **36**, 926.

Noxon, J. F. (1970). *J. Chem. Phys.* **52**, 1852.

Ogren, P. J. (1975). *J. Phys. Chem.* **79**, 1749.

Okabe, H. (1978). "Photochemistry of Small Molecules," Wiley (Interscience), New York.

Ottinger, C., Simonis, J., and Setser, D. W. (1978). *Ber. Bunsenges. Phys. Chem.* **82**, 655.

Petersen, A. B., and Smith, I. W. M. (1978). *Chem. Phys.* **30**, 407.

Phelps, A. V. (1955). *Phys. Rev.* **99**, 1307.

Phelps, A. V. (1959). *Phys. Rev.* **114**, 1011.

Phillips, L. F. (1976). *Chem. Phys. Lett.* **37**, 421.

Piper, L. G., Richardson, W. C., Taylor, G. W., and Setser, D. W. (1972). *Discuss. Faraday Soc.* **53**, 100.

Piper, L. G., Velazco, J. E., and Setser, D. W. (1973). *J. Chem. Phys.* **59**, 3323.

Piper, L. G., Gundel, L. A., Velazco, J. E., and Setser, D. W. (1975a). *J. Chem. Phys.* **62**, 3883.

Piper, L. G., Setser, D. W., and Clyne, M. A. A. (1975b). *J. Chem. Phys.* **63**, 5018.

Piper, L. G., Krech, R. H., and Taylor, R. L. (1978). Interim Progress Report to U.S. Dept. of Energy (COO-4223-4) from Physical Sciences Inc., Woburn, Massachusetts.

Pitchford, L. C., and Deloche, R. (1978). *J. Chem. Phys.* **68**, 1185.

Poirer, R. V., and Carr, R. W. (1971). *J. Phys. Chem.* **75**, 1593.

Powell, H. T., and Ewing, J. J. (1978). *Appl. Phys. Lett.* **33**, 165.

Powell, H. T., Murray, J. R., and Rhodes, C. K. (1975). *Appl. Phys. Lett.* **25**, 730.

Prince, J. F., Collins, C. B., and Robertson, W. W. (1964). *J. Chem. Phys.* **40**, 2619.

Pritt, A. T., and Coombe, R. D. (1976). *J. Chem. Phys.* **65**, 2096.

Richardson, T. H., and Setser, D. W. (1980). *Chem. Phys.* (to be published).

Richardson, W. C., and Setser, D. W. (1973). *J. Chem. Phys.* **58**, 1809.

Roy, C. R., Dreyer, J. W., and Perner, D. (1975). *J. Chem. Phys.* **63**, 2131.

Sadeghi, N., and Nguyen, T. D. (1977). *J. Phys. Lett.* **38**, L283.

Sam, C. L., and Yardley, J. T. (1978). *J. Chem. Phys.* **69**, 4621.

Schaap, A. P. (1976). "Singular Molecular Oxygen," Dowden, Hutchinson & Ross, Strouds-burg, Pennsylvania.

Schearer, L. D. (1974). *Phys. Rev. A* **10**, 1380.

Schmatjko, K. J., and Wolfrum, J. (1978). *Ber. Bunsenges. Phys. Chem.* **82**, 419.

Schmeltekopf, A. L., and Broida, H. P. (1963). *J. Chem. Phys.* **39**, 1261.

Schmeltekopf, A. L., and Fehsenfeld, F. C. (1970). *J. Chem. Phys.* **53**, 3173.

Schofield, K. (1978). *J. Photochem.* **9**, 55.

Setser, D. W., and Richardson, T. H. (1979). (To be published).

Setser, D. W., Stedman, D. H., and Coxon, J. A. (1970). *J. Chem. Phys.* **53**, 1004.

Shaw, M. J., and Jones, J. D. C. (1977). *Int. J. Chem. Kine* **3**, 319.

Shaw, M. J., and Jones, J. D. C. (1977). *J. Appl. Phys.* **14**, 393.

Shemansky, D. (1969). *J. Chem. Phys.* **51**, 689.

Simonaitis, R., and Heicklen, J. (1971). *Int. J. Chem. Kinet.* **3**, 319.

Slafer, W. D., Benard, D. J., and Lee, P. H. (1977). *In* "Electronic Transition Lasers II" (L. E. Wilson, S. N. Suchard, and J. I. Steinfeld, eds.), p. 124. MIT Press, Cambridge, Massachusetts.

Slanger, T. G. (1978). *J. Chem. Phys.* **69**, 4779.

Slanger, T. G., and Black, G. (1974). *J. Chem. Phys.* **60**, 468.

Slanger, T. G., and Black, G. (1976a). *J. Chem. Phys.* **64**, 3763 and 3767.

Slanger, T. G., and Black, G. (1976b). *J. Chem. Phys.* **64**, 4442.

Slanger, T. G., and Black, G. (1978). *J. Chem. Phys.* **68**, 989 and 999.

Slanger, T. G., Wood, B. J., and Black, G. (1973). *J. Photochem.* **2**, 63.

Slanger, T. G., Black, G., and Fournier, J. (1975). *J. Photochem.* **4**, 329.

Small-Warren, N. E., and Chiu, L.-Y. C. (1975). *Phys. Rev. A* **11**, 1177.

Smith, D. J., Setser, D. W., Kim, K. C., and Bogan, D. J. (1977). *J. Phys. Chem.* **80**, 989.

Smith, D. S., and Turner, R. (1963). *Can. J. Phys.* **41**, 1949.

Smith, W. H. (1969). *J. Chem. Phys.* **51**, 521.

Solarz, R. W., Johnson, S. A., and Preston, R. K. (1978). *Chem. Phys. Lett.* **57**, 514.

Stedman, D. H. (1970). *J. Chem. Phys.* **52**, 3966.

Stedman, D. H., and Setser, D. W. (1968). *Chem. Phys. Lett.* **2**, 542.

Stedman, D. H., and Setser, D. W. (1969). *J. Chem. Phys.* **50**, 2256.

Stedman, D. H., and Setser, D. W. (1970). *J. Chem. Phys.* **52**, 3957.

Stedman, D. H., and Setser, D. W. (1971). *Prog. React. Kinet.* **6**,

Stock, M., Smith, E. W., Drullinger, R. E., and Hessel, M. H. (1977). *J. Chem. Phys.* **67**, 2463.

Stock, M., Smith, E. W., Drullinger, R. E., Hessel, M. M., and Pourcin, J. (1978). *J. Chem. Phys.* **68**, 1785 and 4167.

Sung, J. P., and Setser, D. W. (1977). *Chem. Phys. Lett.* **48**, 413.

Sung, J. P., and Setser, D. W. (1978). *Chem. Phys. Lett.* **58**, 98.

Suzuki, K., and Kuchitsu, K. (1978). *Chem. Phys. Lett.* **56**, 50.

Suzuki, K., Nishiyama, I., Ozaki, Y., and Kuchitsu, K. (1978). *Chem. Phys. Lett.* **58**, 145.

Taieb, G., and Broida, H. P. (1976). *J. Chem. Phys.* **65**, 2914.

Tanaka, K., Mackay, G. I., Payzart, J. D., and Bohme, D. K. (1976). *Can. J. Chem.* **54**, 1643.

Taylor, G. W., and Setser, D. W. (1971). *Chem. Phys. Lett.* **8**, 51.

Taylor, G. W., and Setser, D. W. (1973). *J. Chem. Phys.* **58**, 4840.

Taylor, G. W., Coxon, J. A., and Setser, D. W. (1972). *J. Mol. Spectrosc.* **44**, 1087.

Thomas, R. G. O., and Thrush, B. A. (1975). *J. Chem. Soc., Faraday Trans. 2* **71**, 664.

Thomas, R. G. O., and Thrush, B. A. (1977a). *Ber. Bunsenges. Phys. Chem.* **81**, 177.

Thomas, R. G. O., and Thrush, B. A. (1977b). *Proc. R. Soc. London, Ser. A* **356**, 287, 295, and 307.

Thrush, B. A., and Wild, A. H. (1972). *J. Chem. Soc., Faraday Trans. 2*, **68**, 2023.

Touzeau, M., and Pagnon, D. (1978). *Chem. Phys. Lett.* **53**, 355.

Trainor, D. W. (1977a). *J. Chem. Phys.* **66**, 3094.

Trainor, D. W. (1977b). *J. Chem. Phys.* **67**, 1206.

Van Dyck, R. S., Johnson, C. E. and Shugart, H. A. (1971). *Phys. Rev. A* **4**, 1327.

Van Dyck, R. S., Johnson, C. E., and Shugart, H. A. (1972). *Phys. Rev. A* **5**, 991.

Van Itallie, F. J., Doemeny, L. J., and Martin, R. M. (1972). *J. Chem. Phys.* **56**, 3689.

Velazco, J. E., Kolts, J. H., and Setser, D. W. (1976). *J. Chem. Phys.* **65**, 3468.

Velazco, J. E., Kolts, J. H., and Setser, D. W. (1978). *J. Chem. Phys.* **69**, 4357.

Vidaud, P. H., Wayne, R. P., and Yaron, M. (1976a). *Chem. Phys. Lett.* **38**, 306.

Vidaud, P. H., Wayne, R. P., Yaron, M., and von Engle, A. (1976b). *J. Chem. Soc. Faraday Trans.* 2 **72**, 1185.

Vikis, A. C. (1978). *Chem. Phys. Lett.* **57**, 522.

Vikis, A. C., and LeRoy, D. J. (1973a). *Can. J. Chem.* **51**, 1207.

Vikis, A. C., and LeRoy, D. J. (1973b). *Chem. Phys. Lett.* **21**, 103.

Vikis, A. C., and Le Roy, D. J. (1973c). *Chem. Phys. Lett.* **22**, 587.

Wauchop, T. S., and Broida, H. P. (1972a). *J. Quant. Spectrosc. & Radiat. Transfer* **12**, 371.

Wauchop, T. S., and Broida, H. P. (1972b). *J. Chem. Phys.* **56**, 330.

Wayne, R. P. (1969). *Adv. Photochem.* **7**, 311.

Wayne, R. P. (1970). *Ann. N.Y. Acad. Sci.* **171**, 199.

Welge, K. H., and Alkinson, R. (1976). *J. Chem. Phys.* **64**, 531.

West, W. P., Cook, T. B., Dunning, F. B., Rundel R. D., and Stebbings, R. F. (1975). *J. Chem. Phys.* **63**, 1237.

Wieme, W. (1974). *J. Phys. B* **7**, 850.

Wiese, W. L., Smith, M. W., and Glennon, B. N. (1966). "Atomic Transition Probabilities," Vol. 1, NSRDS-NBS-4. U.S. Govt. Printing Office, Washington, D.C.

Wiese, W. L., Smith, M. W., and Miles, B. M. (1969). "Atomic Transition Probabilities," NSRDS-NBS-22. U.S. Govt. Printing Office, Washington, D.C.

Wiesenfeld, J. R., and Wolk, G. L. (1978). *J. Chem. Phys.* **69**, 1805.

Winicur, D. H., and Fraites, J. L. (1974). *J. Chem. Phys.* **61**, 1548.

Winicur, D. H., and Fraites, J. L. (1976). *J. Chem. Phys.* **64**, 89 and 1724.

Woodworth, J. R., and Moos, H. W. (1975). *Phys. Rev. A* **12**, 2455.

Wren, D. J., and Setser, D. W. (1980). *J. Chem. Phys.* (to be published).

Xuan, C. N., Di Stefano, G., Lenzi, M., Margani, A., and Mele, A. (1978a). *Chem. Phys. Lett.* **57**, 207.

Xuan, C. N., Di Stefano, G., Lenz, M., Margani, A., and Mele, A. (1978b). *J. Chem. Phys.* **68**, 959.

Yamada, C., Kawaguchi, K., and Hirota, E. (1978). *J. Chem. Phys.* **69**, 1942.

Yaron, M., and von Engel, A. (1975). *Chem. Phys. Lett.* **33**, 316.

Yee, D. S. C., Steward, W. G., McDowell, C. A., and Brion, C. E. (1975). *J. Electron. Spectrosc. Relat. Phenom.* **7**, 93.

Yokoyama, A., Ueno, T., and Hatano, Y. (1977). *Chem. Phys.* **22**, 459.

Yolles, R. S., and Wise, H. (1968). *J. Chem. Phys.* **48**, 5109.

Young, P. J., Greig, G., and Strausz, O. P. (1970). *J. Am. Chem. Soc.* **92**, 413.

Young, P. J., Hardwidge, E., Tsunashima, S., Grieg, G., and Strausz, O. P. (1974). *J. Am. Chem. Soc.* **96**, 1946.

Young, R. A., and Black, G. (1967). *J. Chem. Phys.* **47**, 2311.

Young, R. A., and St. John, G. A. (1968). *J. Chem. Phys.* **48**, 895 and 898.

Young, R. A., Black, G., and Slanger, T. G. (1968). *J. Chem. Phys.* **49**, 4758.

Zetzsch, C., and Stuhl, F. (1975). *Chem. Phys. Lett.* **33**, 375.

Zetsch, C., and Stuhl, F. (1976a). *Ber. Bunsenges. Phys. Chem.* **80**, 1348.

Zetzsch, C., and Stuhl, F. (1976b). *Ber. Bunsenges. Phys. Chem.* **80**, 1355.

Zetzsch, C., and Stuhl, F. (1977). *J. Chem. Phys.* **66**, 3107.

4

Production and Detection of Reactive Species with Lasers in Static Systems

M. C. LIN and J. R. McDONALD

Chemical Diagnostics Branch
Chemistry Division
Naval Research Laboratory
Washington, D. C.

I. INTRODUCTION

In this chapter the generation and detection of reactive species in static systems will be discussed, with particular emphasis on the use of lasers for this purpose. The reactive species may be atomic, free radicals, or electronically and/or vibrationally excited molecules.

The major difference between static and flow experiments, which have been treated in detail in previous chapters, lies in the manner by which the chemical reactions are initiated. In a flow experiment, the reaction of interest is initiated by means of rapid physical mixing of the reactants. Practically, the time required for mixing should be much shorter than reaction times so that the observed rate data will not be controlled by diffusion or mixing processes. In static experiments, reactants or precursors of reacting species are usually premixed and reactions are initiated externally, either by stationary or by pulsed excitation sources, such as heat, light, and electrical discharge. In many static experiments, reactants are replenished by slowly flowing reaction mixtures through reaction cells. This slow flow, which is kinetically "static," also serves to minimize secondary reactions of reaction products.

In static experiments, the duration of the initiation source used should be short in comparison with the lifetimes of the species of interest. This makes a pulsed initiation source more favorable than a stationary or continuous wave (cw) source because the concentration of the species can be generated at a much higher rate. We will therefore discuss primarily those pulsed techniques which have been proved to be effective for the production of reactive species in static systems. A variety of methods have been used to initiate chemical reactions in static systems. An effective initiation should be intense as well as homogeneous over the whole region excited. The following is a partial list of the techniques which have been effectively used for the initiation of gas phase reactions in static experiments:

(1)   Stationary photolysis with resonance lamps or monochromatized broadband light sources,
(2)   Pulse photolysis with broadband or monochromatized light sources;
(3)   Pulse electric discharge and pulsed microwave excitation;
(4)   Pulse radiolysis and electron beam excitation;
(5)   Pyrolysis by shock waves; and
(6)   Laser photolysis with pulsed or cw, tunable or fixed frequency sources.

Each of these above techniques is characterized by its own literature and many have been discussed in numerous reviews. Because of space limitation, we will discuss in detail only the generation of reactive species by means of stationary and pulsed light sources and the related applications of lasers.

With the exception of photolysis with lasers, resonance lamps, or monochromatized light sources, the techniques listed above are fairly nonspecific in the production of excited and reactive species. As such they are often used to study the emission spectroscopy of relaxed excited state species or, in some cases, for the study of the relaxation of the prepared excited states.

Because of their high electronic excitation efficiencies, pulsed electrical discharge and electron-beam excitation methods have recently been used successfully for pumping various gas laser systems, including chemical lasers, rare gas halide excimer lasers (see Chapter 6 by Djeu) and many other laser systems associated with excited atomic and molecular species, with emissions ranging from the infrared to the vacuum ultraviolet region.

The pulse microwave excitation technique was first employed by Callear and Hedges (1968) for photosensitized reactions of Hg(^{3}P) atoms. This interesting new method could prove to be useful for other metastable atoms. Although the pulse radiolysis technique has been used to study the kinetics of gas phase atomic and radical reactions (Bishop and Dorfman, 1970; Gordon and Mulac, 1975; Hikida et al., 1971), it has been used most success-fully for the kinetics of solvated electrons in solutions (Dorfman and Matheson, 1965).

Rapid homogeneous heating by shock wave compression has been used effectively for the initiation of high temperature gas phase reactions. This method is free from undesirable catalytic effects of vessel surfaces, which have always been a major problem in thermal static experiments. A brief review of the shock tube technique is given in the chapter by Fontijn and Felder.

II. BRIEF REVIEW OF CONVENTIONAL PHOTOLYTIC TECHNIQUES

Inasmuch as the experimental arrangements for the generation and the monitoring of transient species with lasers are very similar to the arrangements for the conventional flash photolysis method, we will begin our discussion with the photolysis method and related experimental techniques.

A. Generation of Reactive Species with Stationary (or Chopped) Light Sources

Photolysis with stationary light sources has long been employed to initiate chemical reactions via electronic excitation in static systems. Since conventional stationary light sources have low powers and thus are less effective than lasers in generating high concentrations of reactive species, experiments are usually carried out over a long period of time. Under these conditions, the rates of reactions of various possible intermediates are in-ferred indirectly from the final product analysis, assuming steady-state conditions for the intermediates involved.

However, by using sensitive detection methods such as modulation and photon counting, photolysis with resonance lamps or monochromatized broadband sources can also be effectively used to follow the production and reaction of transient intermediates. For example, Johnston and co-workers (Morris and Johnston, 1968; Johnston et al., 1969) determined the rates of several reactions involving ClOO and ClO radicals in a photoinitiated Cl_2–O_2 system. Their findings were based on observed phase shifts between

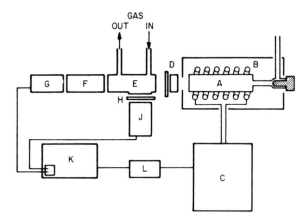

Fig. 1. Block diagram of a phase shift apparatus. A, low-pressure mercury arc; B, six-turn copper coil; C, rf generator, D, composite filter: 4 cm solution of NiSO₄ 210 g/liter and CoSO₄, 60 g/liter, plus 3 mm Corning Glass Filter 7-54; E, Suprasil cell; F, Bausch & Lomb uv monochromator; G, Phillips UPV 150 photomultiplier; H, Corning Glass Filter 3-70; J, Phillips UVP 150 photomultiplier; K, Princeton Applied Research HR8 precision lock-in amplifier; L, Wavetek audio oscillator. [After Atkinson and Cvetanović (1971).]

the square-wave photolyzing light (254 nm) and the radical concentrations monitored by infrared absorption using a Nernst glower. Quinn and co-workers (Parkes *et al.*, 1973; Parkes and Quinn, 1975) applied the same experimental arrangements for the production and detection of CH_3O_2 and $t\text{-}C_4H_9$ radicals.

The phase shift technique is illustrated by the experiments of Cvetanovic and co-workers. The experimental setup shown in Fig. 1 was used for the reaction of oxygen atoms with olefins (Atkinson and Cvetanović, 1971) and for several other series of reactions (Colussi *et al.*, 1975; Singleton *et al.*, 1977). In these experiments, the $O(^3P)$ atom was generated by the Hg-photosensitized decomposition of N_2O and its concentration was monitored by means of the well known NO_2 chemiluminescence ($\lambda > 400$ nm) from the $O + NO$ reaction. The chemical indicator NO was premixed with other reactants and was slowly flowed through the reaction cell. The Hg lamp was powered by a radio frequency (rf) generator similar to that used by Hunziker (1969). The observed phase shift ϕ between the sinusoidal sensitizing radiation (254 nm) and the NO_2 chemiluminescence is related to the chemical removal rate of O atoms by the following equation (Atkinson and Cvetanović, 1971):

$$\tan \phi = 2\pi\omega/(k_2[X] + k_3[NO][M]), \tag{1}$$

where ω, the modulation frequency, is variable up to 8 kHz. k_2 and k_3 are rate constants for O-atom reactions involved in the following simplified but adequate scheme:

$$Hg(^1S_0) + h\nu(254\ nm) \longrightarrow Hg(^3P_1),$$

$$Hg(^3P_1) + N_2O \longrightarrow Hg(^1S_0) + N_2 + O(^3P),$$

$$O(^3P) + X \longrightarrow products,$$

$$O(^3P) + NO + M \longrightarrow NO_2 + M + h\nu\ (\geq 400\ nm).$$

The relationship given by Eq. (1) is nicely demonstrated by the data presented in Fig. 2 for the reaction of O atoms with aldehydes, $X = CH_3CHO$, C_2H_5CHO, $n\text{-}C_3H_7CHO$, and $iso\text{-}C_3H_7CHO$ (Singleton et al., 1977).

This simple modulation technique can be readily extended by using pulsed microwave powered resonance lamps (Phillips, 1971) and chopped cw lasers (Kurylo et al., 1974; Yardley and Moore, 1966, 1968) for studying

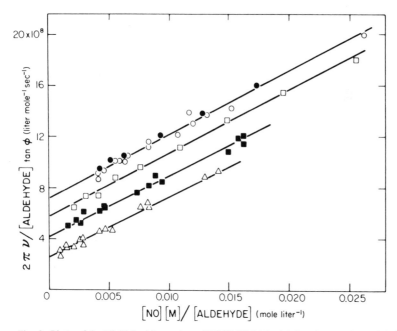

Fig. 2. Plots of $2\pi\Delta$([aldehyde] tan ϕ vs. [NO][M]/[aldehyde] for the reaction of $O(^3P)$ atoms with aldehydes at 298 K. $\triangle$, acetaldehyde; $\blacksquare$, proprionaldehyde; $\square$, butyraldehyde; $\bigcirc$, isobutyraldehyde; $\bullet$, isobutyradelhyde with a 253.7 nm interference filter between the lamp and the reaction cell. [After Singleton et al., (1977). Reproduced with permission of the National Research Council of Canada from Can. J. Chem. **55**, 3321.]

not only chemical reactions, but also the relaxations of electronically and vibrationally excited species. The concentration of the excited species can be monitored by chemiluminescence, fluorescence, or absorption methods previously described for flow experiments. The utility of the modulation technique has also been demonstrated recently by Tsuchiya and co-workers for the study of the relaxation of $Hg(^3P_1)$ atoms by CO, NO, and a number of fluorohydrocarbons. Using this method, they obtained reliable initial product vibrational energy distributions in CO, NO, and HF (Fushiki and Tsuchiya, 1973; S. Tsuchiya, private communication, 1977). Other examples of the application of this useful technique with variations can be found in an excellent review by Phillips (1975).

If the products of photodissociation reactions are formed in their excited states, their formation and decay can be directly monitored by the fluorescence from these states using sensitive photoelectric detectors, usually situated normal to the excitation beam. Since the early fluorescence experiments carried out by Terenin and his co-workers many decades ago (Neuimin and Terenin, 1936; Terenin, 1926; Terenin and Prileshajewa, 1931), numerous systems have been investigated using hydrogen continuum and atomic resonance lamps for the photodissociation of molecules, enabling investigators to study the photochemistry of parent molecules as well as the dynamics of production and relaxation of excited photofragments (Becker and Welge, 1963, 1964, 1965; Carrington, 1964; Dunn et al., 1973; Hampson and Okabe, 1970; Mele and Okabe, 1969; Okabe, 1967; Slanger and Black, 1973, 1974; Slanger et al., 1974; Tanaka et al., 1961; Welge, 1966a,b; Young et al., 1968). The use of various atomic resonance lamps to study the quenching and the sensitized decomposition by excited atomic species has been the subject of numerous reviews (for example, Alkemade and Zeegers, 1971; Calvert and Pitts, 1966; Krause, 1975), and will therefore not be discussed here. Reactions and quenchings of refractory species are discussed elsewhere in this book by Fontijn and Felder, and those species studied by pulse excitation methods will be briefly reviewed in the later part of this chapter.

Okabe (1967) applied the fluorescence technique to measure the threshold incident photon ($\sim 10^6$ photons/sec) energy for the formation of excited photofragments. On the basis of energy conservation, the bond dissociation energy $D_0^0(AB-C)$ in the ABC molecule is related to the threshold incident photon energy $h\nu_0$ and the excess energy $E_0(AB)$ of the excited AB fragment above its ground electronic state by the simple equation: $D_0^0(AB-C) = h\nu_0 - E_0(AB)$, assuming that both reactant and products carry no significant internal and kinetic energies at the threshold dissociation energy. [If the assumption does not hold, only the upper limit $D_0^0(AB) \leq h\nu_0 - E_0(AB)$ can be determined.] Experimentally, the incident photons of varying wavelengths were provided by a hydrogen discharge lamp via a half-meter Seya-

Namioka type vacuum scanning monochromator, and the fluorescence from the excited fragments was measured by a photomultiplier placed normal to the incident light beam. A second photomultiplier aligned colinearly with the incident beam was used to measure the absorption cross sections of the parent molecules at different wavelengths. From this type of experiment, Okabe has obtained valuable information on the photochemistry of many small molecules in the vacuum ultraviolet region and has determined the heats of formation for several important radicals, such as CN (Davis and Okabe, 1968), NH (Okabe, 1970; Okabe and Lenzi, 1967), CS (Okabe, 1972b), SO (Okabe, 1972a), and C_2H (Okabe, 1975; Okabe and Dibeler, 1973).

In addition to the hydrogen continuum and atomic resonance lamps mentioned above, another useful cw and truly continuum source in the extreme uv (5–100 nm) region is synchrotron radiation, emitted by relativistic electrons in a cyclotron storage ring (Gähwiller et al., 1970; Tomboulian and Hartman, 1956). Although the intensity of such radiation is comparatively low,* it has been successfully used to investigate the spectroscopy and photodissociation of a number of molecules because of their large absorption cross sections in this xuv region. For example, the spectroscopy and absorption cross sections of Ar (Carlson et al., 1973), N_2 and O_2 (Lee et al., 1974), and CO and CO_2 (Lee et al., 1975a), N_2O (Lee et al., 1975b), H_2O and D_2O (Ishiguro et al., 1978), NO_2 (Morioka et al., 1976), and a number of other molecules (L. C. Lee et al., 1977a) have been studied with synchotron radiation sources.

B. Generation of Reactive Species with Pulsed Light Sources

Since the first application of the flash photolysis technique by Norrish and Porter (1949; Porter, 1950) and independently by several other groups of investigators (Davidson et al., 1951; Herzberg and Ramsay, 1950; Ramsay, 1952), the kinetics and spectroscopy of numerous transient intermediates occurring in the gaseous and condensed phases have been investigated. Many reviews have already been written for this powerful technique using different versions of flash lamps (McNesby et al., 1971; West, 1976; Willets, 1972). We will briefly describe the four most commonly used experimental arrangements employing flash lamps to generate and, in one setup, to detect reactive transient species. These experimental arrangements are also applicable to the studies that use lasers, instead of flash or resonance lamps.

---

* The number of photons emitted from the 250 MeV electron storage ring of Wisconsin Physical Sciences Laboratory, for example, is about $1–2 \times 10^{10}$ photons sec^{-1} $Å^{-1}$ $mrad^{-1}$ (Gähwiller et al., 1970).

1. Detection by Flash Spectroscopy

Figure 3a shows the schematic diagram of an absorption spectroscopy apparatus first used by Porter (1950) and by Ramsay (1952). The apparatus employs two flash lamps, one of which is used for the high energy (~ 1–$4\,\text{kJ}$) flash photolysis of reaction mixtures, while the other low energy ($< 1\,\text{kJ}$) spectroflash provides a background continuum for probing the transient species generated by the photoflash. The sequence and the delay between the firing of the two flash lamps can be controlled by a suitable delay unit, which may be signaled by the response from a simple photodiode (as shown) or by a pulse generator with variable delay outputs. The spectroscopy of many free radicals has been investigated by means of this technique. The spectroscopic data of many of these free radicals can be found in a review and a monograph by Herzberg (1968, 1971).

More recently, Pimentel, Hand, and their co-workers (Hand et al., 1966; Herr and Pimentel, 1965; Pimentel and Herr, 1964; Tan et al., 1972) have combined the flash photolysis technique with the method of rapid infrared scanning to achieve very fast and highly repetitive scanning in the infrared region. The infrared spectra of many free radicals and transient intermediates (e.g., CF_2, CF_3, CH_3, HOOCl) have been measured by this technique.

By varying the delay of the spectroflash, the system shown in Fig. 3a can also be used to study the kinetics of the formation and decay of transient species. The production and reaction of a number of excited and reactive free radicals that are important to combustion and photochemistry has been investigated using the flash plate–photometry method by Norrish, Porter, and co-workers in their pioneering work. Some of their references related to the production and detection of OH, CN, NH, and ClO are summarized in Table I together with other recent work carried out in a number of laboratories on CH, CH_2, CH_3, NH_2, and several atomic and diatomic species. Using this technique, the relaxation of a number of excited atomic species has also been investigated by Donovan, Husain, and Callear and co-workers. A summary of their work can be found in an extensive review by Donovan and Husain (1970).

2. Detection by Time-Resolved Resonance Absorption

Although the flash plate–photometry technique is very useful for studying the kinetics of many important transient species, the sensitivity and reproducibility of this technique is less than that can be obtained from the time-resolved absorption method, shown schematically in Fig. 3b. This method employs a continuous probing light source (Davidson et al., 1951), particularly a resonance lamp, used in conjunction with a sensitive photoelectric

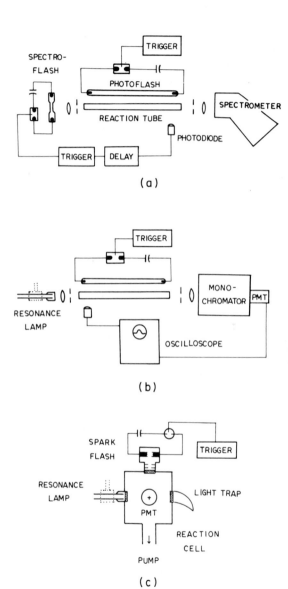

Fig. 3. Generation of reactive species by flash photolysis. (a) Detection by flash spectroscopy; (b) Detection by resonance absorption; (c) Detection by resonance fluorescence.

Table I

Production and Detection of Some Kinetically Important Atomic and Free Radical Species in Static Systems Using Photo-Optical Methods[a]

Species	Production	Detection	References[b]
H	HX(hv, <200 nm), X = Br, I, SH, RCO, alkenyl, —	RA, RF($2^2P_{1/2}-1^2S_{1/2}$, 121.6 nm, Lyman-α), Ch(H + NO)	c
C	$C_3O_2(hv, \sim160$ nm)	PP, RA($3^3P_J-2^3P_J$, 166 nm; $3^1P_1^0-2^1D_2$, 193.1 nm)	d
O(^{3}P)	$NO_2(hv, \geq300$ nm), $O_3(hv, \sim600$ nm), O_2 or $SO_2(hv, \leq200$ nm)	RA, RF($3^3S_1^0-2^3P_J$, 130 nm), Ch(O + CO + NO)	e
O(^{1}D)	$O_3(hv, \approx250$ nm), $N_2O(hv, \approx200$ nm), $CO_2(hv, <160$ nm), $O_2(hv, \leq175$ nm)	RA, RF($3^1D_2^0-2^1D_2$, 115.2 nm), Fl($2^1D_2-2^3P_J$, 630 nm)	f
O(^{1}S)	$N_2O(hv, \sim120$ nm), $CO_2(hv, \sim110$ nm), $O_2(hv, <130$ nm)	Fl($3^1S_0-2^1D_2$, 557.7 nm)	g
N(^{4}S)	$N(^2D) + M$	RF($3^4P_{1/2}-2^4S_{3/2}^0$, 120.1 nm)	h
N(^{2}D)	$N_2O(hv, \sim120-140$ nm)	RA($3^2P_J-2^2D_J^0$, 149 nm)	i
S(^{3}P)	$CS_2(hv, \leq220$ nm), $COS(hv, \leq250$ nm), $S(^1D) + M$	PP, RA, RF($4^3S_1^0-3^3P_J$, 182 nm)	j
F	$F_2(hv, \sim280$ nm), $F_2O(hv, \geq200$ nm), $CF_3I(hv, \leq200$ nm), $SF_6(mhv, 10\ \mu m)$, $O + CF_n(n = 1, 2, 3)$	RF($3^4F_{5/2}-2^2P_{3/2}$, 97.65 nm), Ch(F + RH, F + NO)	k
Cl	$Cl_2(hv, \sim330$ nm), $ClNO(hv, \leq600$ nm), CF_2Cl_2 or $CFCl_3(hv, <200$ nm; nhv, $10\ \mu m$)	RA, RF($4^4P_{5/2}-3^2P_{3/2}^0$, 138.0 nm; $4^4P_{1/2}-3^2P_{1/2}^0$, 139.0 nm)	l
Br	$Br_2(hv, \sim420$ nm), $RBr(hv, \sim200$ nm; nhv, $10\ \mu m$; R = alkyl or perfluoroalkyl)	RA, RF($5^4P_{5/2}-4^2P_{3/2}^0$, 157.7 nm; $5^2P_{1/2}-4^2P_{1/2}^0$, 153.2 nm), Fl($4^2P_{1/2}^0-4^2P_{3/2}^0$, 2.7 μm)	m
I	$I_2(hv, \leq500$ nm), $RI(hv, \leq300$ nm; nhv, $10\ \mu m$; R = alkyl or perfluoroalkyl)	PP, RA($6^2P_{3/2}-5^2P_{1/2}^0$, 206.2 nm; $6^4P_{5/2}-5^2P_{3/2}^0$, 183.0 nm), Fl($5^2P_{1/2}^0-5^2P_{3/2}^0$, 1.3 μm)	n

		...H, LIF (A³Π$_u$–X³Σ$_g^-$, 6?–?758 nm, ? nm/ps band, d³Π$_g$–a³Π$_u$, 340–785 nm, Swan band)	
CH	CH$_4$(hv, ≥105 nm), CHBr$_3$(hv, ≥165 nm), CH$_3$Br, C$_2$H$_2$(nhv, 193 nm)	PP, Fl, LIF(A²Δ–X²Π, 431–490 nm; C²Σ⁺–X²Π, 314 nm)	p
CF	CFBr$_3$(hv, ≥165 nm)	PP, Fl(A²Σ⁺–X²Π, 220–300 nm)	q
CN	XCN(hv, X = Cl, ≤227 nm; Br, ≤246 nm; I, ≤310 nm), RCN or RNC(hv, ≥165 nm; nhv, 10 μm; R = alkyl)	PP, Fl, LIF(A²Π–X²Σ⁺, 440–1500 nm; B²Σ⁺–X²Σ⁺, 340–460 nm)	r
CS	CS$_2$(hv, ≤220 nm)	PP, Fl, LIF(A¹Π–X¹Σ⁺, 240–330 nm)	s
NH	HN$_3$(hv, ≤226 nm), HNCO(hv, <200 nm), NH$_3$(hv, <155 nm)	PP, Fl, LIF(A³Π–X³Σ⁻, 302–368 nm; b¹Σ⁺–X³Σ⁻, 460–408 nm; c¹Π–a¹Δ, 303–365 nm)	t
OH	H$_2$O(hv, ≤186 nm), H$_2$O$_2$(hv, <350 nm), O(¹D) + HX(X = H, Cl, OH, CH$_3$)	RA, RF, PP, LIF(A²Σ⁺–X²Π, 260–410 nm)	u
SO	SO$_2$(hv, <220 nm), SOCl$_2$(hv, <200 nm); O + COS, CS$_2$; S + O$_2$	PP, Fl(A³Π–X³Σ⁻, 240–263 nm; B³Σ⁻–X³Σ⁻, 190–457 nm, etc.)	v
ClO	Cl$_2$O(hv, 220–850 nm), ClO$_2$(hv, 270–400 nm), Cl$_2$/O$_2$(hv, >200 nm), Cl + O$_3$	PP(A²Π–X²Π, 260–310 nm)	w
BrO	Br$_2$/O$_2$(hv, <200 nm), Br + O$_3$	PP(A²Π–X²Π, 290–355 nm)	x
CH$_2$	CH$_2$N$_2$(hv, <500 nm), CH$_2$CO(hv, <230 nm), CH$_2$Br$_2$, CH$_2$I$_2$(hv, ≥165 nm)	PP(b̃–X̃, 141 nm; b̃¹B$_1$–ã¹A$_1$, 500–900 nm; LIF(b̃–ã, 589–596 nm)	y
CHF	CHFBr$_2$(hv, ≤200 nm)	PP(Ã¹A″–X̃¹A′, 450–600 nm)	z
CF$_2$	CF$_2$Br$_2$(hv, ≥165 nm; nhv, 10 μm), CF$_2$Cl$_2$, CF$_2$HCl(nhv, 10 μm)	PP, LIF(Ã¹A$_1$–X̃¹A$_1$, 220–325 nm)	aa
HCO	RCHO(hv, <340 nm, R = H, CH$_3$, C$_2$H$_5$)	PP(Ã²A′–X̃²A′, 460–860 nm) ILA(Ã²A″(0, 9, 0)–X̃²A′(0, 0, 0), 614.1 nm)	bb

(continued)

Table I *(continued)*

Species	Production	Detection	References[b]
C_2O	$C_3O_2(h\nu, <300$ nm$)$	PP($\tilde{A}^3\Pi - \tilde{X}^3\Sigma^-$, 500–900 nm)	cc
HO_2	$X + H_2O_2(X = F, Cl, HO)$, $H + O_2 + M$	A, Fl($\tilde{A}^2A' - \tilde{X}^2A''$, 1.2–2.3 μm; uv A, 220–280 nm)	dd
NH_2	$NH_3(h\nu, \leq210$ nm; $n h\nu, 10\ \mu$m$)$, $NH(^1\Delta) + HX$	PP, Fl, LIF($\tilde{A}^2A_1 - \tilde{X}^2B_1$, 430–900 nm)	ee
CH_3	$CH_3I(h\nu, <300$ nm$)$, $CH_3N_2CH_3(h\nu, \leq400$ nm$)$, $CH_3HgCH_3(h\nu, \geq200$ nm$)$	PP($\tilde{B}^2A_1' - \tilde{X}^2A_2'$, 216 nm; $\tilde{C}^2E'' - \tilde{X}^2A_2'$, 150 nm)	ff

[a] $h\nu$ = single photon dissociation, $n h\nu$ = multiphoton dissociation, RF = resonance fluorescence, Ch = chemiluminescence, Fl = fluorescence, LIF = laser-induced fluorescence, PP = plate photometry (usually obtained from spectroscopic flashes), ILA = intracavity laser absorption.

[b] Due to space limitation, the references quoted here are not inclusive and they are cited mainly for different detection methods used in static experiments. For further references related to the generation and detection of various species, the reader is referred to Calvert and Pitts (1966), Clyne and Nip (Chapter 1), Herzberg (1966, 1971), Kinsey (1977), and Rosen (1970).

[c] Braun and Lenzi (1967); Klemm et al. (1975a); Kurylo (1972a); Kurylo et al. (1970, 1971); Ridley et al. (1972).

[d] Braun et al. (1969); Husain and Kirsch (1971a,b,c).

[e] Atkinson and Pitts (1974, 1977); Davis et al. (1972a,b,c); Donovan et al. (1970c, 1971); Klemm and Stief (1974); Klemm et al. (1975b); Kurylo (1972b); Lee et al. (1976); Slanger and Black (1970); Stuhl and Niki (1971).

[f] Reported works used mainly RA and Fl. Davidson et al. (1976); Donovan et al. (1970d, 1971); Gauthier and Snelling (1975); Gilpin et al. (1971); Heidner et al. (1973); Heidner and Husain (1973, 1974); L. C. Lee et al. (1977b); Streit et al. (1976).

[g] Atkinson and Welge (1972); Black et al. (1975c); Filseth et al. (1970a); Filseth and Welge (1969); Koyano et al. (1975); Welge and Atkinson (1976).

[h] J. H. Lee et al. (1978).

[i] Black et al. (1969); Black et al. (1975a); Husain et al. (1972); Husain et al. (1974); Slanger and Black (1976); Slanger et al. (1971).

[j] Davis et al. (1972a,b); Davis and Klemm (1973); Donovan et al. (1970c,d); Fair et al. (1971); Little et al. (1972).

[k] Rate measurements were made mainly in flow systems: See Clyne and Nip (Chapter 1). To our knowledge, no published work has used the RF method. References on the use of CF_3I and other F-atom sources can be found in Berry (1972), Hsu and Lin (1978), and Wurzberg et al. (1978).

[l] Davis et al. (1970); J. H. Lee et al. (1977); Michael et al. (1977). Other examples of $n h\nu$ generation of Cl atoms can be found in Table III.

Houston (1977); Husain and Wisenfeld (1967).

[a] Most measurements were made in flames, carbon furnaces, and shock tubes. Ballik and Ramsay (1963a,b); Baronavski and McDonald (1977c); Becker et al. (1974a); Matsuda et al. (1972); Norrish et al. (1953, 1954); Tatarczyk et al. (1976); Tsang et al. (1962); Vear et al. (1972).

[b] Most measurements were made in flames and discharge flow systems. The following references are relevant to PP, FI, and LIF detection methods. Anderson et al. (1975); Baronavski and McDonald (1978); Barnes et al. (1973); Bosnali and Perner (1971); Braun et al. (1967); Butler et al. (1979); Clerc and Schmidt (1972); McDonald et al. (1978a); Matsuda et al. (1972); Norrish et al. (1953); Rebbert and Ausloos (1972).

[q] Hsu and Lin (1978); Hsu et al. (1978b); Porter et al. (1965); Thrush and Zwolenik (1963).

[r] Baronavski and McDonald (1977a); Basco (1965); Basco et al. (1963); Basco and Norrish (1965); Cody et al. (1977); Davis and Okabe (1968); Jackson and Cody (1974); Kinsey (1977); Knudtson and Berry (1978); Lesiecki and Guillory (1977); Luk and Bersohn (1973); Mele and Okabe (1969); Sabety-Dzvonik and Cody (1976); Schacke et al. (1973); Tokue et al. (1974-1975).

[s] Bell et al. (1972); G. Black et al. (1977); Donovan et al. (1970a); Hynes and Brophy (1979); Lee and Judge (1975); Okabe (1972b); Smith (1967).

[t] Baronavski et al. (1978); Becker and Welge (1964); Hansen et al. (1976); Husain and Norrish (1963); McDonald et al. (1978b); Masanet et al. (1974-1975); Okabe and Lenzi (1967); Zetzsch and Stuhl (1975).

[u] Atkinson et al. (1975, 1977); Basco and Norrish (1961); Baardsen and Terhune (1972); Davis et al. (1974); Greiner (1967, 1968, 1969, 1970); Kinsey (1977); Lengel and Crosley (1978); Morley and Smith (1972); Norrish and Porter (1952); Ravishankara et al. (1978); Stuhl and Niki (1972); Wang et al. (1975).

[v] Donovan et al. (1969a); Okabe (1972); Smith (1967).

[w] Basco and Morse (1973); Coxon (1976, 1977); Coxon et al. (1976); Coxon and Ramsay (1976); Durie and Ramsay (1958); Edgecombe et al. (1957); Lipscomb et al. (1956); Nicholas and Norrish (1968); Porter and Wright (1953); Watson et al. (1977).

[x] Burn and Norrish (1963); Durie and Ramsay (1958).

[y] Braun et al. (1970); Danon et al. (1978); Herzberg (1961); Herzberg and Johns (1966).

[z] Hsu et al. (1978b); Merer and Travis (1966); Lin (1978).

[aa] Hsu et al. (1978b); King and Stephenson (1977); Mathews (1976); Tyerman (1969); Venkateswarlu (1950).

[bb] Clark et al. (1978); Herzberg and Ramsay (1955); Johns et al. (1963); Reilly et al. (1978).

[cc] Devillers and Ramsay (1971).

[dd] Becker et al. (1974b); Giachardi et al. (1975); Hunziker and Wendt (1974); Kijewski and Troe (1971).

[ee] Baronavski et al. (1978); Campbell et al. (1976a,b); Dressler and Ramsay (1959); Halpern et al. (1975); Hancock et al. (1975); Kroll (1975); Lesclaux et al. (1975, 1976); McDonald et al. (1978b); Vervloet and Merienne-Lafore (1978).

[ff] Basco et al. (1970, 1972); Herzberg (1961); James and Simons (1974); van den Bergh and Callear (1971).

245

detector instead of a photographic plate. Although the method of resonance absorption is not new (Michell and Zemansky, 1934), the first use of this method to monitor kinetically important species such as H, O, Cl, OH, etc., was made rather recently (Tanaka and McNesby, 1962; Myerson et al., 1965; Michael and Weston, 1966; Bishop and Dorfman, 1970; Del Greco and Kaufman, 1962; Donovan et al., 1970c).

The flash photolysis–resonance absorption arrangement has since been used extensively by Donovan, Husain, and co-workers to study the relaxation and the reaction of a number of atomic species in both the ground electronic and the excited electronic states, including $O(^3P_J)$ (Donovan et al., 1970c, 1971), $N(^2D_J, {}^2P_J)$ (Husain et al., 1972), $C(2^3P_J, 2^1D_2)$ (Husain and Kirsch, 1971a,b,c), $I(5^2P_{1/2})$ (Deakin and Husain, 1972; Deakin et al., 1971), $O(2^1D_2)$ (Donovan et al., 1970d, 1971; Heidner et al., 1973; Heidner and Husain, 1974) and many other metallic and nonmetallic species. These and other studies (by Davidovito, Breckenridge, Callear, Trainor and Eywing, and their co-workers), particularly on refractory species, have been summarized by Fontijn and Felder in Section IV,D of their review. More recent work of Husain and co-workers can be found in one of their latest papers on P atoms (Husain and Norris, 1977). Smith and co-workers have recently used the resonance absorption technique to investigate the kinetics of several reactions involving the OH radical, employing an OH-resonance lamp operated by a microwave discharge of a trace amount of H_2O in Ar (Morley and Smith, 1972; Smith and Zellner, 1973, 1974, 1975, 1976). A similar study has also been made by Trainor and von Rosenberg (1974). A more detailed description of the theory and operation of various resonance lamps is given by Clyne and Nip.

The advantage of employing continuous probing sources in a flash photolytic system lies in the fact that the pseudo-first-order kinetics of the probed species, in the presence of an excess amount of a substrate, can be obtained in a single flash and thus reduces the effect of intensity fluctuation in both the spectro- and photoflashes that occurs in the dual flash arrangement. Additionally, if the photoflash can be operated at a reasonably high repetition rate (≥ 1 Hz), signal averaging (by means of a multichannel scalar or a waveform eductor) can significantly enhance the signal-to-noise ratio and thus improve the accuracy of measured kinetic data.

3. Detection by Resonance Fluorescence

Another important application of resonance lamps, used in conjunction with the flash photolysis method, is for the method of induced fluorescence (or photoluminescence). This technique, which was developed many decades ago by Wood (1905), has been used extensively to study the relaxation of electronically excited atoms (Michell and Zemansky, 1934; Krause, 1975). The theory and application of the resonance fluorescence technique using different resonance lamps has been treated in some detail by Clyne and Nip in an earlier chapter. It can be directly applied to the flash-initiated static experiments. Since the technique is extremely sensitive, the energy of the photoflash can be lowered considerably from the values used in the previous two flash setups. Because of this lower energy requirement and the continuous monitoring of species concentration, signal averaging can now be carried out readily, thanks to the enhanced repetition rates. The photoflash is usually generated with spark discharge lamps, powered by < 100 J energy.

The application of this technique to study the reactions of many kinetically important species such as H, OH, O, Cl, etc., was made rather recently. Braun and Lenzi (1967) first investigated the reactions of H atoms with C_2H_4 using a Lyman-α resonance lamp (Okabe, 1965; Davis and Braun, 1968), based on the experimental arrangement shown in Fig. 3c. In this setup, the fluorescence detector (photomultiplier or electron magnetic multiplier used by Braun and Lenzi, 1967) is usually positioned perpendicularly to both the photoflash and the resonance lamps. Extreme care should be taken to eliminate scattered light from both lamps by using light baffles and traps (Wood's horn). As indicated earlier by Clyne and Nip, the resonance fluorescence technique, unlike the resonance absorption method, usually employs highly reversed resonance lamps to obtain greater fluorescence intensity. Since it is unnecessary to make any corrections for the line shape [see comments given earlier by Fontijn and Felder in section II,C of their review on the use of the modified Beer–Lambert law: $\log(I_0/I) = \varepsilon(lc)^\gamma$], the data determined by the resonance fluorescence technique are probably more reliable and accurate than the data obtained by the resonance absorption technique.

Since 1967, many reactions involving H, Cl, O, S, and OH have been investigated in static systems by means of this technique. References for some of these studies are given in Table I. In these studies, the pseudo-first-order decay rates of the species of interest (which were usually generated by highly repetitive low energy flash photolysis as mentioned before) were measured in the presence of an excess amount of a substrate by employing a suitable signal averaging device or other automated data processing system.

4. Detection by Fluorescence

If the reactive intermediates are formed in vibrationally or electronically excited states which emit strongly, their production and decay can be monitored directly by observing their fluorescences from those excited states with sensitive photoelectric detecting systems. Reactive species which are formed in the ground state or other nonemitting states can be sometimes followed indirectly by their chemiluminescent reactions with other reagents, such as $O + NO$, $H + NO$, etc. The experimental arrangement employed in those experiments is essentially the same as that given in Fig. 3c, with no resonance lamp. Stuhl and Niki (1971), for example, used the $O + NO$ chemiluminescence method to study the reactions of $O(^3P)$ atoms with a number of reactants: NO, O_2, and CO. For the $O + CO$ reaction, the CO blue flame emission (~ 300–400 nm) was used directly to monitor the decay of the O atom or the production of CO_2 in the system. This method was recently adopted by Atkinson and Pitts (1974, 1977) to measure the rates of many $O(^3P)$-atom reactions important to atmospheric chemistry. Another example of the use of this technique is von Rosenberg and Trainor's study (1974) of the recombination of O atoms with O_2 in which they measured the infrared fluorescence of O_3 using several bandpass filters to separate emissions from different vibrational modes.

As discussed earlier in conjunction with the application of cw vacuum uv resonance lamps, the fluorescence measurements from excited photofragments can provide important kinetic information not only about their reaction and relaxation processes, but also about the photochemistry of dissociating molecules. By using improved short pulsed light sources (Hundley et al., 1967; Ware, 1971) and fluorescence detecting methods (Bennett, 1960; Bollinger and Thomas, 1961; Yguerabide, 1965), lifetime and relaxation measurements can both be directly made on the nanosecond scale (see, for example, Ware, 1971). Several studies have been made of the relaxation and/or reaction of electronically excited alkali atoms using these improved flash photolysis and fluorescence detection methods (Barker and Weston, 1976; Brus, 1970; Lin and Weston, 1976). Recently, Luk and Bersohn (1973) investigated the relaxation and reaction of the B state of CN by various gases, using a nanosecond vacuum uv flash lamp to dissociate ICN and a photoncounting method similar to that described by Brus (1970) for the detection of the CN (B → X) fluorescence.

Welge and co-workers employed a simple spark discharge lamp to study the photodissociation of simple molecules (O_2, O_3, CO_2, N_2O) and the relaxation (radiative and collisional) of their electronically excited photofragments by a simple fluorescence sampling method (Filseth et al., 1970b). Some of their work is given as follows: the production and quenching of $O(^1S_0)$ (Atkinson and Welge, 1972; Filseth et al., 1970a; Filseth and Welge, 1969; Koyano et al., 1975; Welge and Atkinson 1976), $O(^1D_2)$ (Gilpin et al., 1971), and $O_2(^1\Sigma_g^+)$ (Filseth et al., 1970b). For the relaxation of the $O(^1S_0)$ atom, they observed the 558 nm emission enhancement by Xe, Kr, Ar, N_2, and H_2 in order of decreasing efficiency (Filseth et al., 1970a). Their results

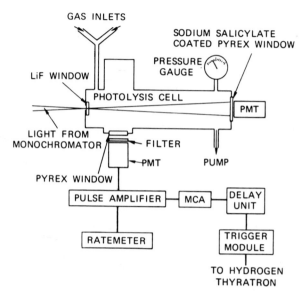

Fig. 4. Schematic diagram of the vacuum uv photolysis–fluorescence apparatus of Black *et al.* (1975a). The light source was a McPherson Model 630 capillary discharge lamp which operated with ~0.5 Torr H_2 at a pulse rate of <400 Hz with a half-width of ~1 μsec.

were similar to those obtained by Hampson and Okabe (1970) using the stationary photolysis method described earlier. This emission enhancement is known to be caused by the formation of excimers which promote the strongly forbidden $^1S_0 \rightarrow {}^1D_2$ transition. In fact, stimulated emissions from both XeO and KrO have been found to occur in the electron-beam excitation of high pressure $Xe-O_2$ and $Kr-O_2$ mixtures (Powell *et al.*, 1974).

Similar studies have been made by Black, Slanger, and co-workers for the quantum yields of production of several excited species using a pulse H_2-discharge lamp. Their experimental arrangement, shown in Fig. 4 is similar to that used by Okabe (1967) for the measurement of incident photon threshold energy employing a cw H_2-discharge lamp, which we described earlier. The measured quantum yields for the production of $O(^1S)$, $N(^2D)$, and $N_2(A^3\Sigma_u^+)$ from the vacuum uv (110–150 nm) photolysis of N_2O (Black *et al.*, 1975a) are summarized in Fig. 5. The production of $O(^1S)$ was monitored by the $^1S_0-{}^1D_2$ emission at 558 nm. The $N(^2D)$ yield was determined by the intensity of the NO β-band emission from NO $(B^2\Pi_r)$ generated by the subsequent reaction,

$$N(^2D) + N_2O \longrightarrow N_2 + NO(B^2\Pi),$$

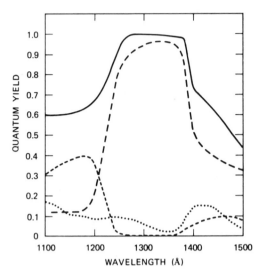

Fig. 5. Quantum yields of $O(^1S)$, $N(^2D)$, and $N_2(A^3\Sigma_u^+)$ from the photodissociation of N_2O between 110 and 150 nm. --, $O(^1S)$; ---, $N(^2D)$; ···, $N_2(A^3\Sigma_u^+)$; —, $O(^1S) + N(^2D) + N_2(A^3\Sigma_u^+)$. [After Black *et al.* (1975a).]

according to an earlier work of Black *et al.* (1969). Similarly, formation of $N_2(A^3\Sigma_u^+)$ was measured by the intensity of the NO γ-band emission from $NO(A^2\Sigma^+)$ generated by the $E \rightarrow E$ transfer process (Callear and Smith, 1963; Callear and Wood, 1971):

$$N_2(A^3\Sigma_u^+) + NO \longrightarrow N_2 + NO(A^2\Sigma^+),$$

where NO was premixed with N_2O and N_2 (buffer gas). Similar measurements have also been carried out for the production of $S(^1S_0)$ from the photodissociation of COS at 110–170 nm (Black *et al.*, 1975b), $CS(A^1\Pi)$ from CS_2 at 105–210 nm (G. Black *et al.*, 1977), and $O(^1D_2)$ from O_2 at 116–177 nm (L. C. Lee *et al.*, 1977b). Additionally, the quenching of the $O(^1S_0)$ and $S(^1S_0)$ atoms by rare gases has been investigated by monitoring the time-resolved fluorescence intensity at 558 and 773 nm, respectively (Black *et al.*, 1975c,d), using a method similar to that used by Welge and co-workers mentioned above.

Recently, a series of elegant experiments on the photodissociation and absolute absorption cross sections of a number of small molecules have been carried out in the vuv region by Judge, Lee, and co-workers. They employed the pulse photolysis and fluorescence detection system (Judge and Lee, 1972) shown in Fig. 6. In this system, the vacuum uv light was generated by

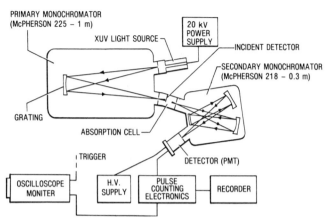

Fig. 6. Schematic diagram of the far uv photolysis–fluorescence apparatus of Judge and Lee (1972). (See the text for a more detailed description of the system.)

a condensed spark discharge through a boron nitride capillary containing argon, air, or nitrogen at a pressure of about 0.02 Torr. The discharge lamp was windowless and could be operated at a rate of 40 Hz, with a pulse duration of about 5 μsec. The output of the lamp system covered a broad range of 40–200 nm. The incident photolysis wavelength was selected by a 1 m McPherson model 225 monochromator. Typical average outputs at the exit slit of the monochromator were on the order of 10^{11} photons/sec. Through differential pumping at the entrance and exit slits, the average pressure in the main chamber of the monochromator could be maintained at about 7×10^{-5} Torr. The fluorescence from the excited photofragments was analyzed by a normal incidence 0.3 m McPherson model 218 monochromator, which viewed the fluorescence through a Suprasil window at a position 5° off the primary dissociation beam to avoid direct interference. The absorption cross sections of parent molecules and the cross sections for the production of various photofragments, including ionic species, have been measured at different incident dissociation wavelengths within the range of 40–200 nm. The molecules studied by Judge, Lee, and co-workers include CO (Judge and Lee, 1972), H_2 and D_2 (Lee and Judge, 1976), CO_2 (Judge and Lee, 1973; Lee and Judge, 1973; Phillips et al., 1976, 1977), COS and CS_2 (Lee and Judge, 1975), and H_2O (L. C. Lee et al., 1978). Figure 7 shows the $CS(A^1\Pi \rightarrow X^1\Sigma^+)$ fluorescence spectra from the photodissociation of COS and CS_2 at 92.3 nm. The observed $CS(A^1\Pi)$ population distributions are approximately Poissonian, indicating the impulsive nature of both dissociation reactions at this wavelength. For the CS_2 dissociation at 124 nm, the $CS(A^1\Pi)$ vibrational population distribution is similar, but vibrationally colder.

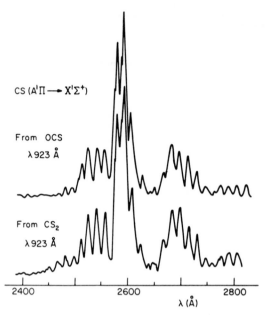

Fig. 7. The CS($A^1\Pi \rightarrow X^1\Sigma^+$) fluorescence spectra produced from the photodissociation of OCS and CS_2 at 92.3 nm. (After Lee and Judge (1975).]

Undoubtedly, this simple but versatile experimental setup can be used very effectively for studying not only the spectroscopy of electronic transitions, but also the dynamics of photodissociation of important molecules such as HCN, ICN, H_2O, CH_4, etc. The studies can be carried out over a broad range of this high energy region for which continuously tunable lasers are still not commercially available.

III. LASER SOURCES

As indicated earlier, the most useful excitation sources for generating specific species (atoms, radicals, or excited states of specific interest) for gas phase kinetic studies are light sources. These sources should have some combination of the following properties:

(1)   a monochromatic output at a desired resonance frequency,

(2)   a high spectral brightness (i.e., a large photon flux per unit spectral bandwidth),

(3)   a reasonable range of tunability to allow wavelength selection,

(4)   for a pulsed excitation, the pulse duration should be shorter than the decay times of the reactive and excited species of interest.

Table II

Summary of Some Chemically Useful Laser Sources[a]

Laser	Major lasing wavelength (μm)	Pumping method	Lasing mechanism	Power (Watt)[b] CW	Power (Watt)[b] Pulse
H_2	0.12 0.16	Electron beam, Blumlein discharge	$C^1\Pi_u - X^1\Sigma_g^+$ $B^1\Sigma_u^+ \to X^1\Sigma_g^+$		10^3 $10^4 - 10^6$
Xe_2	~0.17	Electron beam	$^3\Sigma_u^+ - {}^1\Sigma_g^+$		10^6
ArF^c	0.193	Electron beam, Blumlein, or preionized pulsed discharge $(Ar^* + F_2, NF_3)$	$B^2\Sigma_{1/2}^+ \to X^2\Sigma_{1/2}^+$		10^7
KrF^c	0.249	Initiation methods same as ArF $(Kr^* + F_2, NF_3)$	$B \to X$		10^8
$XeCl^c$	0.308	Initiation methods same as ArF $(Xe^* + HCl)$	$B \to X$		10^7
He–Cd	0.325 0.442	Electrical discharge	$Cd^+, 5s^2(^2D_{3/2}) \to (^1S)5p^2P_{1/2}^0$ $Cd^+, 5s^2(^2D_{5/2}) \to (^1S)5p^2P_{3/2}^0$	0.1 0.1	10^3
N_2	0.337	Initiation methods same as ArF	$C^3\Pi_u \to B^3\Pi_g(0, 0)$		$10^5 - 10^6$
$Ar-N_2$	0.358		$C \to B(0, 1)$		10^7
XeF^c	0.351	Initiation methods same as ArF $(Xe^* + F_2, NF_3)$	$B \to X$		10^7
$He_2^+ - N_2$	0.428	Electron beam, pulsed electrical discharge	$N_2^+, B^2\Sigma_u \to X^2\Sigma_g$		10^7
Coumarin dye	0.4–0.5	Flash photolysis or other optical pumping methods	$S_1 \to S_0$	0.1–1	$10^5 - 10^6$
Ar^+	0.488 0.515	Electrical discharge	$Ar^+, (^3P)4p^2D_{5/2}^0 \to (^3P)4s^2P_{3/2}$ $Ar^+, (^3P)4p^4D_{5/2}^0 \to (^3P)^4s^2P_{3/2}$	1–20	10^2

(continued)

253

Table II (*continued*)

Laser	Major lasing wavelength (μm)	Pumping method	Lasing mechanism	Power (Watt)[b] CW	Pulse
Kr^+	0.568 0.647	Electrical discharge	$Kr^+, (^3P)5p\,^4D^0_{5/2} \rightarrow (^3P)4s\,^2P_{3/2}$ $Kr^+, (^3P)5p\,^4D^0_{5/2} \rightarrow (^3P)5s\,^2P_{3/2}$	1–10	
Rhodamine-6G dye	0.56–0.61	Flash photolysis or other optical pumping methods	$S_1 \rightarrow S_0$	0.1–5	$<10^7$
He–Ne	0.633 3.391	Electrical discharge	$Ne, 5s'[1/2]^0_1 \rightarrow 3p'[3/2]_2$ $Ne, 5s'[1/2]^0_1 \rightarrow 4p'[3/2]_2$	0.01 0.01	10^7
Ruby	0.694	Flash photolysis	$Cr^{3+}/Al_2O_3, {}^2E(\bar{E}) \rightarrow {}^4A_2$		10^9–10^{10}
Nd–glass	1.06	Flash photolysis	$Nd^{3+}/glass, {}^4F_{3/2} \rightarrow {}^4I_{11/2}$		$>10^{11}$
Nd–YAG	1.064	Flash photolysis	$Nd^{3+}/YAG, {}^4F_{3/2} \rightarrow {}^4I_{11/2}$	10^{-2}	10^7
Nd–YAG, freq. doubled	0.532			10–10^2	10^5
Nd–YAG, freq. tripled	0.353				10^4
Nd–YAG, freq. quad.	0.266				10^4
I*	1.315	Flash photolysis, $R_F I + h\nu$ (>200 nm),	$5\,^2P^0_{1/2} \rightarrow 5\,^2P^0_{3/2}$		$>10^9$

		Flash photolysis, electron beam, electrical or microwave discharge, $F + RH$; $R = H$, $CH_3, \cdots$	Vibration–rotation transitions	<10	10^6–10
HCl/DCl	3.6–4.0/4.9–5.2	Initiation methods same as HF, $H + Cl_2$; $Cl + RH(R = I, Br)$	Vibration–rotation transitions	<10	10^3
HBr/DBr	4.0–4.4/5.6–5.9	Initiation methods same as HF, $H + Br_2$; $Br + HI$	Vibration–rotation transitions	<10	10^3
CO	4.8–6.0	Electron beam or electrical discharge of CO–N_2–He mixtures chemical pumping: $O + CX(X = S, H, N, Se)$, $CH + O_2$, NO^d	Vibration–rotation transitions	<10	10^3–10^6
CO₂	9.6, 10.6	Electron beam or electrical discharge of CO_2–N_2–He mixtures	9.6 μm: $CO_2(001) \rightarrow CO_2(020)$ 10.6 μm: $CO_2(001) \rightarrow CO_2(100)$	$<10^2$	10^9

[a] For a more complete list of laser lines and the operational principles of various laser systems see "Handbook of Lasers" (Pressley, 1971) and "Handbook of Chemical Lasers" (Gross and Bott, 1976). There are different nonlinear optical mixing techniques which have been successfully used to generate wavelengths ranging from the near-ir to the vacuum uv region.

[b] Approximate power levels that are readily available for chemical applications under different modes of operation. Informations on commercial laser systems can be found in the 1978 "Buyer's Guide" published by "Laser Focus."

[c] For the operation of these table-top rare gas halide lasers see the following chapter by Djeu.

[d] A summary of different chemical CO lasers is given by Lin (1974).

The development of the laser over the past fifteen years has at least partially provided light sources with the above requirements and has profoundly affected the development of experimental science in the areas of optics, spectroscopy, chemical kinetics, and excited state chemistry. The laser, whether it is a pulsed or a continuously operating device, is a highly directional, monochromatic point source of unparalleled spectral brightness. Continuous wave tunable laser sources are currently available for the visible region, whereas tunable pulsed sources available now cover the range between ~ 215 nm and ~ 1 μm with pulse widths as short as 5×10^{-13} sec (in the visible region) and peak powers greater than 10^{12} W (for the iodine atom laser operating at 1.315 μm and the CO_2 laser at 10.6 μm). Fixed frequency pulsed lasers cover a broad range from the far ir (via optical pumping with ir lasers) to the xuv region (through nonlinear multifrequency mixing). Various optical phenomena and photochemical processes, which would not have been possible without the special properties of the laser, have now been discovered and have been employed for a variety of optical, spectroscopic and chemical applications, such as the conversion of uv and visible photons to vuv photons, and the use of multiple-photon dissociation for isotope enrichment. Some of the most common and important laser sources are summarized in Table II. Further references on lasers and their applications in different areas can be found in many recent topical reviews (Brewer and Moorradian, 1974; Kimel and Speiser, 1977; Moore, 1974; Shimoda, 1976).

IV. GENERATION OF REACTIVE SPECIES WITH LASERS

A. Visible and Ultraviolet Photodissociation Processes

High energy photodissociations can give rise to electronically excited or metastable fragments, depending on the energetics and group theoretical symmetry and spin conservation restrictions on the processes. Vacuum uv photolysis with continuum sources and monochromators, as discussed earlier, can yield information about the threshold energy for the appearance of emissive fragments, in turn yielding information about bond strengths, etc. However, because of the high energies required to break bonds and produce energetic fragments, relatively few of these studies have been carried out with laser photolysis.

On the other hand, there are several questions that are difficult or impossible to answer in static systems using classical photodissociation sources, for example:

(1)   In a given photodissociation process, how is the excess energy partitioned among the various internal energy states (electronic, vibrational, and rotational) of the fragments?

(2)   How does this energy partitioning vary as a function of photolysis energy?

(3)   What are the excited state lifetimes of the radicals in the absence of collisions and what are the rates of important bimolecular quenching processes?

(4)   What is the dependence of secondary reaction rates on the internal energy of the primary fragments?

To address these questions, one needs to produce the primary fragments with short, intense, monochromatic, and tunable pulsed sources. The systems must be studied at pressures approaching collision free conditions. Ultraviolet and vacuum ultraviolet laser sources are ideal for generating these primary species. Studies in static systems employing molecular laser photolysis include: HCHO by Clark *et al.* (1978) and Houston and Moore (1976), glyoxal by Atkinson *et al.* (1978a,b), NO_2 by Creel and Ross (1976), CF_3NO by Spears and Hoffland (1977), ICN by Baronavski and McDonald (1977a), and HN_3 by McDonald and co-workers (1978).

Rare gas halide excimer lasers, particularly the ArF laser, which is commercially available, have provided us with intense photolysis laser sources in the vacuum ultraviolet region. The generation of electronically excited fragments from a variety of parent molecules is straightforward with the 193 nm ArF source. Figure 8 shows a portion of the fluorescence spectrum of the $NH_2(^2A_1 \rightarrow {}^2B_1)$ radical produced by the 193 nm photolysis of NH_3 (Donnelly *et al.*, 1978). These spectra, taken at very low pressures, yield information about the initial partitioning of the excess photodissociation energy into vibrational and rotational excitation in the $NH_2(^2A_1)$ fragment. Figure 9 shows a block diagram of the experimental setup used to measure photofragment fluorescence emission spectra, with the excimer laser source used for photolysis.

Many of the photodissociation reactions we have investigated using intense 193 nm laser photolysis are dominated by multiphoton processes for collimated or mildly focused beams. The NH_2 spectrum in Fig. 8 results from the absorption of a single laser photon by ammonia. At higher laser powers, emission from $NH(A^3\Pi \rightarrow X^3\Sigma)$ is strong. This process results from the absorption of two 193 nm photons. These multiphoton processes are

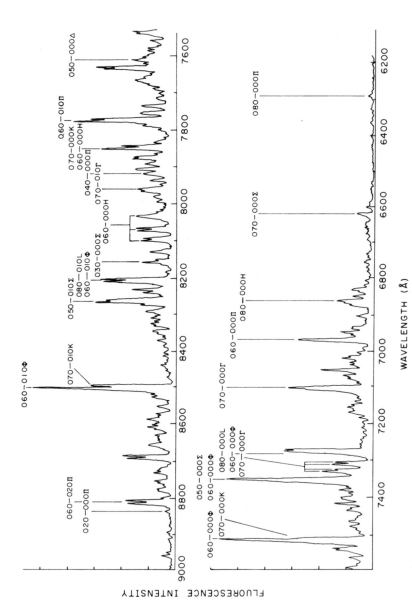

Fig. 8. A portion of the $NH_2(^2A_1 \rightarrow ^2B_1)$ photofragment fluorescence spectrum resulting from dissociation of NH_3 with 193 nm excitation with an ArF excimer laser. The photolysis was carried out under slowly flowing conditions at pressures sufficiently low that the $NH_2(^2A_1)$ fragment was monitored under near collision free conditions during its radiative lifetime of 20–40 μsec. The experimental arrangement is that shown in Fig. 9. [After Donnelly *et al.* (1979).]

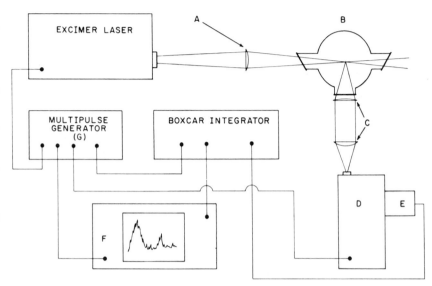

Fig. 9. Block diagram of the experimental arrangement typically used to measure spectra and emission lifetimes of excited state fragments resulting from laser photodissociation of parent species. The laser for these studies, ArF, was characterized by 20 nsec, pulses at 193 nm of up to ≈ 50 mJ/pulse at a repetition rate of <5 Hz. The laser beam was collimated or focused by lens A into the photolysis cell B where it dissociated the gas under study. The resulting fragment emission were focused by a telescope C onto the input slit of 0.6 m monochromator D. The photoelectric detector E presented the dispersed fluorescence signal to a boxcar integrator for signal averaging and ultimate recording on a strip chart recorder F. A multipulse generator G controlled the firing of the laser, the scanning of the monochromator, the time synchronization of the boxcar integrator gate, and the scanning of the strip chart recorder.

extremely efficient and dominant for most photodissociations we have investigated in which the parent molecule has a single photon resonance at 193 nm.

One further example will illustrate the multiphoton dissociation mechanisms which are common in these experiments. Figures 10 and 11 show partial scans of the photofragment fluorescence spectra resulting from the photolysis of acetylene with a mildly focused ArF laser beam. These emissions result from primary fragments and do not represent measurements of collision induced processes. The intensity of the fluorescence from the C_2 Phillips band system $(A^1\Pi_u \to {}^1\Sigma_g^+)$, part of which is shown in Fig. 10, depends upon the square of the ArF laser power. The most likely mechanism involves a two step sequential process with a C_2H intermediate:

$$C_2H_2 + h\nu_1(193\ nm) \longrightarrow H + C_2H,$$
$$C_2H + h\nu_2(193\ nm) \longrightarrow H + C_2({}^1\Pi_u).$$

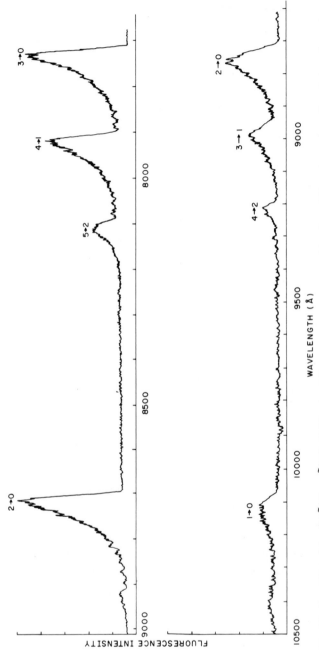

Fig. 10. A portion of the $C_2(\tilde{A}^1\Pi_u \to \tilde{X}^1\Sigma_g^+)$ Phillips band fluorescence spectrum resulting from 193 nm laser photodissociation of acetylene. The experimental arrangement is that shown in Fig. 9. In this case the ArF laser was mildly focused ($\sim 0.5 \times 4$ mm²) in the photolysis cell becuase the $C_2(\tilde{A}^1\Pi_u)$ state requires absorption of two laser photons. [After Donnelly *et al.* (1978).]

260

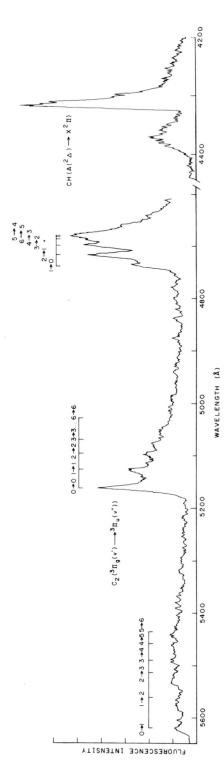

Fig. 11. A continuation of the fragment emission spectra of acetylene in the visible is shown. The conditions are described in Fig. 10. These spectra were measured under pressure conditions such that the fragments were collision free during their radiative lifetimes. In this figure the emissions of $CH(\tilde{A}^2\Delta \rightarrow \tilde{X}^2\Pi)$ fragment and the swan band $C_2(\tilde{d}^3\Pi_g \rightarrow a^3\Pi_u)$ systems are shown.

However, the sequential two-photon absorption mechanism via electronically excited states of C_2H_2 cannot be ruled out. The C_2 fragment shows significant vibrational populations up to $v' = 6$, which is the approximate thermodynamic cutoff based upon the available energy for the above mechanism.

The $CH(A^2\Delta \to X^2\Pi)$ emission shown in Fig. 11 results also from absorption of two laser photons. In this case, the mechanism must involve the absorption of two laser photons by acetylene to produce one $^2\Delta$ and one $^2\Pi$ CH fragement. As shown in Fig. 11, the C_2 Swan Band System ($d^3\Pi_g \to a^3\Pi_u$) is also produced by primary photolysis. Again, the C_2 fragment is produced with a large excess of vibrational energy. The mechanism producing ($d^3\Pi_g$) fragments probably involves three photons of ArF laser as the intensity of the fluorescence depends approximately on the cube of the laser power. Jackson and co-workers (1978), in similar experiments with an ArF photodissociation source, have measured fragment emissions from products of the dissociation of such species as C_2N_2, C_2H_2, C_2H_4, CH_3OH, C_2H_5OH, H_2O, NO_2, and I_2.

Because of the complexity of photofragment processes generated with excimer lasers, one must carefully analyze the results of experiments before postulating mechanisms. The analysis is further complicated by the fact that the absorption spectroscopy of most molecules below 100 nm is unknown and one cannot be clearly guided by state symmetry correlations in predicting the products of these photodissociation processes. Given these precautions, however, it is clear that excimer lasers represent a powerful new tool for the production of very high concentrations of high energy atoms, radicals, and molecules.

B. Infrared Multiphoton Dissociation of Polyatomic Molecules

One of the most interesting and significant discoveries after the advent of the laser was the dissociation of polyatomic molecules by intense infrared lasers (≥ 10 MW/cm^2) via successive absorption of infrared photons. This ir multiphoton dissociation phenomenon was first observed by Isenor and Richardson (1971) in their study of the sparks produced in molecular gases by a high-power pulsed CO_2 laser. Surprisingly, they detected the appearance of luminescence from electronically excited photofragments in the early portion of the laser pulse, within a period shorter than that required for molecular collisions (Isenor et al., 1973). Subsequently, groups of scientists in Russia (Ambartzumian et al., 1974, 1975a,b) and in the United States (Lyman et al., 1975) demonstrated isotopic selectivity in these high power ir laser dissociation processes. This observation clearly indicated that such a dissociation process occurring under the influence of intense electromagnetic radiation is indeed collisionless and nonthermal (Isenor et al., 1973). The most interesting theoretical question concerns how it is possible for molecules to overcome large anharmonicities, often greater than 10 cm^{-1}, and absorb successively many photons of equal energy in the absence of collisions. A considerable amount of experimental (Ambartzumian et al., 1974, 1975a,b, 1976; Bauer and Chien, 1977; Brunner et al., 1977; Campbell et al., 1976a,b; Coggiola et al., 1977; Gandini et al., 1977; Grant et al., 1977; Hudgens, 1978; King and Stephenson, 1977; Koren et al., 1976; Lesiecki and Guillory, 1977; Ritter and Freund, 1976; Sudbø et al., 1978) and theoretical work (Bloembergen, 1975; Cantrell and Galbraith, 1976; Hodgkinson and Briggs, 1976; Larson, 1976; Larson and Bloembergen, 1976; Letokhov and Makarov, 1976; Mukamel and Jortner, 1976; Speiser and Jortner, 1976; Stone et al., 1976) has been carried out to elucidate the mechanisms and the dynamics of these new photochemical reactions. Presently, the most plausible interpretation invokes the assumption that the successive absorption of photons occurs via three vaguely defined regions (Larson and Bloembergen, 1976; Mukamel and Jortner, 1976). In the first region, the intense electromagnetic radiation couples with the mode excited (e.g., the v_3 mode of SF_6) almost resonantly in the first few vibrational levels, aided by rotational compensation (Ambartzumian et al., 1976), the anharmonic splitting present in degenerate overtones (Cantrell and Galbraith, 1976), and to some extent, by the broadening of the absorbed lines due to the strong oscillating electric field of the laser (Bloembergen, 1975; Larson and Bloembergen, 1976; Mukamel and Jortner, 1976; Stone et al., 1976). Once the molecules are excited beyond this region, which is isotopically selective and is the bottleneck of the multiphoton dissociation process, the excited mode starts to communicate with others because of the increasing density of states. The absorption in this second quasi-continuum region, therefore, becomes easier and less selective. In the third region, which lies near the dissociation limit, the density of states is so great that the absorption is essentially nonselective. This mechanism has been corroborated by the results of the SF_6 decomposition reaction, which was studied with two lasers (Ambartzumian et al., 1976). The first laser (942.2 cm^{-1}) was used to excite the v_3 mode of SF_6 with a low constant power density of 4 MW/cm^2. The second laser, which had a higher power density of 58 MW/cm^2, was used to pump the molecule in a broad range of wavelengths between 970 and 1084 cm^{-1}, far out of resonance with the absorbing v_3 mode. Decomposition was detected with the second laser tuned as far as 1084 cm^{-1}. This clearly indicates the presence of the quasi-continuum, onsetting at several levels above the ground vibrational state.

The occurrence of strong state mixings at higher energy levels is also manifested by the results of a series of elegant molecular beam experiments by Y. T. Lee and co-workers (Coggiola et al., 1977; Grant et al., 1977;

Sudbø *et al.*, 1978). Their recoil velocity measurements for the photofragments of a number of molecules show that the observed recoil energy distributions can be quantitatively accounted for by a statistical model based on the RRKM theory of unimolecular reactions (Forst, 1973). For the decomposition of SF_6, the theoretical distribution corresponding to approximately nine excess CO_2 photons (i.e., nine photons in excess of the dissociation threshold) agrees closely with the observed translational energy distribution (Grant *et al.*, 1977). A nice experiment on the multiphoton CO_2 laser dissociation of CF_2Cl_2 and CF_2Br_2 has recently been done by King and Stephenson (1977), who employed a tunable uv laser to measure the vibrational energy distribution of the CF_2 fragment. Their data on the v_2 (bending) mode of CF_2, obtained from the dissociation of CF_2Cl_2 by 929 cm^{-1} CO_2 laser pulses, are shown in Fig. 12. It is worth noting that the observed $CF_2(0, v_2, 0)$

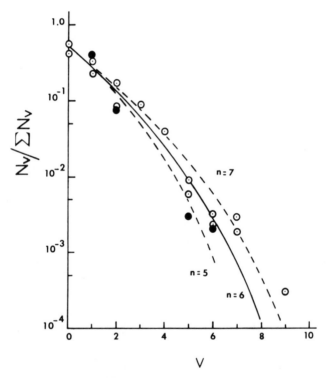

Fig. 12. Normalized vibrational energy distribution for the CF_2 bending mode as a function of excess number of photon n observed in the multiphoton dissociation of CF_2Cl_2. A pulse CO_2 laser operating at the P(34) line of the 10.6 μm band was used. $\bigcirc$, Fluence = 5.7 J/cm^2; $\bullet$, Fluence = 2.7 J/cm^2. Theoretical curves are the results of calculations based on Eq. (2). [After King and Stephenson (1977).]

Table III

Generation of Atomic and Free Radical Species by Multiphoton Dissociation with 10 μm CO_2 Lasers[a]

Molecule	λ_{ex}[b]	Major primary products
BCl_3	10.2–10.9	$BCl_2 + Cl$
NH_3	10.72	$NH_2 + H$
N_2F_4	10.25	$NF_2 + NF_2$
SiF_4	10.6	$SiF_3 + F$
SF_6	10.59	$SF_5 + F$
SF_5Cl	11.05	$SF_5 + Cl$
CH_3OH	9.67	$CH_3 + OH$
CH_3CN	10.89	$CH_3 + CN$
CF_3Cl	9.17	$CF_3 + Cl$
CF_3Br	9.22	$CF_3 + Br$
CF_2Cl_2	9.18, 10.81	$CF_2Cl + Cl$
CF_2Br_2	9.22	$CF_2Br + Br$
CHF_2Cl	9.24	$CF_2 + HCl$
$CHClCF_2$	10.33	$C_2F_2 + HCl$
		$CHCF_2 + Cl$
C_2HCl_3	10.76	$C_2HCl_2 + Cl$
CF_3CF_2Cl	10.22	$CF_3CF_2 + Cl$
		$CF_3 + CF_2Cl$

[a] Most examples listed in this table are the results of Y. T. Lee and co-workers (Sudbo *et al.*, 1978) obtained from molecular beam experiments which are free from secondary product reactions (except in some cases where secondary photodissociation reactions may occur; e.g., $SF_5 + mhv \rightarrow SF_4 + F$). Lee's results show convincingly that these ir laser multiphoton excitations usually result in decomposition via thermokinetically lowest free energy paths.

[b] Approximate excitation wavelengths in micrometers. These molecules usually can absorb many CO_2 laser lines. NH_3, for example, has been shown to undergo dissociation by more than half-dozen CO_2 lines (see, for example, Campbell *et al.*, 1976b).

vibrational population distribution (N_v) can be accounted for by a statistical distribution evaluated from the expression (Lin *et al.*, 1976; Shortridge and Lin, 1976):

$$N_v \propto \Sigma P(nhv - E_v), \tag{2}$$

where n is the number of photons in excess of the dissociation energy, E_v is the energy of the v_2 vibration at the vth level, and $\Sigma P(E)$ the total energy level sums for all vibrational modes of the CF_2Cl_2 molecule at energy $E = nhv - E_v$, except those of the CCl_2 and CF_2 scissoring vibrations. The former vibration becomes the reaction coordinate (for producing Cl_2) and the latter corresponds to the excited v_2 mode of the CF_2 fragment. The theoretical curves with $n = 6$ and 7 agree closely with the experimental data. The results of similar calculations for the $CF_2(0, v_2, 0)$ distributions observed in the decomposition of CF_2Br_2 and CF_2Cl_2 with $1082\ cm^{-1}\ CO_2$ laser pulses indicate that about three to four and five excess photons, respectively, were absorbed by these two dissociating molecules. These results support the conclusion drawn from Lee's molecular beam experiments that the ir multiphoton collisionless dissociation processes occur via the lowest thermokinetic free energy paths, with the total excitation energies $(Nhv = nhn + \Delta H°)$ randomly distributed among all internal degrees of freedom (J. G. Black et al., 1977). Recent results seem to substantiate this conclusion (Braun and Tsang, 1976; Hudgens, 1978; Lussier and Steinfeld, 1977; Richardson and Setser, 1977).

The ir multiphoton dissociation method can therefore be used, instead of uv lasers, for the generation of various free radicals and excited species for low pressure gas phase kinetic studies. Table III lists some of the reactions that produce free radicals which were identified mainly in Lee's molecular beam system (Sudbø et al., 1978).

V. MONITORING OF REACTIVE SPECIES AND REACTIONS WITH LASERS

A. Resonance Absorption Techniques

1. Infrared Laser Absorption Measurements

Although semiconductor (diode) lasers cover a relatively broad range of ir wavelengths (~ 1–$10\ \mu m$), most ir laser resonance absorption measurements made to date have employed molecular gas lasers. This stems partly from the inherent shortcomings of diode lasers, which have relatively low power ($\sim 1\ mW$ for cw operation) and require low operation temperatures ($\lesssim 10\ K$), and partly from the difficulty in making quantitative concentration measurements for transient species, particularly for time-resolved kinetic spectroscopy studies. For the detection of stable molecular species, however, these lasers have been used successfully for applications such as the measurement of trace amounts of pollutants in the atmosphere (see, for example, Ku et al., 1975). However, in this section, we will discuss mainly the use of ir molecular gas lasers for monitoring the concentration of the corresponding molecular species at various vibrational levels. Molecules that have been studied using this method are CO_2 (Djeu et al., 1968), CO (Djeu, 1974; Djeu et al., 1971; Hancock et al., 1971; Houston and Moore, 1976;

Lin and Shortridge, 1974; Powell and Kelley, 1974; Schmatjko and Wolfrum, 1975, 1977), HCl (Menard-Bourcin *et al.*, 1972, 1975), and HF (Hinchen and Hobbs, 1976).

Experimentally, it is more convenient to use a cw than a pulsed laser for the resonance absorption measurement because a cw laser provides a time-resolved measurement in a single shot. We have employed the cw CO laser absorption method to study the dynamics and mechanisms of CO formation in a number of chemical lasers and combustion and energy transfer reactions (Hsu and Lin, 1976; 1977a,b; Lin *et al.*, 1976; Lin and Shortridge, 1974; Shortridge and Lin, 1975, 1976; Umstead and Lin, 1977).

A typical experimental arrangement designed for investigating the relaxation of electronically excited sodium atoms $Na(3^2P)$ by CO is shown in Fig. 13. The cw CO laser (1.7 m active length, sealed by a pair of BaF_2 windows) was powered by the high voltage discharge of a slowly flowing gas mixture containing ~ 0.1 Torr of CO, and 5 Torr each of N_2 and He with a trace amount of NO (used to remove carbon deposit and also to generate an additional amount of vibrationally excited N_2 via the $N + NO$ reaction). To lower the translational/rotational temperature of the laser medium, the tube was continuously cooled with liquid nitrogen. The CO laser output could

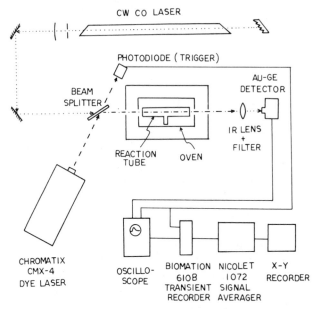

Fig. 13. Continuous wave CO laser resonance absorption apparatus for the study of the $Na(^2P) + CO$ relaxation reaction. [After D.S.Y. Hus and M.C. Lin (unpublished work).]

be line tuned by means of a grating from $\Delta v = 1 \to 0$ to as high as $26 \to 25$, practically covering the exothermicities of most elementary reactions that may generate CO.

Since the extent of absorption depends on the length of the reaction cell, the absorption measurement was usually made in an arrangement similar to that shown in Fig. 3b for a conventional resonance lamp, carried out along the axis of the cell as indicated in the figure. The most important factor that affects the reproducibility of laser resonance absorption measurements is the stability of the laser used. To achieve stable single mode oscillation, considerable care was taken to stabilize the laser at the line center (i.e., the peak of the Doppler profile) by employing an appropriate lock-in stabilizer used in conjunction with a parallel ir detector (Lin and Shortridge, 1974).

The absorption measurement was made by presetting the laser at a desired vibration–rotation transition, measuring first the incident laser intensity (I_0) by chopping the beam at an appropriate rate before reaction, and then the transmitted beam intensity (I) during the course of the reaction, in a time-resolved manner. A typical set of data taken from the relaxation of $Na(3^2P)$ atoms by CO is shown in Fig. 14 and the evaluated relative CO vibrational population distribution at different times in Fig. 15. The vibrational population was evaluated from the following set of simultaneous equations by means of a computer:

$$k_0(v, J) = \frac{8\pi^3}{3kT} \left(\frac{Mc^2}{2\pi kT}\right)^{1/2} |R_v|^2 J F_{vr} \{B_{v+1} N_{v+1} \exp[-B_{v+1} J(J-1)hc/kT]$$

$$- B_v N_v \exp[-B_v J(J+1)hc/kT]\} \tag{3}$$

with $v_{max} > v > 0$; where $k_0(v, J)$, the absorption coefficient at the line center for the P branch transition ($v \to v + 1, J \to J - 1$), is experimentally derived from $\ln(I/I_0)/l$, with l being the cell length. B_v is the rotational constant for the vth vibrational level, $|R_v|^2$ is the rotationless matrix element, and F_{vr} the Herman–Wallis vibration–rotation interaction factor for the $v \to v + 1$ transition. For the CO molecule, $F_{vr} = 1.00$ (Toth et al., 1969), $R_0 = 0.104$ D (Young and Eachus, 1966) and $|R_v|^2/|R_0|^2$ for $v = 1$–11 have been measured by Djeu and Searles (1972). v_{max} is the highest vibrational level observed in the experiment. Other constants used above have the usual meanings. In the above equations, the linewidths for all transitions involved were assumed to be Doppler broadened. For CO, this assumption is generally valid for pressures < 10 Torr for such buffer gases as Ar, He, or SF_6. However, corrections for pressure broadening can be made readily since the Lorentz widths for various gases are known (Williams et al., 1971).

The CO laser probing system shown in Fig. 13 has also been used to study

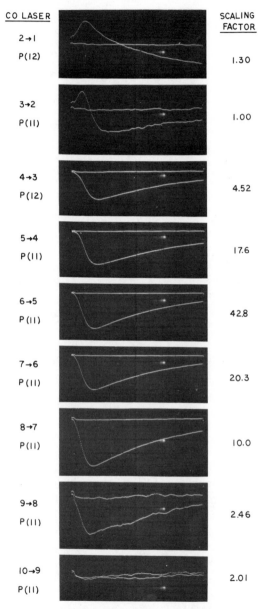

CO LASER		SCALING FACTOR
2→1 P(12)		1.30
3→2 P(11)		1.00
4→3 P(12)		4.52
5→4 P(11)		17.6
6→5 P(11)		42.8
7→6 P(11)		20.3
8→7 P(11)		10.0
9→8 P(11)		2.46
10→9 P(11)		2.01

Fig. 14. Typical set of CO laser absorption traces for the CO ($v \geq 1$) formed in the $Na(^2P_{1/2})$ + CO reaction at 528 K. The CO pressure was 0.10 Torr diluted with 5.05 Torr of Ar. The full recorded time range was 25.5 μsec. The probing transitions are shown in the left-hand side of the traces and the incident laser intensity (I_0) for each transition was proportional to the scaling factor shown in the right.

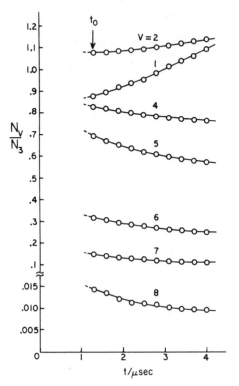

Fig. 15. Time-resolved vibrational population of CO(v) from the Na($^2P_{1/2}$) + CO relaxation reaction at 528 K evaluated from the data shown in Fig. 14. t_0 shown in the figure is the time at which the initial relative population was evaluated using a least squares fitting method. [See Shortridge and Lin (1976).]

the dynamics of the quenching of O($^1D_2$) atoms by CO and the reactions of O(^3P) atoms with a number of radicals and unsaturated hydrocarbon molecules. In these studies, the excitation dye laser and the reaction cell shown in Fig. 13 were replaced by a coaxial flash reaction tube constructed with appropriate brass or stainless joints sealed with O-rings, similar to that described by Smith and Zellner (1973). The inner reaction tube can be replaced readily for suitable optical cutoffs, such as Pyrex ($\lambda \geq 300$ nm), Vycor ($\lambda \geq 210$ nm), quartz ($\lambda \geq 195$ nm), or Suprasil ($\lambda \geq 165$ nm).

Henry and co-workers (Menard-Bourcin et al., 1972, 1975) have employed a pulsed chemical HCl laser to probe the production of vibrationally excited HCl from H + Cl$_2$ and H + ClNO reactions. In these pulsed experiments, absorptions were measured at various time intervals through appropriate delays between the chemical initiation and the probing laser pulses. Similar experiments can be readily carried out for the production of vibrationally excited hydrogen (or deuterium) halide molecules (HBr, HF, DCl, etc.) involved in chemical and relaxation reactions. Data obtained from these measurements are believed to be more reliable than those from fluorescence measurements using cold gas and/or other narrow bandpass filters.

The ir laser resonance absorption measurements can typically detect $\sim 10^{13}$ molecules/cm^3 in a path length of about 1 m in a single flash experiment. Multiple-path measurements aided by signal averaging using repetitive flashes should readily extend the detectability down to 10^{10}–10^{11} molecules/cm^3.

2. Intracavity Absorption Measurements

The intracavity laser absorption method is significantly more sensitive than the extracavity laser resonance absorption technique described in the preceding section. This is because the output of a laser operating near oscillation threshold is very sensitive to small variations in its gain. The power of such a laser is approximately proportional to the net gain of its medium. The sensitivity of intracavity laser absorption has been shown to be many orders of magnitude higher than that of extracavity laser absorption. For example, Hänsch et al. (1972) demonstrated that the intracavity absorption of I_2 using a cw dye laser operating near 600 nm was at least 10^5 times more sensitive than the single-pass absorption measurement made under similar conditions outside the laser cavity.

Experimentally, roughly two types of intracavity measurements have been made: (i) those employing gas lasers in the ir region by Djeu and co-workers (1971; Djeu, 1974; Searles and Djeu, 1973) and by Pimentel and co-workers (Molina and Pimentel, 1972; Poole and Pimentel, 1975; Tablas and Pimentel, 1970), and (ii) those using broadband dye lasers in the uv and visible regions (Hänsch et al., 1972; Keller et al., 1972; Peterson et al., 1971; Schröder et al., 1975). These measurements, although similar in principle, differ considerably in the methods of data acquisition and interpretation.

The ir intracavity absorption technique of Djeu, based on measurements of a quantity known as the laser oscillation range, has been used successfully for the quantitative determination of absolute concentrations of the CO formed at various vibrational levels in the CS_2/O_2 (Djeu et al., 1971) and C_2H_2/O_2 (Searles and Djeu, 1973) low pressure diffusive flames, and in the O + CS reaction occurring under low pressure, fast-flow conditions (Djeu, 1974). The theory of this method has been described in detail by Djeu (1974). Since the method can only be conveniently employed in flow experiments, it will not be discussed in further detail here. As has been demonstrated by Djeu and co-workers (1971, 1973), this technique can be effectively used to probe the production of molecules such as CO and CO_2 in various hydro-carbon flames for detailed kinetic modeling of combustion chemistry.

The technique of Pimentel and co-workers used two identical flash laser tubes operating in tandem in the same optical cavity. A known HCl or HF chemical laser system was employed to probe the vibrational population distribution of an unknown system, based on the method of equal gain (Parker and Pimentel, 1969). Experimentally, two different rotational lines of a selected vibrational transition (say, $v + 1 \rightarrow v$) were made to reach the same oscillation threshold time (i.e., equal gain) by adjusting the pressure of the reaction mixture in one of the laser tubes. Similar experiments were then repeated for a different pair of rotational lines of the same vibrational transition. By use of the gain (or absorption) equation shown in the preceding section, assuming that the rotational and translational temperatures of both systems are thermalized, the data obtained from the above two sets of experiments can then be used to evaluate the population ratio (N_{v+1}/N_v) of the unknown system. This method has been used to study the production of HCl from photodissociation of chloroethylenes (Molina and Pimentel, 1972) and HF from several exothermic reactions.

The theory of intracavity dye laser absorption experiments has been discussed by a number of investigators (Brunner and Paul, 1974; Hänsch et al., 1972; Keller et al., 1972; Tohma, 1975). According to the work of Tohma (1975), who took into account the effects of spatial relaxation and mode competition, intracavity absorption enhancement with multimode operation could be quite large not only in the oscillation threshold region but also in the high gain region so long as mode competition is significant. Measurements with broadband dye lasers have been made for several ions (Eu^{3+}, Ho^{3+}, Pr^{3+}), atoms (Na, Sr, Ba), diatomic (I_2), and polyatomic molecules (HCO, NO_2, NH_2, and other larger molecules). References on these examples can be found in a recent review by Kimel and Speiser (1977).

More recently, Moore, Pimentel and co-workers (Clark et al., 1978; Reilly et al., 1978) employed the intracavity dye laser absorption method to study the reaction of HCO with O_2 and NO. The HCO radical was generated by the photodissociation of H_2CO with a frequency-doubled flashlamp-pumped dye laser (0.6 mJ at 294 nm). The rate of formation and disappearance of HCO was monitored by the intensity of intracavity dye laser absorption of the 614.4 nm [$^2A'(0, 0, 0) \rightarrow {}^2A''(0, 9, 0)$] band at different time intervals between the photodissociation and probing laser pulses. (The wavelength of the probing laser was adjusted by using different concentrations of appropriate laser dyes.) Their experimental arrangement is shown in Fig. 16 and a typical absorption spectrum of the HCO(0, 0, 0) $\rightarrow$ (0, 9, 0) band is shown in Fig. 17.

In order to obtain quantitative measurements, care must be taken to

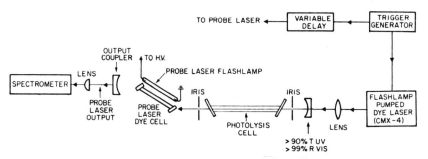

Fig. 16. Apparatus for a laser photolysis–intracavity dye laser absorption experiment. [After Reilly *et al.* (1978).]

avoid the saturation of both intracavity absorption and photographic emulsion. This was done by selecting those lines with weaker absorption (such as the ones bracketed in Fig. 17 with optical densities less than 0.54). The rate constants for the HCO reaction with NO and O_2 were measured to be $(1.4 \pm 0.2) \times 10^{-11}$ and $(4.0 \pm 0.8) \times 10^{-12}$ cm³/sec, respectively, from experiments using 10 Torr of H_2CO and varying amounts of each gas.

The work of Moore, Pimentel, and co-workers described above clearly demonstrates the usefulness of the intracavity broadband laser absorption technique for monitoring transient species which do not fluoresce readily

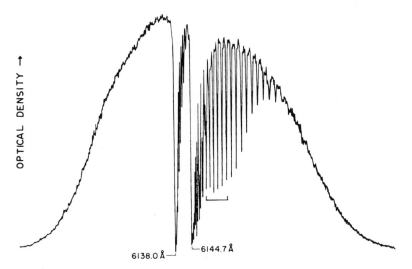

Fig. 17. Microdensitometer tracing of an intracavity dye laser absoroption spectrum of HCO $(0,0,0) \rightarrow (0,9,0)$ produced by laser photolysis of H_2CO. Optical densities of the six single rotational lines bracketed were used for kinetic studies. [After Reilly *et al.* (1978).]

because of either predissociation or small Einstein A coefficients. Radicals such as CH_2, C_3, NCO, C_2O, C_2H, C_2N, CHF, CHCl, NO_3, etc., can be followed with the existing uv/visible dye lasers.

3. Optoacoustic Methods

Almost a hundred years ago, Alexander Graham Bell (1880) discovered the optoacoustic effect by observing that focused intensity modulated light falling on a solid material produces an audible sound. It was soon demonstrated that sound could also be produced by radiation absorption in gases in an enclosed sample cell. Many years later it was shown that the amplitude of the pressure wave produced by optical excitation is proportional to the concentration of the absorbing gases. Subsequently, the optoacoustic effect was applied in the development of the spectrophone. In this application, an arc source was dispersed with a monochromator to scan optical resonances in the absorbing gas (Viengerov, 1940). The optoacoustic effect was shown to be applicable to the measurement of vibrational-to-translational energy transfer rates and the optical frequencies which could be used to generate signals were extended into the microwave region (Gorelik, 1946; Hershberger et al., 1946; Slobodskaya, 1948). Commerical gas analyzers based upon this technique were soon developed. Among the applications of optoacoustic detectors is the measurement of collisional deactivation rates (Cottrell et al., 1966). De Groot and co-workers (1972) demonstrated that the optoacoustic effect can be also used to study photochemical reactions.

Optoacoustic detection has gas phase applications in the area of concentration measurement in mixed gas systems like polluted atmospheres and as a laboratory research device to measure concentrations of species and energy transfer rates. Allen et al. (1977) have also demonstrated that the technique can be used in open seeded turbulent flames to monitor temperature fluctuations and electronic-to-translational energy transfer. Molecular number density measurements on the order of $1.5 \times 10^{14}/cm^3$ have been reported by Patel et al. (1974); Schnell and Fischer (1978) have published detection limits on the order of 0.1 ppm in 760 Torr N_2 using ir laser excitation for such species as phosgene, H_2O, D_2O, HDO, and NO.

While the use of optoacoustic detectors is conceptually quite simple, the design of a system for a specific application involves the consideration of several factors. Visible, ultraviolet, infrared, and even microwave sources may be satisfactory for excitation sources depending upon the resonance transitions to be pumped. Continuous wave or modulated laser sources probably provide the best sensitivity. A decision must be made about whether to intensity modulate or frequency modulate the source, depending on the phenomena one wishes to detect. Careful consideration should be given to the design of the sample cavity. The cell, pumping beam profile, and detector configuration can be designed for nonresonant detection, optically, or acoustically resonant detection.

These considerations may seem to raise confusing engineering questions especially for spectroscopists and photochemists unfamiliar with the particular problems involved. Fortunately, these design parameters have been treated in considerable detail by Rosengren (1975). This paper treats all the design parameters pertinent to the fabrication of an optoacoustic detector for widely varied applications. Rosengren also evaluates the responsivity, noise problems, and overall detector sensitivities which can be expected from optoacoustic detectors. In addition, there is a recently published text edited by Pao (1977) which is devoted entirely to optoacoustic spectroscopy and detection. While the ultimate sensitivity of optoacoustic detection falls short of laser induced fluorescence, long path absorption, and mass spectrometric techniques, the technique is quite sensitive and is applicable in many situations in which these other techniques are not.

B. Resonance Fluorescence Processes

Fluorescence studies of radicals and chemiluminescent reaction processes in static systems is not a separate topic from similar studies in flow systems and molecular beams. The use of laser induced fluorescence techniques in molecular beams, effusive beams, and crossed beams has been recently discussed by Moore (1974), Ezekiel (1974), Zare and Dagdigian (1974), and more recently by Kinsey (1977). However, for the purposes of the present discussion, we will limit ourselves to studies in static systems and in slowly flowing reactors.

1. Detection of Photodissociation Products

Because the primary fragments of photodissociation reactions in most cases are not emissive, probing techniques are required to determine details about the production and energy distributions of these species. We have found it convenient to study the ground state and metastable fragments by laser induced fluorescence techniques. Figure 18 shows a block diagram of the basic photofragment fluorescence excitation spectroscopy instrumentation which we have employed in the study of several systems (Baronavski and McDonald, 1977a, 1978; Baronavski et al., 1978; Donnelly et al., 1978).

In these experiments, primary fragments for study are produced by a photolysis laser. The photolysis beam is set up colinearly with the probing beam from a tunable, flash lamp-pumped dye laser. These two lasers are synchronized in time by pulse generators with the effect that the pumping

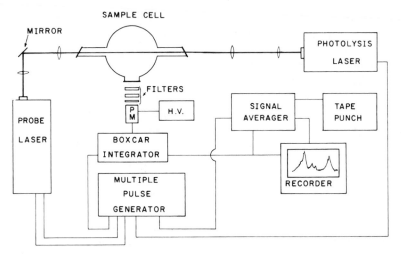

Fig. 18. A block diagram of the experimental arrangement used to measure ground state or metastable photo fragments resulting from primary laser photodissociation processes. The details of the experimental arrangement are given in the text. [After Baronavski and McDonald (1977a).]

and probing pulses can be made simultaneous or can be delayed by any arbitrarily chosen interval. The probe laser wavelength is scanned across electronic resonances of the primary photofragments, thus inducing fluorescence of the fragments. The fluorescence spectral region is isolated with filters to block scattered light from the lasers; the fluorescence light pulse generated by the probe laser is monitored with a photomultiplier and captured by a boxcar integrator operating as a gated integrator. The fluorescence signal is then made available to a 4000 channel signal averager which is stepped from channel to channel by the pulse generators. The scan rates of the probe laser and multichannel analyzer can be arbitrarily set so that each channel corresponds to a set wavelength interval and so that a given number of laser shots are averaged in each channel. This gives a 4000 channel digital display of the photofragment fluorescence excitation spectrum, which can be corrected for variations in the laser power from the photolysis and probe sources.

The laser photodissociation of ICN illustrates the use of fluorescence probes to detect reactive fragments. The photolysis laser at 266 nm can produce several possible fragment combinations, i.e.;

$$\text{ICN} + h\nu \,(266\,\text{nm}) \quad\longrightarrow\quad \text{I}(^2\text{P}_{3/2}) + \text{CN}\,(\tilde{X}^2\Sigma^+, v \le 3)$$
$$\longrightarrow\quad \text{I}(^2\text{P}_{1/2}) + \text{CN}\,(\tilde{X}^2\Sigma^+, v \le 1)$$
$$\longrightarrow\quad \text{I}(^2\text{P}_{3/2}) + \text{CN}\,(\tilde{A}^2\Pi, v = 0).$$

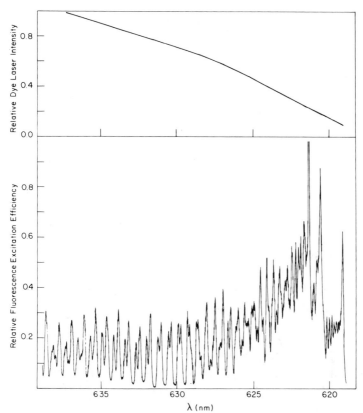

Fig. 19. A portion of the laser induced fluorescence excitation spectrum of the $CN(\tilde{X}^2\Sigma^+)$ fragment produced by the 266 nm photodissociation of ICN. This spectrum is of a region near the band head of the $v'' = 0 \rightarrow v' = 4$ level of the $CN(\tilde{X}^2\Sigma^+ \rightarrow \tilde{A}^2\Pi)$ transition. The ICN pressure in the static cell was ~ 15 mTorr. The photodissociation laser energy was ~ 100 $\mu J/$pulse. The probe laser energy (shown at the top of the figure) varied from ~ 100 $\mu J/$pulse at 620 nm to ~ 1 mJ/pulse at 640 nm. [After Baronavski and McDonald (1977a).]

Prompt fluorescence from $I(^2P_{1/2})$ at 1.315 μm and $CN(\tilde{A}^2\Pi)$ at 600–800 nm are measured to determine yields of electronically excited products. Resonance fluorescence from $I(^2P_{1/2})$ shows that $\sim 60\%$ of the iodine fragments are created in the excited state. The lack of prompt fluorescence from $CN(\tilde{A}^2\Pi)$ state indicates that all CN fragments are created in their electronic ground state.

The $CN(\tilde{X}^2\Sigma^+)$ fragments are probed by laser induced fluorescence using a scanning dye laser. Figure 19 shows a low resolution scan of the $v'' = 0 \rightarrow v' = 4$ region of the spectrum. By computer fitting with synthetic spectra one can determine the lower state ($v'' = 0$) rotational populations. Addition of a buffer gas was found to cool the very hot rotational distribution. This scan reveals excitation of the $v'' = 1 \to v' = 5$ as well as the $v'' = 0 \to v' = 4$ transitions. These measurements, combined with the measurements of the $CN(\tilde{A}^2\Pi)$ and $I(^2P_{1/2})$ state prompt emissions, completely determine the electronic, vibrational, and rotational distributions of all the primary fragments of this dissociation process.

In a variation of this two laser technique, other workers have successfully employed short pulse flash lamp photodissociation combined with laser induced fluorescence excitation spectroscopy to study the fragment energy distributions of related photodissociation processes. For instance, using this technique, Jackson and co-workers (Cody et al., 1977; Jackson and Cody, 1974; Sabety-Dzvonik et al., 1976, 1977; Sabety-Dzvonik and Cody, 1976) have studied the dissociation of CN-containing species such as $NC-C\equiv C-CN$, ICN, BrCN, and C_2N_2. With the use of flash lamps and an appropriate choice of filters, one can vary the photodissociation energy and study its effect on product distributions. However, the use of laser photodissociation sources have advantages over the use of these flash lamps:

(1) The flash lamp sources supply a broad wavelength distribution of photolysis energies;

(2) The flash pulses are so long that primary rotational distributions often are scrambled by collisions.

(3) The long tail of flash lamp emission can often obscure prompt emission from electronically excited primary fragments or cause secondary absorption by the photofragments.

One further example demonstrating laser fluorescence detection of primary metastable photofragments involves the photodissociation of HN_3 at 266 nm. This photodissociation can potentially give rise to NH fragments in four different electronic states, $\tilde{X}(^3\Sigma^-)$, $\tilde{A}(^3\Pi)$, $\tilde{a}(^1\Delta)$, and $\tilde{b}(^1\Sigma^+)$. [The photolysis energy is insufficient to populate the $\tilde{c}(^1\Pi)$ state, as is shown in Fig. 20.] The absence of prompt fluorescence from the $\tilde{A}(^3\Pi)$ and $\tilde{c}(^1\Pi)$ states confirms that these states are not populated by the primary dissociation. Probing the $\tilde{c}(\Pi) \leftarrow b(^1\Sigma^+)$ transition by laser induced fluorescence reveals that this state is not a primary product. Figure 21 shows the laser induced fluorescence excitation spectrum of the $NH\tilde{c}(^1\Pi) \leftarrow \tilde{a}(^1\Delta)$ system, confirming the $(^1\Delta)$ state as a primary fragment. The analysis of the unrelaxed spectrum shows that the rotational distribution in the $v'' = 0$ level is Boltzmann and corresponds to a rotational temperature of ~ 1200 K. Thus, the experiments completely determine the electronic, vibrational, and rotational energy distribution of the primary NH fragment in this dissociation reaction. These experiments demonstrate that it is feasible to make measurements of elec-

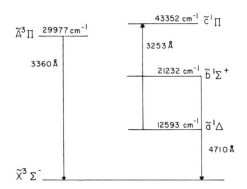

Fig. 20. An energy level diagram of the NH radical. NH radiative lifetimes: $\tilde{A}^3\Pi \to \tilde{X}^3\Sigma^-$, $\tau = 455 \pm 90$ nsec; $\tilde{c}^1\Pi \to \tilde{a}^1\Delta$, $\tau = 480 \pm 90$ nsec; $\tilde{c}^1\Pi \to \tilde{b}^1\Sigma^+$, $\tau = 485 \pm 90$ nsec; $\tilde{d}^1\Sigma \to \tilde{c}^1\Pi$, $\tau = 18 \pm 3$ nsec.

tronically excited fragments by laser induced fluorescence techniques when these species are created in metastable levels.

A few comments should be made concerning the sensitivity of experiments involving laser induced fluorescence probing of primary fragments. When one is working under collision free conditions and probing electronically allowed fragment transitions in the visible region with a laser such as the Chromatix CMX-4, good signal-to-noise ratios can be obtained with fragment concentrations in the laser beam of $\sim 1 \times 10^9$ cm^{-3}. Slightly higher concentrations are required for forbidden transitions such as CN(A–X) or for fragments which absorb in the uv such as OH. The photolysis laser used at 266 nm for the CN and NH work delivers ~ 100 μJ/pulse. For forbidden transitions such as those of HN$_3$ or ICN, sufficient fragment densities to carry out the experiments we have described can be created at parent molecule pressures of 1 – 20 mTorr.

2. Detection of Chemical Reaction Products

By using pulsed lasers for excitation of reactants or for creation of reactive fragments it is possible to study many reaction systems under essentially single collision conditions. For many years chemists have been studying single collision reactions in beams and crossed beams. We will discuss two examples investigated in our laboratory which demonstrate that it is feasible to study primary reaction processes in slowly flowing cells and under static

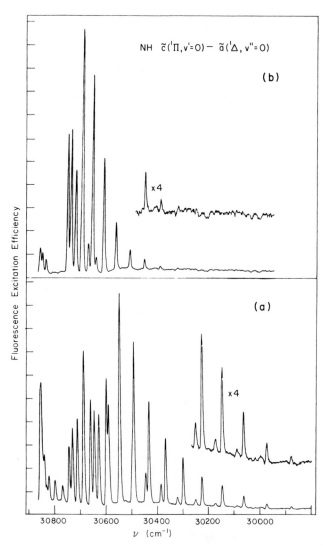

Fig. 21. The fluorescence excitation spectrum of the $HN(a^1\Delta \rightarrow c^1\Pi)$ transition. The fragment resulted from the 266 nm photolysis of HN_3 as described in the text. The experimental arrangement is that shown in Fig. 18. In (a) the photolysis cell contained 18 m Torr of HN_3. In (b) the experimental conditions were identical except that ~ 3 Torr of the buffer gas were added to the photolysis cell. [After Baronavski *et al.* (1978).]

cell conditions. Other groups are carrying out studies of a similar type which involve variations of these applications.

a. *Slowly Flowing Cells* G. K. Smith *et al.* (1979) are studying electronically excited oxygen atom reactions represented by the following equations:

$$O_3 + h\nu_1 \ (266 \text{ nm}) \longrightarrow O(^1D_2) + O_2(^1\Delta_g),$$
$$O(^1D_2) + H_2 \longrightarrow OH(X^2\Pi, v, J) + H, \qquad v \leq 4,$$
$$OH(X^2\Pi, v, J) + h\nu_2 \longrightarrow OH(A^2\Sigma, v', J'),$$
$$OH(^2\Sigma, v', J') \longrightarrow OH(X^2\Pi, v'', J'') + h\nu_3.$$

Localized high concentrations of $O(^1D_2)$ atoms are produced by the laser dissociation of ozone. The OH produced by collisions with H_2 in the electronic ground state can contain up to about four quanta of vibrational energy. The OH concentrations are probed by laser induced fluorescence measurements on the $OH(A \leftarrow X)$ vibronic levels. Figure 22 shows a part of the laser induced fluorescence excitation spectrum of the $v' = 0 \leftarrow v'' = 0$ levels of this transition. These excitation spectra yield a measurement of the internal energy distribution of the OH product. By monitoring the intensity of the OH laser induced fluorescence signal as a function of the delay time between the photolysis and probe lasers, the production of OH at different internal states and therefore the dynamics of $O(^1D_2) + H_2$ can be studied. The experimental setup used in this experiment is very similar to that shown in Fig. 18.

b. *Static Cells* As we have discussed earlier, the 266 nm photolysis of HN_3 gives $>99.8\%$ $NH(^1\Delta)$ in the $v = 0$ vibrational level (Baronavski *et al.*, 1978; McDonald *et al.*, 1977). This is a convenient and very clean source of $NH(^1\Delta)$ radicals. Under single collision conditions, when the radical is allowed to react with the bath gas of HN_3, the following reaction takes place (McDonald *et al.*, 1978b):

$$NH(^1\Delta) + HN_3(A'') \longrightarrow N_3 + NH_2(^2A_1, v \leq 5)$$
$$NH_2(^2A_1) \longrightarrow NH_2(^2B_1) + h\nu \qquad \text{(fluorescence)}.$$

Figure 23 shows the time-resolved production of chemiluminescence from $NH_2(^2A_1)$, produced by the reaction of the primary $NH(^1\Delta)$ fragment. A kinetic analysis of the rise and fall of the chemiluminescence as a function of HN_3 pressure yields a measure of the reaction rate constant.

In variations of this experiment, Baronavski and McDonald studied the reaction of $NH(^1\Delta)$ with various other species, such as HCl, CH_4, C_2H_6, cyclopropane, and cyclohexane. The reactions of these species with $NH(^1\Delta)$ yield nonchemiluminescent products. By monitoring the time-resolved chemiluminescence of the $NH(^1\Delta) + NH_3$ reaction in competition with the

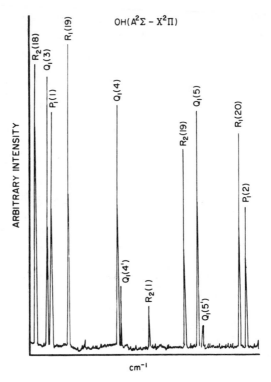

Fig. 22. A portion of a laser induced fluorescence spectrum for the OH radical formed in the $O(^1D) + H_2$ reaction. The $O(^1D)$ atom was generated by the laser photodissociation of O_3 at 266 nm and the OH radical probed by a flash lamp-pumped dye laser (Chromatix CMX-4) doubled to ~ 300 nm. Typical experimental conditions are O_3: 50 mTorr; H_2: 100 mTorr with the delay time between photolysis and probing pulses varying from 1 to 10 μsec. [After Smith et al. (1979).]

dark reactions of NH with these other molecules, we determined the reaction rates of $NH(^1\Delta)$ with the molecules listed. All these experiments are easily carried out in large static bulbs. While the photolysis laser creates relatively high local concentrations of $NH(^1\Delta)$, it takes several hours of irradiation to create significant concentrations of secondary products. Using photolysis laser pulses of ~ 100 μJ, chemiluminescence signals can easily be measured at HN_3 pressures down to about 1 mTorr.

While we have cited specific examples of reaction measurements for both flowing gas–laser induced fluorescence probing and static gas-chemiluminescence probing, the techniques are general and have a wide range of applicability for many types of systems involving both dark reactions and chemi-

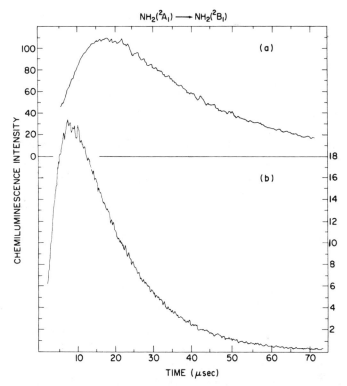

Fig. 23. The time-resolved chemiluminescence of the $NH_2(^2A_1)$ emission resulting from the reaction of $NH(^1\Delta)$ fragments with HN_3. The $NH(^1\Delta)$ fragments were produced by the 266 nm pulsed laser photodissociation of HN_3 at time zero. The HN_3 pressures were (a) 9.90 mTorr and (b) 47.0 mTorr. See the text for details. [After McDonald *et al.* (1978).]

luminescent reactions. Thus, it is plain that single collision reactions can be studied using lasers to determine product energy distributions and reaction rate constants in static and slowly flowing cells.

3. *Resonance Excitation for Relaxation and Spectroscopic Studies*

a. Excitation by ir Lasers   Infrared lasers have been used extensively for studying the dynamics of vibrational energy transfer processes and the reactions of vibrational excited species. Extensive literature related to energy transfer have recently been reviewed by Moore (1973) and Weiz and Flynn (1974). Accordingly, we will discuss only briefly the fundamental aspects of common experimental arrangements.

Most vibrational energy transfer experiments were carried out with a single frequency based on the principle of induced resonance fluorescence shown in Fig. 3c. Since in most of these experiments the excitation was achieved by optical pumping of molecules in the ground vibrational state, care should be taken to avoid absorption of the induced fluorescence by the ground state molecules as well as the saturation of the detecting system (detector and the associated electronic system) by the strong incident excitation light. The former problem can be circumvented by using lower concentrations of reactant molecules and, at the same time, trying to focus the excitation beam closer to the observation window. The problem of saturation can be most readily avoided by the use of proper narrowband interference filters and/or suitable electronic gating and delay devices. For example, in their study of the relaxation of HF ($v = 1$) by various gases, Hancock and Green (1973) employed a pulsed HF laser operating at the $P_{10}(7)$ line and observed HF($1 \rightarrow 0$) fluorescence from the R branch using a narrowband interference filter to avoid the strong scatter light of the incident beam.

b. Excitation by Visible and uv Lasers Prelaser flash experiments were mostly carried out on the microsecond time scale. Such experiments limited the secondary probing techniques to the study of metastable triplet levels. The use of nanosecond flash lamps extended these studies to processes which are about a hundred times faster. Because the pumping intensities are very low, however, the studies are mostly limited to excited state dynamic processes investigated by monitoring the fluorescence of the prepared upper states. Many very elegant and important studies of excited state dynamic processes have been carried out using these fast discharge lamp techniques. Given below are a few typical examples which will acquaint the reader with the use and limitations of the technique. Lee (1977) has recently reviewed the studies on formaldehyde. Benzene has been studied by Selinger and Ware (1970), by Parmenter and Schuyler (1970), and by Spears and Rice (1971), among others. Studies of single vibronic levels of aniline have been reported by Scheps et al. (1974) and by Ware and Garcia (1974). Studies of small ketones are reported by Halpern and Ware (1971) and by Hansen and Lee (1975), and svl (single vibronic level) studies of diketones are reported by Parmenter and Poland (1969) and by Anderson et al. (1973).

The use of dye laser excitation sources rather than nanosecond flash lamps has allowed narrower bandwidth excitation with higher photon fluxes. The study of the excited state dynamics of single vibronic or single rovibronic levels using laser excitation sources has become routine. Dozens of papers have appeared on a wide variety of molecules, significantly advancing the development of the theory of radiationless transition processes. The experimental results of many of these studies have recently been reviewed in a com-

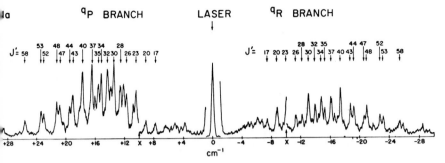

Fig. 24. The fluorescence spectrum of tetrazine resulting from $6a_0^1$ excitation with a single mode Argon ion laser at 530.8 nm. Excitation was in the ^qQ branch. The positive abscissa indicates red shift from the laser. Intensity discontinuities occur at X points when the scan is interrupted to pump off photolysis products. [After Dworetsky *et al.* (1974).]

plete and current report by Avouris *et al.* (1977). The extremely narrow bandwidths ($\lesssim 10^{-4}$ cm^{-1}) available using single mode gas lasers, tunable dye lasers, and diode lasers have created a revolution in ultrahigh resolution excitation studies. In fluorescence experiments, in which it is necessary to excite single rotational levels of single vibrational states of intermediate and large molecules, such sources are indispensable. Figure 24 shows a fluorescence spectrum measured by Dworetsky *et al.* (1974) from a single rovibronic level of a large molecule, tetrazine, excited with a single mode Argon ion laser at 530.8 nm.

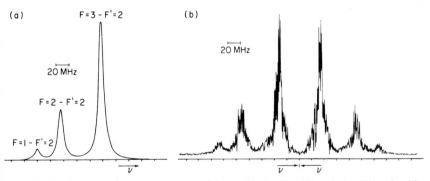

Fig. 25. Part of the hyperfine structure of the sodium D_2 line. The calculated signals with the natural linewidth are shown in (a). The experimental spectrum is shown in (b). The experimental conditions involved an atomic beam collimation ratio of 1 : 400. A triangular sweep mode was used in the averaging system, thus the right part of the experimental curve should be a mirror image of the left. [After Lange *et al.* (1973).]

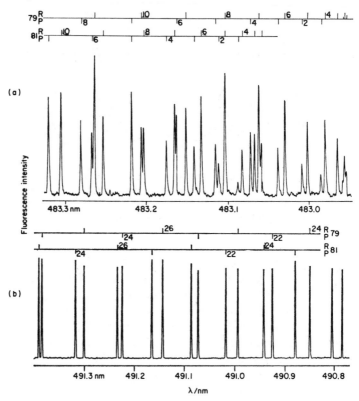

Fig. 26. Fluorescence excitation spectra of BrF showing the total fluorescence intensity as a function of laser wavelength. The excitation laser bandwidth was 0.001 nm. Spectrum (a) shows the 8–0 band near the head. Note the complete resolution of most rotational lines even near the band head. Spectrum (b) shows the middle section of the 7–0 band. Note the simple PR structure of the bands, which with the two equally abundant species ^{79}BrF and ^{81}BrF, gives characteristic groups of four lines. [After Clyne and McDermid (1977).]

Figure 25 shows the fluorescence excitation spectrum measured by Lange *et al.* (1973) of hyperfine components of the Na D_2 line with a 15-MHz-bandwidth tunable dye laser. Here, the spectroscopic resolution is limited only by the natural line width of the resonance absorption processes. There are numerous applications of such high resolution sources in the areas of absorption spectroscopy, fluorescence spectroscopy, svl radiationless transition processes, and photochemical processes such as isotope enrichment. Figure 26, from Clyne and McDermid (1977), shows the application of a relatively narrowband laser (0.001 nm) to the measurement of the fluorescence excitation spectrum near 500 nm of a portion of the 8–0 and 7–0 bandheads of the B → X transition of the interhalogen molecule BrF. Such spectra clearly resolve resonances of the ^{79}Br and ^{81}Br isotopic species. Fluorescence excitation spectra of this quality are equivalent to or better than the best quality absorption spectra obtainable with grating instruments.

Because of the high photon fluxes of nanosecond laser sources, it has become feasible to prepare molecular excited states by the simultaneous absorption of two or more photons without the need for resonant intermediate single photon states. The information derived from such measurements differs from single photon absorption processes because of polarization effects and because the two photon selection rules differ from those of single photon resonances. The development of these techniques has been reported by Monson and McClain (1970, 1972), by McClain (1971, 1973), and by Bray and Hochstrasser (1976). Figure 27 shows typical low resolution one and two photon absorption spectra of biphenyl by Drucker and McClain (1974). The spectroscopic information to be derived from these two processes is different and complementary. The photon resonances can be monitored

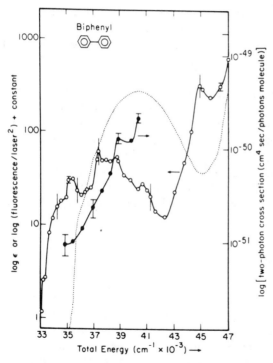

Fig. 27. Contrast between absolute two-photon spectrum and fluorescence-detected two-photon spectrum for biphenyl. Although different in detail, both spectra show significant intensity below the one-photon threshold. ○, fluorescence detection; ●, direct absorption; · · ·, one-photon. [After Drucker and McClain (1974).]

either by absorption or by fluorescence emission occurring from the states prepared by the two photon absorption. For a more extensive discussion of two photon resonance processes, the reader is referred to the recent review by Kimel and Speiser (1977). This review article and the references contained therein will acquaint the experimenter with techniques used in the set up of two photon experiments.

Two photon absorption processes which proceed via resonant intermediate states have a wide range of applications in the preparation of high energy excited states. In atomic spectroscopy, two-step excitation has several potential advantages:

(1)   It is possible to reach states which are not accessible from the ground state because of selection rule prohibitions;

(2)   very high energy states can be reached using high resolution laser sources at energies higher than those available from conventional laser sources;

(3)   one can carry out Doppler-free excitations without the use of molecular beams by using colinear laser beams;

(4)   one can study with high resolution photochemical processes extending into the ionization energy regime.

The study of excited state condensed phase absorption processes with nanosecond lasers has proceeded along the same lines as the much slower flash photolysis techniques. With nanosecond lasers, direct absorption processes of thermalized excited singlet states can be probed. Excited state transitions in aromatic molecules, charge transfer complexes, laser dyes, excimers, and exciplexes have been probed by these stepwise excitation techniques. The review articles by West (1976) and by Kimel and Speiser (1977) have covered recent progress in the field of nanosecond flash photolysis using laser sources.

4. Saturation Spectroscopy

One of the major difficulties involved in probing transient species with all spectroscopic methods lies in the determination of their absolute concentrations. This difficulty has recently been overcome for selected species which have relatively large Franck–Condon factors and known radiative lifetimes. Take the C_2 radical, for example. Baronavski and McDonald (1977b,c) have recently demonstrated that the concentration of C_2 in an C_2H_2/O_2 flame can be measured with reasonable certainty using the technique of laser fluorescence saturation spectroscopy. Their experiment was designed on the basis of Daily's theoretical development of two-level scheme

(Daily, 1976, 1977) for use in atmospheric pressure flames. Baronavski and McDonald extended this model to the intensity region where partial saturation occurred.

For a two-level system, the rates of concentration changes in the upper (N_2) and lower (N_1) levels under the influence of resonance radiation (with intensity I_v) is given by

$$dN_2/dt = -(Q + A_{21} + B_{21}I_v)N_2 + B_{12}I_vN_1,$$
$$dN_1/dt = -N_1B_{12}I_v + (Q_{21} + A_{21} + B_{21}I_v)N_2, \tag{4}$$

where A_{21}, B_{21}, and B_{12} are the Einstein coefficients for levels 1 and 2. Q is the total quenching rate of molecules from level 2 which includes predissociation and other relaxation processes. For an excitation pulse with a duration much longer than $1/Q$, the steady-state approximation leads to the following expression for the fluorescence signal (S) collected 90° to the excitation beam under the near saturation conditions (Baronavski and McDonald, 1977c):

$$S = hv\frac{A_{21}}{8\pi}L\Omega_c A_c\left[N_1^0 - \frac{N_1^0(Q + A_{21})}{B_{21} + B_{12}I_v}\right]. \tag{5}$$

In the above expression, L is the length of the excited region observed, A_c the cross-sectional area of the probe laser beam waist, Ω_c the solid angle of the collection optics, and N_1^0 the total concentration ($N_1 + N_2$) of the species at $t = 0$, assuming $N_1^0 \gg N_2^0$. Using this technique they measured a C_2 concentration of $\sim 5 \times 10^{15}$ cm^{-3} in the reactive zone of an acetylene flame. These experiments give a total upper state quenching rate of $\sim 1.2 \times 10^{12}$ sec^{-1}. In a more recent application, Pasternack et al. (1978) have used aspirating slot burners to probe the concentrations of Na and MgO in an atmospheric acetylene flame. The concentrations of these species were independently measured by classical absorption techniques and the saturated fluorescence measurements and the results are in agreement to within experimental errors. The laser saturated fluorescence signals were shown to vary linearly with concentration for sodium between 10^9 and 10^{12} atoms/cm^3. The practical limits of sensitivity presently are in the region of 10 parts per trillion for atomic species and about 100 parts per trillion for molecular systems with reasonably short radiative lifetimes (i.e., $t_f \leq 500$ nsec.) This technique, in addition to flame systems, should have immediate applicability in a wide range of experimental situations varying from static photolysis cells to molecular beams. The latter case represents a particularly difficult situation for measuring absolute concentrations.

VI. ULTRAFAST PROCESSES

Many primary processes occur on a subnanosecond time scale:

(1) photophysical and photochemical processes in the condensed phase:

(a) vibrational rotational, and orientational relaxation of excited states;

(b) internal conversion among electronic excited states of like spin multiplicity;

(c) resonant energy transfer processes;

(d) recovery of ground state populations in mode-locking dyes;

(e) primary photodissociations and photochemical rearrangements;

(2) dynamic and photochemical processes in the gas phase:

(a) unimolecular photodissociations and intra-molecular energy transfers,

(b) statistical limit internal conversion and intersystem crossing processes,

(c) coherence effects involved in molecules demonstrating statistical limit radiationless transition phenomena.

The term "ultrafast process" is reserved for those phenomena which one must produce or probe on a picosecond or subpicosecond time scale. To obtain this time resolution one must employ mode locked lasers, which are of two types. The first type of laser source employs ion gas lasers to pump dye lasers. The mode-locked devices operating in their most advantageous mode give picosecond or slightly subpicosecond pulses which are typically spaced ~ 10 nsec apart, this separation being the round trip time of the optical cavity. Such laser sources can be tunable across the visible region of the spectrum. For certain applications, the visible pulses can be frequency doubled to give a low energy ultraviolet excitation capability.

The second type of picosecond sources is the more commonly used solid state laser such as the ruby or Nd–glass laser. These devices put out a train of picosecond pulses for each firing of the laser. These pulses can be either used to synchronously pump a dye laser or a single pulse can be extracted from the pulse train for excitation purposes. Because such pumping techniques are easily amenable to amplification, it is possible to obtain picosecond pulses of very high energy. Such high energy pulses are then easily used for frequency doubling or higher order frequency generation processes, making excitation sources available at all wavelengths in the visible and ultraviolet. The repetition rates of the solid state devices are typically 0.1–10 Hz. Picosecond laser sources represent a high level of technological development. While components for construction of picosecond systems are commercially available, most experiments require such a level of instrumental sophistication and associated diagnostic electronics that the study of ultrafast processes has been limited to a relatively small number of laboratories. A secondary problem associated with carrying out picosecond experiments is the problem of detection techniques on this time scale. Often the unambiguous measurement of picosecond events is more difficult than the excitation processes required to produce them. The descriptive review of experimental techniques given by West (1976) is still very close to the current state-of-the-art. The most recent advances in technology in this field have been made in the development of new laser glasses for neodymium and in the beginnings of production of second generation streak cameras. The new experimenter in the field should carefully review the current technology before making economic commitments.

Kimel and Speiser have reviewed experimental picosecond work recently. In the work cited below, we accentuate the most recent experimental developments in the field.

A. Gas Phase Applications

The application of picosecond techniques to the study of excited state processes as yet has not found extensive applications with gas phase. Electronic relaxation processes in POPOP and perylene vapors have been studied by Shapiro et al. (1974). Ground state vibrational relaxation has been studied by Maier et al. (1977) for coumarin 6 in the vapor phase, using crossed beam techniques. Intersystem crossing rates in acridine and phenazine vapors have been reported by Hirata and Tanaka (1977a,b). Excited state energy transfer processes, unimolecular photodissociation reactions, and statistical limit relaxation processes are areas in which the gas phase applications of picosecond techniques may prove important in the near future.

B. Studies in The Condensed Phase

Picosecond experimenters have been, of necessity, occupied with the measurement of fast processes involved in the condensed media used in the practice of their trade. Fluorescence rates, excited state and ground state vibrational recovery rates, intersystem crossing rates, and sequential multiphoton pumping rates control the lasing efficiency of dyes and their usefulness as mode lock and Q-switch materials. Many measurements of these excited state processes have been carried out using picosecond excitation techniques (Bush et al., 1975; Fleming et al., 1977a,b; Formosinho, 1976; Fouassier et al., 1975, 1976; Heritage and Penzkofer, 1976; Jaraudias et al., 1977; Kobayashi and Nagakura, 1977; Laubereau et al., 1975; Lill et al., 1977; Mialocq et al.,

1976, 1977a,b; Mialocq and Goujon, 1977; Mourou and Malley, 1975; Yu *et al.*, 1977).

In the condensed phase, picosecond techniques have been used to study a wide range of radiative and radiationless transition processes of excited state molecules. Among the species studied are azulene (Heritage and Penzkofer, 1976; Ippen *et al.*, 1977), benzophenone (R. W. Anderson *et al.*, 1974a,b, 1975; Rentzepis and Bush, 1971), iodine (Chuang *et al.*, 1974a), anthracene (Hochstrasser and Wessel, 1974; Nakashima and Mataga, 1975), tetrazine (Campillo *et al.*, 1977; Hochstrasser *et al.*, 1976), tetraphenyl-hydrazine (Anderson and Hochstrasser, 1976), and tetramethyldioxetane (Smith *et al.*, 1977).

Picosecond laser excitation techniques have also been used to study electronic energy transfer, charge transfer process, and rotational and diffusional relaxation processes (Chuang and Eisenthal, 1975; Chuang *et al.*, 1974b; Fujiwara *et al.*, 1977; Gnädig and Eisenthal, 1977; Griffiths *et al.*, 1974; Haar *et al.*, 1977; Hochstrasser and Nelson, 1976; Kinney-

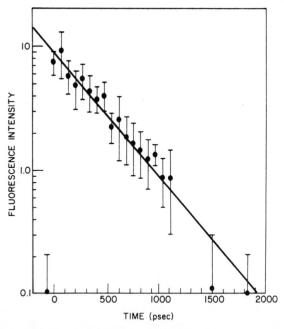

Fig. 28. The fluorescence decay of s-tetrazine in benzene solution (300 K, $\tau_f = 450 \pm 55$ psec). Excitation was with an 8 psec pulse at 530 nm from a frequency doubled Nd/glass laser. The experimental data points were the photoelectrically measured fluorescence intensity passing a gated optical Kerr shutter with a gate width (FWHM) of 26 psec. [After Hochstrasser *et al.* (1976).]

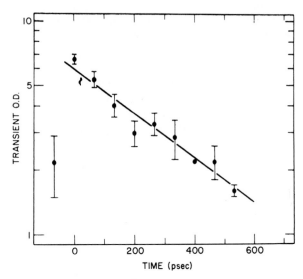

Fig. 29. The transient absorption of the S_1 state of s-tetrazine in benzene solution (300 K, $\tau_{^1B_{3u}(n\pi^*)} = 424 \pm 75$ psec). The transient state was produced by a 8 psec pulse from a frequency doubled Nd/glass laser. The transient absorption was photographed against an infrared generated CCl_4 continuum. Kinetic data were obtained by passing the continuum through a variable spatial delay line observing changes in the transient absorption. [After Hochrasser *et al.* (1976).]

Wallace and Jonah, 1977; Nishimura *et al.*, 1977; Ostertag and Grun, 1977; Porter *et al.*, 1977).

Figures 28 and 29 show two measurements of the $^1B_{3u}$ excited state relaxation of tetrazine. The first plot is a direct measure of the fluorescence decay rate of the prepared electronic state and the second shows the time resolved decay behavior of a transient absorption originating from the prepared excited state. Measurements of subnanosecond excited state phenomena were not possible prior to the application of picosecond techniques.

VII. CONCLUDING REMARKS

In this chapter, we have briefly described the use of lasers for generation and detection of reactive species in the gas phase. Emphasis was placed mainly on studies carried out in static systems using fluorescence, resonance absorption, and resonance fluorescence methods. Since these detection methods have been previously used extensively with conventional light sources, a brief review of the methods was made in conjunction with the flash photolysis technique (see Section II).

For reference purposes, we have summarized in Table I various methods that have been used successfully for the production as well as the detection of some kinetically important atomic and free radical species, with the exception of metallic and refractory species, which are discussed in detail elsewhere by Fontijn and Felder. Because of space limitation, we have excluded such techniques as laser Raman scattering and coherent anti-Stokes Raman Spectroscopy (CARS). The usefulness of these techniques for detecting reactive species at low concentration levels, however, is still not yet established.

ACKNOWLEDGMENTS

We are grateful to many colleagues for permitting us to use their published and unpublished work. We also thank Ms. Laura Colcord and Mr. Timothy Creamer for their assistance in preparing this manuscript. Many constructive comments by Ms. Colcord are particularly appreciated.

REFERENCES

Alkemade, C. T., and Zeegers, P. J. (1971). "Spectrochemical Methods of Analysis: Analysis of Atoms and Molecules." Wiley, New York.

Allen, J. E., Jr., Anderson, W. R., and Crosley, D. R. (1977). *Opt. Lett.* **1**, 118.

Ambartzumian, R. V., Letokhov, V. S., Ryabov, E. A., and Chekalin, N. V. (1974). *Pis'ma Zh. Eksp. Teor. Fiz.* **20**, 597.

Ambartzumian, R. V., Gorokhov, Yu. A., Letokhov, V. S., and Makarov, G. N. (1975a). *Pis'ma Zh. Eksp. Teor. Fiz.* **21**, 375.

Ambartzumian, R. V., Gorokhov, Yu. A., Letokhov, V. S., and Makarov, G. N. (1975b). *Pis'ma Zh. Eksp. Teor. Fiz.* **22**, 96.

Ambartzumian, R. V., Furzikov, N. P., Gorokhov, Yu. A., Letokhov, V. S., Makarov, G. N., and Puretzky, A. A. (1976). *Opt. Commun.* **18**, 517.

Anderson, L. G., Parmenter, C. S., and Poland, H. M. (1973). *Chem. Phys.* **1**, 401.

Anderson, R. A., Peacher, J., and Wilcox, D. M. (1975). *J. Chem. Phys.* **63**, 5287.

Anderson, R. W., Jr., and Hochstrasser, R. M. (1976). *J. Phys. Chem.* **80**, 2155.

Anderson, R. W., Jr., Hochstrasser, R. M., Lutz, H., and Scott, G. W. (1974a). *Chem. Phys. Lett.* **24**, 162.

Anderson, R. W., Jr., Hochstrasser, R. M., Lutz, H., and Scott, G. W. (1974b). *Chem. Phys. Lett.* **28**, 153.

Anderson, R. W., Jr., Hochstrasser, R. M., Lutz, H., and Scott, G. W. (1975). *Chem. Phys. Lett.* **32**, 204.

Atkinson, G. H., McIlwain, M. E., Venkatesh, C. G., and Chapman, D. M. (1978a). *J. Photochem.* **8**, 307.

Atkinson, G. H., McIlwain, M. E., and Venkatesh, C. G. (1978b). *J. Chem. Phys.* **68**, 726.

Atkinson, R., and Cvetanović, R. J. (1971). *J. Chem. Phys.* **55**, 659.

Atkinson, R., and Pitts, J. N., Jr., (1974). *Chem. Phys. Lett.* **29**, 28.

Atkinson, R., and Pitts, J. N., Jr. (1977). *J. Chem. Phys.* **67**, 38.

Atkinson, R., and Welge, K. H. (1972). *J. Chem. Phys.* **57**, 3689.

Atkinson, R., Hansen, D. A., and Pitts, J. N., Jr. (1975). *J. Chem. Phys.* **62**, 3284.

Atkinson, R., Perry, R. A., and Pitts, J. N., Jr. (1977). *J. Chem. Phys.* **66**, 1197.

Avouris, P., Gilbert, W. M., and El-Sayed, M. A. (1977). *Chem. Rev.* **77**, 793.

Baardsen, E. L., and Terhune, R. W. (1972). *Appl. Phys. Lett.* **21**, 209.

Ballik, E. A., and Ramsay, D. A. (1963a). *Astrophys. J.* **137**, 61.

Ballik, E. A., and Ramsay, D. A. (1963b). *Astrophys. J.* **137**, 84.

Barker, J. R., and Weston, R. E., Jr., (1976). *J. Chem. Phys.* **65**, 1427.

Barnes, R. H. Moeller, C. E., Kircher, J. F., and Verber, C. M. (1973). *Appl. Opt.* **12**, 2531.

Baronavski, A. P., and McDonald, J. R. (1977a). *Chem. Phys. Lett.* **45**, 172.

Baronavski, A. P., and McDonald, J. R. (1977b). *J. Chem. Phys.* **66**, 3300.

Baronavski, A. P., and McDonald, J. R. (1977c). *Appl. Opt.* **16**, 1897.

Baronavski, A. P., and McDonald, J. R. (1978). *Chem. Phys. Lett.* **56**, 369.

Baronavski, A. P., Miller, R. G., and McDonald, J. R. (1978). *Chem. Phys.* **30**, 119.

Basco, N. (1965). *Proc. R. Soc. London, Ser. A* **283**, 302.

Basco, N., and Morse, R. D. (1973). *J. Mol. Spectrosc.* **45**, 35.

Basco, N., and Norrish, R. G. W. (1961). *Proc. R. Soc. London, Ser. A*, **260**, 293.

Basco, N., and Norrish, R. G. W. (1965). *Proc. R. Soc. London, Ser. A* **283**, 291.

Basco, N., Nicholas, J. E., and Norrish, R. G. W. (1963). *Proc. R. Soc. London, Ser. A* **272**, 147.

Basco, N., James, D. G. L., and Stuart, R. D. (1970). *Int. J. Chem. Kinet.* **2**, 215.

Basco, N., James, D. G. L., and James, F. C. (1972). *Int. J. Chem. Kinet.* **4**, 129.

Bauer, S. H., and Chien, K. R. (1977). *Chem. Phys. Lett.* **45**, 529.

Becker, K. H., and Welge, K. H. (1963). *Z. Naturforsch, Teil A* **18**, 600.

Becker, K. H., and Welge, K. H. (1964). *Z. Naturforsch, Teil A* **19**, 1006.

Becker, K. H., and Welge, K. H. (1965). *Z. Naturforsch, Teil A* **20**, 442.

Becker, K. H., Haaks, D., and Tatarczyk, T. (1974a). *Z. Naturforsch., Teil A* **29**, 829.

Becker, K. H., Fink, E. H., Langen, P., and Schurath, U. (1974b). *J. Chem. Phys.* **60**, 4623.

Bell, A. G. (1880). *Proc. Am. Assoc. Adventures Sci.* **29**, 115.

Bell, S., Ng, T. L., and Suggitt, C. (1972). *J. Mol. Spectrosc.* **44**, 267.

Bennett, R. G. (1960). *Rev. Sci. Intrum.* **31**, 1275.

Berry, M. J. (1972). *Chem. Phys. Lett.* **15**, 269.

Bishop, W. P., and Dorfman, L. M. (1970). *J. Chem. Phys.* **52**, 3210.

Black, G., Slanger, T. G., St. John, G. A., and Young, R. A. (1969). *J. Chem. Phys.* **51**, 116.

Black, G., Sharpless, R. L., Slanger, T. G., and Lorents, D. C. (1975a). *J. Chem. Phys.* **62**, 4266.

Black, G., Sharpless, R. L., Slanger, T. G., and Lorents, D. C. (1975b). *J. Chem. Phys.* **62**, 4274.

Black, G., Sharpless, R. L., and Slanger, T. G. (1975c). *J. Chem. Phys.* **63**, 4546.

Black, G., Sharpless, R. L., and Slanger, T. G. (1975d). *J. Chem. Phys.* **63**, 4551.

Black, G., Sharpless, R. L., and Slanger, T. G. (1977). *J. Chem. Phys.* **66**, 2113.

Black, J. G., Yablonovitch, E., Bloembergen, N., and Mukamel, S. (1977). *Phys. Rev. Lett.* **38**, 1131.

Bloembergen, N. (1975). *Opt. Commun.* **15**, 416.

Bollinger, L. M., and Thomas, G. E. (1961). *Rev. Sci. Instrum.* **32**, 1044.

Bosnali, M. W., and Perner, D. (1971). *Z. Naturforsch., Teil A* **26**, 1768.

Braun, W., and Lenzi, M. (1967). *Discuss. Faraday Soc.* **44**, 252.

Braun, W., and Tsang, W. (1976). *Chem. Phys. Lett.* **44**, 354.

Braun, W., McNesby, J. R., and Bass, A. M. (1967). *J. Chem. Phys.* **46**, 2071.

Braun, W., Bass, A. M., Davis, D. D., and Simons, J. D. (1969). *Proc. R. Soc. London, Ser. A* **312**, 417.

Braun, W., Bass, A. M., and Pilling, M. (1970). *J. Chem. Phys.* **52**, 5131.

Bray, R. G., and Hochstrasser, R. M. (1976). *Mol. Phys.* **31**, 1199.

Brewer, R. G., and Mooradian, A. (1974). "Laser Spectroscopy." Plenum, New York.

Brunner, F., Cotter, T. P., Kompa, K. L., and Proch, D. (1977). *J. Chem. Phys.* **67**, 1547.

Brunner, W., and Paul, H. (1977). *Opt. Commun.* **12**, 252.

Brus, L. E. (1970). *J. Chem. Phys.* **52**, 1716.

Burn, G., and Norrish, R. G. W. (1963). *Proc. R. Soc. London, A* **271**, 289.

Bush, G. E., Greve, K. S., Olson, G. L., Jones, R. P., and Rentzepis, P. M. (1975). *Chem. Phys. Lett.* **33**, 412 and 417.

Butler, J. E., Goss, L. P., Lin, M. C., and Hudgens, J. W. (1979). *Chem. Phys. Lett.* **63**, 104.

Callear, A. B., and Hedges, R. E. M. (1968). *Nature (London)* **218**, 163.

Callear, A. B., and Smith, I. W. M. (1963). *Trans. Faraday Soc.* **59**, 1720.

Callear, A. B., and Wood, P. M. (1971). *Trans. Faraday Soc.* **67**, 272.

Calvert, J. G., and Pitts, J. N., Jr. (1966). "Photochemistry." Wiley, New York.

Campbell, J. D., Hancock, G., and Welge, K. H. (1976a). *Chem. Phys. Lett.* **43**, 581.

Campbell, J. D., Hancock, G., Halpern, J. B., and Welge, K. H. (1976b). *Chem. Phys. Lett.* **44**, 404.

Campillo, A. J., Hyer, R. C., Shapiro, S. L., and Swenberg, E. (1977). *Chem. Phys. Lett.* **48**, 495.

Cantrell, C. D., and Galbraith, H. W. (1976). *Opt. Commun.* **18**, 513.

Carlson, R. W., Judge, D. L., Ogawa, M., and Lee, L. C. (1973). *Appl. Opt.* **12**, 409.

Carrington, T. (1964). *J. Chem. Phys.* **41**, 2012.

Chuang, T. J., and Eisenthal, K. B. (1975). *J. Chem. Phys.* **62**, 2213.

Chuang, T. J., Cox, R. J., and Eisenthal, K. B. (1974a). *J. Am. Chem. Soc.* **96**, 6828.

Chuang, T. J., Hoffman, G. W., and Eisenthal, K. B. (1974b). *Chem. Phys. Lett.* **25**, 201.

Clark, J. H., Moore, C. B., and Reilly, J. P. (1978). *Int. J. Chem. Kinet.* **10**, 427.

Clerc, M., and Schmidt, M. (1972). *Faraday Discuss. Chem. Soc.* **53**, 217.

Clyne, M. A. A., and McDermid, J. (1977). *J. Chem. Soc. Faraday Trans. 2* **73**, 1094.

Cody, R., Sabety-Dzvonik, M., and Jackson, W. M. (1977). *J. Chem. Phys.* **66**, 2145.

Coggiola, M. J., Schulz, P. A., Lee, Y. T., and Shen, Y. R. (1977). *Phys. Rev. Lett.* **38**, 17.

Colussi, A. J., Singleton, D. L., Irwin, R. S., and Cvetanović, R. J. (1975). *J. Phys. Chem.* **79**, 1900.

Cottrell, T. A., McFarlane, I. M., Read, A. W., and Young, A. H. (1966). *Trans. Faraday Soc.* **62**, 2655.

Coxon, J. A. (1976). *J. Photochem.* **5**, 337.

Coxon, J. A. (1977). *J. Photochem.* **6**, 439.

Coxon, J. A., and Ramsay, D. A. (1976). *Can. J. Phys.* **54**, 1034.

Coxon, J. A., Jones, W. E., and Skolnik, E. G. (1976). *Can. J. Phys.* **54**, 1043.

Creel, C. L., and Ross, J. (1976). *J. Chem. Phys.* **64**, 3560.

Daily, J. W. (1976). *Appl. Opt.* **15**, 955.

Daily, J. W. (1977). *Appl. Opt.* **16**, 568.

Danon, J., Filseth, S. V., Feldmann, D., Zacharias, H., Dugan, C. H., and Welge, K. H. (1978). *Chem. Phys.* **29**, 345.

Davidson, J. A., Sadowski, C. M., Schiff, H. I., Streit, G. E., Howard, C. J., Jennings, D. A., and Schmeltekopf, A. L. (1976). *J. Chem. Phys.* **64**, 57.

Davidson, N., Marshall, R., Larsh, A. E., Jr., and Carrington, T. (1951). *J. Chem. Phys.* **29**, 345.

Davis, D. D., and Braun, W. (1968). *Appl. Opt.* **7**, 2071.

Davis, D. D., and Klemm, R. B. (1973). *Int. J. Chem. Kinet.* **5**, 841.

Davis, D. D., and Okabe, H. (1968). *J. Chem. Phys.* **49**, 5526.

Davis, D. D., Braun., and Bass, A. M. (1970). *Int. J. Chem. Kinet.* **2**, 101.

Davis, D. D., Braun, W., and Bass, A. M. (1970). *Int. J. Chem. Kinet.* **2**, 101.

Davis, D. D., Klemm, R. B., and Pilling, M. (1972a). *Int. J. Chem. Kinet.* **4**, 367.

Davis, D. D., Klemm, R. B., Braun, W., and Pilling, M. (1972b). *Int. J. Chem. Kinet.* **4**, 383.

Davis, D. D., Huie, R. E., Herron, J. T., Kurylo, M. J., and Braun, W. (1972c). *J. Chem. Phys.* **56**, 4868.

Davis, D. D., Schiff, R., and Fischer, S. (1974). *J. Chem. Phys.* **61**, 2213.

Deakin, J. J., and Husain, D. (1972). *J. Chem. Soc. Faraday Trans. 2* **68**, 41.

Deakin, J. J., Husain, D., and Wiesenfeld, J. R. (1971). *Chem. Phys. Lett.* **10**, 146.

De Groot, M. S., Emeis, C. A., Hesselman, I. A. M., Dreut, E., and Farenhorst, E. (1972). *Chem. Phys. Lett.* **17**, 332.

Del Greco, F. P., and Kaufman, F. (1962). *Discuss. Faraday Soc.* **33**, 128.

Devillers, C., and Ramsay, D. A. (1971). *Can. J. Phys.* **49**, 2839.

Djeu, N. (1974). *J. Chem. Phys.* **60**, 4109.

Djeu, N., and Searles, S. K. (1972). *J. Chem. Phys.* **57**, 4681.

Djeu, N., Kan, T., and Wolga, G. T. (1968). *IEEE J. Quantum Electron.* **QE-4**, 256.

Djeu, N., Pilloff, H. S., and Searles, S. K. (1971). *Appl. Phys. Lett.* **18**, 538.

Donnelly, V. M., McDonald, J. R., and Baronavski, A. P. (1978). *Chem. Phys. Lett.* (in press).

Donohue, T., and Wiesenfeld, J. R. (1975). *J. Chem. Phys.* **63**, 3130.

Donovan, R. J. and Husain, D. (1966a). *Trans. Faraday Soc.* **62**, 11.

Donovan, R. J., and Husain, D. (1966b). *Trans. Faraday Soc.* **62**, 2023.

Donovan, R. J., and Husain, D. (1966c). *Trans. Faraday Soc.* **62**, 2643.

Donovan, R. J., and Husain, D. (1966d). *Trans. Faraday Soc.*, **62**, 2987.

Donovan, R. J., and Husain, D. (1970). *Chem. Rev.* **70**, 489.

Donovan, R. J., Husain, D., and Jackson, P. T. (1969a). *Trans. Faraday Soc.* **65**, 2930.

Donovan, R. J., Husain, D., and Stevenson, C. D. (1969b). *Trans. Faraday Soc.* **65**, 2941.

Donovan, R. J., Husain, D., and Stevenson, C. D. (1970a). *Trans. Faraday Soc.* **66**, 1.

Donovan, R. J., Husain, D., Fair, R. W., Strausz, O. P., and Gunning, H. E. (1970b). *Trans. Faraday Soc.* **66**, 1635.

Donovan, R. J., Husain, D., and Kirsch, L. J. (1970c). *Trans. Faraday Soc.* **66**, 2551.

Donovan, R. J., Husain, D., and Kirsch, L. J. (1970d). *Chem. Phys. Lett.* **6**, 488.

Donovan, R. J., Husain, D., and Kirsch, L. J. (1971). *Trans. Faraday Soc.* **67**, 375.

Dorfman, L. M., and Matheson, M. S. (1965). *Prog. React. Kinet.* **3**, 237.

Dressler, K., and Ramsay, D. A. (1959). *Philos. Trans. R. Soc. London, Ser. A* **251**, 553.

Drucker, R. P., and McClain, W. M. (1974). *J. Chem. Phys.* **61**, 2609.

Dunn, O. J., Filseth, S. V., and Young, R. A. (1973). *J. Chem. Phys.* **59**, 2892.

Durie, R. A., and Ramsay, D. A. (1958). *Can. J. Phys.* **36**, 35.

Dworetsky, S. H., Brus, L. E., and Hozak, R. S. (1974). *J. Chem. Phys.* **61**, 1581.

Edgecombe, F. H. C., Norrish, R. G. W., and Thrust, B. A. (1957). *Proc. R. Soc. London, Ser. A* **243**, 24.

Ezekiel, S. (1974). *In* " Laser Spectroscopy, Molecular Beam Spectroscopy with Argon and Dye Lasers" (R. G. Brewer, and A. Mooradian, eds.), p. 361, Plenum, New York.

Fair, R. W., Van Roodselaar, A., and Strausz, O. P. (1971). *Can. J. Chem.* **49**, 1659.

Filseth, S. V., and Welge, K. H. (1969). *J. Chem. Phys.* **51**, 839.

Filseth, S. V., Stuhl, F., and Welge, K. H. (1970a). *J. Chem. Phys.* **52**, 239.

Filseth, S. V., Zia, Z., and Welge, K. H. (1970b). *J. Chem. Phys.* **52**, 5502.

Fleming, R., Knight, A. E. W., Morris, J. M., Robbins, R. J., and Robinson, G. W. (1977a). *Chem. Phys.* **23**, 61.

Fleming, R., Knight, A. E. W., Morris, J. M., Robbins, R. J., and Robinson, G. W. (1977b). *Chem. Phys. Lett.* **49**, 1.

Formosinho, S. J. (1976). *Mol. Photochem.* **7**, 41.

Forst, W. (1973). "Theory of Unimolecular Reactions." Academic Press, New York.

Fouassier, J.-P., Lougnot, D.-J., and Faure, J. (1975). *Chem. Phys. Lett.* **35**, 189.

Fouassier, J.-P., Lougnot, D.-J., and Faure, J. (1976). *Opt. Commun.* **18**, 263.

Fujiwara, H., Nakashima, N., and Montaga, N. (1977). *Chem. Phys. Lett.* **47**, 185.

Fushiki, Y., and Tsuchiya, S. (1973). *Chem. Phys. Lett.* **22**, 47.

Gähwiller, C., Brown, F. C., and Fujita, H. (1970). *Rev. Sci. Instrum.* **41**, 1275.

Gandini, A., Willis, C., and Back, R. A. (1977). *Can. J. Chem.* **55**, 4156.

Gauthier, M. J. E., and Snelling, D. R. (1975). *J. Photochem.* **4**, 27.

Giachardi, D. J., Harris, G. W., and Wayne, R. P. (1975). *Chem. Phys. Lett.* **32**, 586.

Gilpin, R., Schiff. H. I., and Welge, K. H. (1971). *J. Chem. Phys.* **55**, 1081.

Gnädig, K., and Eisenthal, K. B. (1977). *Chem. Phys. Lett.* **46**, 339.

Gordon, S., and Mulac, W. A. (1975). *Int. J. Chem. Kinet.* **7**, Suppl. 289.

Gorelik, G. (1946). *Dokl. Akad. Nauk. SSSR* **54**, 779.

Grant, E. R., Coggiola, M. J., Lee, Y. T., Schulz, P. A., Sudbø, Aa.S., and Shen, Y. R. (1977). *Chem. Phys. Lett.* **52**, 595.

Greiner, N. R. (1967). *J. Chem. Phys.* **46**, 2795.

Greiner, N. R. (1968). *J. Phys. Chem.* **72**, 406.

Greiner, N. R. (1969). *J. Chem. Phys.* **48**, 1413.

Greiner, N. R. (1970). *J. Chem. Phys.* **53**, 1070.

Griffiths, J. E., Clerc, M., and Rentzepis, P. M. (1974). *J. Chem. Phys.* **60**, 3824.

Gross, R. W. F., and Bott, J. F. (1976). "Handbook of Chemical Lasers." Wiley, New York.

Haar, H. P., Klein, V. K. A., Hafner, F. W., and Hauser, M. (1977). *Chem. Phys. Lett.* **49**, 563.

Halpern, A. M., and Ware, W. R. (1971). *J. Chem. Phys.* **54**, 1271.

Halpern, J. B., Hancock, G., Lenzi, M., and Welge, K. H. (1975). *J. Chem. Phys.* **63**, 4808.

Hampson, R. F., Jr., and Okabe, H. (1970). *J. Chem. Phys.* **52**, 1930.

Hancock, G., Lange, W., Lenzi, M., and Welge, K. H. (1975). *Chem. Phys. Lett.* **33**, 168.

Hancock, G., Morley, C., and Smith, I. W. M. (1971). *Chem. Phys. Lett.* **12**, 193.

Hancock, J. K., and Green, W. H. (1973). *J. Chem. Phys.* **59**, 6350.

Hand, C. W. Kaufman, P. Z., and Hexter, R. M. (1966). *Appl. Opt.* **5**, 1097.

Hänsch, T. W., Schawlow, A. L., and Toschek, P. E. (1972). *IEEE J. Quantum Electron.* **qe-** 802.

Hansen, D. A., and Lee, E. K. C. (1975). *J. Chem. Phys.* **62**, 183.

Hansen, I., Höinghaus, K., Zetzsch, C., and Stuhl, F. (1976). *Chem. Phys. Lett.* **42**, 370.

Heidner, R. F., III, and Husain, D. (1973). *Int. J. Chem. Kinet.* **5**, 819.

Heidner, R. F., III, and Husain, D. (1974). *Int. J. Chem. Kinet.* **6**, 77.

Heidner, R. F., III, Husain, D., and Wiesenfeld, J. R. (1973). *J. Chem. Soc. Faraday Trans. 2* **69**, 927.

Heritage, J. P., and Penzkofer, D. (1976). *Chem. Phys. Lett.* **44**, 76.

Herr, K. C., and Pimentel, G. C. (1965). *Appl. Opt.* **4**, 25.

Hershberger, W. D., Bush, E. T., and Leck, G. W. (1946). *RCA Rev.* **7**, 422.

Herzberg, G. (1961). *Proc. R. Soc. London, Ser. A* **262**, 291.

Herzberg, G. (1966). "Molecular Spectra and Molecular Structure II." Van Nostrand, Reinhold, Princeton, New Jersey.

Herzberg, G. (1968). *Adv. Photochem.* **5**, 1.

Herzberg, G. (1971). "The Spectra and Structures of Simple Free Radicals." Cornell Univ. Press, Ithaca, New York.

Herzberg, G., and Johns, J. W. C. (1966). *Proc. R. Soc. London, Ser. A* **295**, 107.

Herzberg, G., and Ramsay, D. A. (1950). *Discuss. Faraday Soc.* **9**, 80.

Herzberg, G., and Ramsay, D. A. (1955). *Proc. R. Soc. London, Ser. A* **233**, 34.

Hesser, J. E. (1970). *Astrophys. J.* **159**, 703.

Hikida, T., Eyre, J. A., and Dorfman, L. M. (1971). *J. Chem. Phys.* **54**, 3422.

Hinchen, J. J., and Hobbs, R. H. (1976). *J. Chem. Phys.* **65**, 2732.

Hirata, Y., and Tanaka, I. (1977a). *Chem. Phys. Lett.* **41**, 336.

Hirata, Y., and Tanaka, I. (1977b). *Chem. Phys. Lett.* **43**, 568.

Hochstrasser, R. M., and Nelson, A. C. (1976). *Opt. Commun.* **18**, 361.

Hochstrasser, R. M., and Wessel, J. E. (1974). *Chem. Phys.* **6**, 19.

Hochstrasser, R. M., King, D. S., and Nelson, A. C. (1976). *Chem. Phys. Lett.* **42**, 8.

Hodgkinson, D. P., and Briggs, J. S. (1976). *Chem. Phys. Lett.* **43**, 451.
Hofmann, H., and Leone, S. R. (1978). *Chem. Phys. Lett.* **54**, 314.
Houston, P. L. (1977). *Chem. Phys. Lett.* **47**, 137.
Houston, P. L., and Moore, C. B. (1976). *J. Chem. Phys.* **65**, 757.
Hsu, D. S. Y., and Lin, M. C. (1976). *Chem. Phys. Lett.* **42**, 78.
Hsu, D. S. Y., and Lin, M. C. (1977a). *Int. J. Chem. Kinet.* **9**, 507.
Hsu, D. S. Y., and Lin, M. C. (1977b). *Chem. Phys.* **21**, 235.
Hsu, D. S. Y., and Lin, M. C. (1978). *Int. J. Chem. Kinet.* **10**, 839.
Hsu, D. S. Y., Colcord, L. J., and Lin, M. C. (1978a). *J. Phys. Chem.* **82**, 121.
Hsu, D. S. Y., Umstead, M. E., and Lin, M. C. (1978b). *ACS Symp. Ser.* **66**, 128.
Hudgens, J. W. (1978). *J. Chem. Phys.* **68**, 777.
Hundley, L., Coburn, T., Gorwin, E., and Stryer, L. (1967). *Rev. Sci. Instrum.* **38**, 488.
Hunziker, H. E. (1969). *Chem. Phys. Lett.* **3**, 504.
Hunziker, H. E., and Wendt, H. R. (1974). *J. Chem. Phys.* **60**, 4622.
Husain, D., and Kirsch, L. J. (1971a). *Trans. Faraday Soc.* **67**, 2025.
Husain, D., and Kirsch, L. J. (1971b). *Trans. Faraday Soc.* **67**, 2886.
Husain, D., and Kirsch, L. J. (1971c). *Trans. Faraday Soc.* **67**, 3166.
Husain, D., and Norris, P. E. (1977). *J. Chem. Soc. Faraday Trans. 2* **73**, 415.
Husain, D., and Norrish, R. G. W. (1963). *Proc. R. Soc. London, Ser. A* **273**, 145.
Husain, D., and Wiesenfeld, J. R. (1967). *Trans. Faraday Soc.* **63**, 1349.
Husain, D., Kirsch, L. J., and Wiesenfeld, J. R. (1972). *Faraday Discuss. Chem. Soc.* **53**, 201.
Husain, D., Mitra, S. K., and Young, A. N. (1974). *J. Chem. Soc. Faraday Trans. 2* **70**, 1721.
Hynes, A. J., and Brophy, J. H. (1979). *Chem. Phys. Lett.* **63**, 93.
Ippen, E. P., Schzuk, C. V., and Woevner, R. L. (1977). *Chem. Phys. Lett.* **46**, 20.
Isenor, N. R., and Richardson, M. C. (1971). *Appl. Phys. Lett.* **18**, 224.
Isenor, N. R., Merchant, V., Hallsworth, R. S., and Richardson, M. C. (1973). *Can. J. Phys.* **51**, 1281.
Ishiguro, E., Sasanuma, M., Masuko, H., Morioka, Y., and Nakamura, M. (1978). *J. Phys. B* **11**, 993.
Jackson, W. M., and Cody, R. (1974). *J. Chem. Phys.* **61**, 4183.
Jackson, W. M., Halpern, J. B., and Lin, C. S. (1978). *Chem. Phys. Lett.* **45**, 107.
James, F. C., and Simons, J. P. (1974). *Int. J. Chem. Kinet.* **6**, 887.
Jaraudias, J., Goujon, P., and Mialocq, J. C. (1977). *Chem. Phys. Lett.* **45**, 107.
Johns, J. W. C., Priddle, S. H., and Ramsay, D. A. (1963). *Discuss. Faraday Soc.* **35**, 90.
Johnston, H. S., Morris, E. D., Jr., and Van den Bogaerde, J. (1969). *J. Am. Chem. Soc.* **91**, 7712.
Judge, D. L., and Lee, L. C. (1972). *J. Chem. Phys.* **57**, 455.
Judge, D. L., and Lee, L. C. (1973). *J. Chem. Phys.* **58**, 104.
Keller, R. A., Zalewski, E. F., and Peterson, N. C. (1972). *J. Opt. Soc. Am.* **62**, 319.
Kijewski, H., and Troe, J. (1971). *Int. J. Chem. Kinet.* **3**, 223.
Kimel, S., and Speiser, S. (1977). *Chem. Rev.* **77**, 437.
King, D. S., and Stephenson, J. C. (1977). *Chem. Phys. Lett.* **51**, 48.
Kinney-Wallace, G. A., and Jonah, C. D. (1977). *Chem. Phys. Lett.* **47**, 362.
Kinsey, J. L. (1977). *Annu. Rev. Phys. Chem.* **28**, 349.
Klemm, R. B., and Stief, L. J. (1974). *J. Chem. Phys.* **61**, 4900.
Klemm, R. B., Payne, W. A., and Stief, L. J. (1975a). *Int. J. Chem. Kinet.* **7**, Suppl., 61.
Klemm, R. B., Glicker, S., and Stief, L. J. (1975b). *Chem. Phys. Lett.* **33**, 512.
Knudtson, J. T., and Berry, M. J. (1978). *J. Chem. Phys.* **68**, 4419.
Kobayashi, T., and Nagakura, S. (1977). *Chem. Phys.* **23**, 153.
Koren, G., Oppenheim, U. P., Tal, D., Okom, M., and Weil, R. (1976). *Appl. Phys. Lett.* **29**, 40.

Koyano, I., Wauchop, T. S., and Welge, K. H. (1975). *J. Chem. Phys.* **63**, 110.

Krause, L. (1975). *Adv. Chem. Phys.* **28**, 267.

Kroll, M. (1975). *J. Chem. Phys.* **63**, 319.

Ku, R. T., Hinckley, E. D., and Sample, J. O. (1975). *Appl. Opt.* **14**, 854.

Kurylo, M. J. (1972a). *J. Phys. Chem.* **76**, 3518.

Kurylo, M. J. (1972b). *Chem. Phys. Lett.* **14**, 117.

Kurylo, M. J., Peterson, N. C., and Braun, W. (1970). *J. Chem. Phys.* **53**, 2776.

Kurylo, M. J., Peterson, N. C., and Braun, W. (1971). *J. Chem. Phys.* **54**, 4662.

Kurylo, M. J., Braun, W., Kaldor, A., Freund, S. M., and Wayne, R. P. (1974). *J. Photochem.* **3**, 71.

Lange, W., Luther, J., Nottleck, B., and Schröder, H. W. (1973). *Opt. Commun.* **8**, 157.

Larson, D. M. (1976). *Opt. Commun.* **19**, 404.

Larson, D. M., and Bloembergen, N. (1976). *Opt. Commun.* **17**, 254.

Laubereau, A., Seilmeier, A., and Kaiser, W. (1975). *Chem. Phys. Lett.* **36**, 232.

Lee, E. K. C. (1977). *Acc. Chem. Res.* **10**, 319.

Lee, L. C., and Judge, D. L. (1973). *Can. J. Phys.* **51**, 378.

Lee, L. C., and Judge, D. L. (1975). *J. Chem. Phys.* **63**, 2782.

Lee, L. C., and Judge, D. L. (1976). *Phys. Rev. A*, **14**, 1094.

Lee, L. C., Carlson, R. W., Judge, D. L., and Ogawa, M. (1974). *J. Chem. Phys.* **61**, 3261.

Lee, L. C., Carlson, R. W., Judge, D. L., and Ogawa, M. (1975a). *J. Chem. Phys.* **63**, 3987.

Lee, L. C., Carlson, R. W., Judge, D. L., and Ogawa, M. (1975b). *J. Phys. B*, **8**, 977.

Lee, J. H., Timmons, R. B., and Stief, L. J. (1976). *J. Chem. Phys.* **64**, 300.

Lee, J. H., Michael, J. V., Payne, W. A., Stief, L. J., and Whytock, D. A. (1977). *J. Chem. Soc. Faraday Trans. 1* **73**, 1530.

Lee, J. H., Michael, J. V., Payne, W. A., and Stief, L. J. (1978). *J. Chem. Phys.* **69**, 3069.

Lee, L. C., Phillips, E., and Judge, D. L. (1977a). *J. Chem. Phys.* **67**, 1237.

Lee, L. C., Slanger, T. G., Black, G., and Sharpless, R. L. (1977b). *J. Chem. Phys.* **67**, 5602.

Lee, L. C., Oren, L., Phillips, E., and Judge, D. L. (1978). *J. Phys. B* **11**, 47.

Lengel, R. K., and Crosley, D. R. (1978). *J. Chem. Phys.* **68**, 5309.

Lesclaux, R., Khe, P. V., Dezauzier, P., and Soulignac, J. C. (1975). *Chem. Phys. Lett.* **35**, 493.

Lesclaux, R., Soulignac, J. C., and Khe, P. V. (1976). *Chem. Phys. Lett.* **43**, 520.

Lesiecki, M. L., and Guillory, W. A. (1977). *J. Chem. Phys.* **66**, 4239.

Letokhov, V. S., and Makarov, A. A. (1976). *Opt. Commun.* **17**, 250.

Lill, E., Schneider, S., and Don, F. (1977), *Opt. Commun.* **22**, 107.

Lin, M. C. (1974). *J. Chem. Phys.* **61**, 1835.

Lin, M. C. (1978). *J. Chem. Phys.* **68**, 2004.

Lin, M. C., and Shortridge, R. G. (1974). *Chem. Phys. Lett.* **29**, 42.

Lin, M. C., Shortridge, R. G., and Umstead, M. E. (1976). *Chem. Phys. Lett.* **37**, 279.

Lin, S.-M., and Weston, R. E., Jr. (1976). *J. Chem. Phys.* **65**, 1443.

Lipscombe, F. J., Norrish, R. G. W., and Thrush, B. A. (1956). *Proc. R. Soc. London, Ser. A* **233**, 455.

Little, D. J., Dalgleisch, A., and Donovan, R. J. (1972). *Faraday Discuss. Chem. Soc.* **53**, 211.

Luk, C. K., and Bersohn, R. (1973). *J. Chem. Phys.* **58**, 2153.

Lussier, F. M., and Steinfeld, J. J. (1977). *Chem. Phys. Lett.* **50**, 175.

Lyman, J. L., Jensen, R. J., Rink, J., Robinson, C. P., and Rockwood, S. D. (1975). *Appl. Phys. Lett.* **27**, 87.

McClain, W. M. (1971). *J. Chem. Phys.* **55**, 2789.

McClain, W. M. (1973). *J. Chem. Phys.* **58**, 324.

McDonald, J. R., Baronavski, A. P., and Donnelly, V. M. (1978a). *Chem. Phys.* **33**, 161.

McDonald, J. R., Miller, R. G., and Baronavski, A. P. (1977). *Chem. Phys. Lett.* **51**, 57.

McDonald, J. R., Miller, R. G., and Baronavski, A. P. (1978b). *Chem. Phys.* **30**, 133.
McNesby, J. R., Braun, W., and Ball, J. (1971). *In* "Creation and Detection of the Excited State" (L. A. Lamola, ed.), Vol. 1, p. 503. Dekker, New York.
Maier, J. P., Seilmeier, A., Laubereau, A., and Kaiser, W. (1977). *Chem. Phys. Lett.* **46**, 527.
Masanet, J., Gilles, A., and Vermeil, C. (1974–1975). *J. Photochem.* **3**, 417.
Mathews, C. W. (1967). *Can. J. Phys.* **45**, 2355.
Matsuda, S., Slagle, I. R., Fife, D. J., Marquart, J. R., and Gutman, D. (1972). *J. Chem. Phys.* **57**, 5277.
Mele, A., and Okabe, H. (1969). *J. Chem. Phys.* **51**, 4798.
Menard-Bourcin, F., Menard, J., and Henry, L. (1972). *C. R. Hebd. Seances Acad. Sci., Ser. B* **274**, 24 and 1134.
Menard-Bourcin, F., Menard, J., and Henry, L. (1975). *J. Chem. Phys.* **63**, 1479.
Merer, A. J., and Travis, D. N. (1966). *Can. J. Phys.* **44**, 1541.
Mialocq, J. C., and Goujon, P. (1977). *Opt. Commun.* **20**, 342.
Mialocq, J. C., Boyd, A. W., Jaraudias, J., and Sutton, J. (1976). *Chem. Phys. Lett.* **37**, 236.
Mialocq, J. C., Boyd, A. W., Jaraudias, J., and Sutton, J. (1977a). *Chem. Phys. Lett.* **45**, 107.
Mialocq, J. C., Jaraudias, J., and Goujon, P. (1977b). *Chem. Phys. Lett.* **47**, 123.
Michael, J. V., and Weston, R. E., Jr. (1966). *J. Chem. Phys.* **45**, 3632.
Michael, J. V., Whytock, D. A., Lee, J. H., Payne, W. A., and Stief, L. J. (1977). *J. Chem. Phys.* **67**, 3533.
Michael, J. V., Lee, J. H., Payne, W. A., and Stief, L. J. (1978). *J. Chem. Phys.* **68**, 4093.
Michell, A. C. G., and Zemansky, M. W. (1934). "Resonance Radiation and Excited Atoms." Cambridge Univ. Press, London and New York.
Molina, M. J., and Pimentel, G. C. (1972). *J. Chem. Phys.* **56**, 3988.
Monson, P. R., and McClain, W. M. (1970). *J. Chem. Phys.* **53**, 29.
Monson, P. R., and McClain, W. M. (1972). *J. Chem. Phys.* **56**, 4817.
Moore, C. B. (1973). *Adv. Chem. Phys.* **23**, 41.
Moore, C. B. (1974). "Chemical and Biochemical Application of Lasers," Vol. 1. Academic Press, New York.
Morioka, Y., Masuko, H., Nakamura, M., Ishiguro, E., and Sasanuma, M. (1976). *J. Phys. B* **9**, 2321.
Morley, C., and Smith, I. W. M. (1972). *J. Chem. Soc., Faraday Trans. 2* 1016.
Morris, E. D., Jr., and Johnston, H. S. (1968). *Rev. Sci. Instrum.* **39**, 620.
Mourou, G., and Malley, M. M. (1975). *Chem. Phys. Lett.* **32**, 476.
Mukamel, S., and Jortner, J. (1976). *Chem. Phys. Lett.* **40**, 150.
Myerson, A. L., Thomas, H. M., and Joseph, P. J. (1965). *J. Chem. Phys.* **42**, 3331.
Nakashima, N., and Mataga, N. (1975). *Chem. Phys. Lett.* **35**, 487.
Neuimin, H., and Terenin, A. (1936). *Acta Physicochim. URSS* **5**, 465.
Nicholas, J. E., and Norrish, R. G. W. (1968). *Proc. R. Soc. London, Ser. A* **307**, 391.
Nishimura, T., Nakashima, N., and Mataga, N. (1977). *Chem. Phys. Lett.* **46**, 334.
Norrish, R. G. W., and Porter, G. (1949). *Nature (London)* **164**, 658.
Norrish, R. G. W., and Porter, G. (1952). *Proc. R. Soc. London, Ser. A* **210**, 439.
Norrish, R. G. W., Porter, G., and Thrush, B. A. (1953). *Proc. R. Soc. London, Ser. A* **216**, 165.
Norrish, R. G. W., Porter, G., and Thrush, B. A. (1954). *Proc. R. Soc. London, Ser. A* **227**, 723.
Okabe, H. (1965). *J. Opt. Soc. Am.* **54**, 478.
Okabe, H. (1967). *J. Chem. Phys.* **47**, 101.
Okabe, H. (1970). *J. Chem. Phys.* **53**, 3507.
Okabe, H. (1972a). *J. Chem. Phys.* **56**, 3378.
Okabe, H., (1972b). *J. Chem. Phys.* **56**, 4381.
Okabe, H. (1975). *J. Chem. Phys.* **62**, 2782.

Okabe, H., and Dibeler, D. H. (1973). *J. Chem. Phys.* **59**, 2430.

Okabe, H., and Lenzi, M. (1967). *J. Chem. Phys.* **47**, 5241.

Ostertag, E., and Grun, J. B. (1977). *Appl. Phys. Lett.* **31**, 509.

Pao, Y.-H., ed. (1977). "Optoacoustic Spectroscopy and Detection." Academic Press, New York.

Parker, J. H., and Pimentel, G. C. (1969). *J. Chem. Phys.* **51**, 91.

Parkes, D. A., and Quinn, C. P. (1975). *Chem. Phys. Lett.* **33**, 483.

Parkes, D. A., Paul, D. M., Quinn, C. P., and Robson, R. C. (1973). *Chem. Phys. Lett.* **23**, 425.

Parmenter, C. S., and Poland, H. M. (1969). *J. Chem. Phys.* **51**, 1551.

Parmenter, C. S., and Schuyler, M. W. (1970). *Chem. Phys. Lett.* **6**, 339.

Pasternack, L., Baronavski, A. P., and McDonald, J. R. (1978). *J. Chem. Phys.* **69**, 4830.

Patel, C. K. N., Burkhardt, E. G., and Lambert, C. A. (1974). *Science* **184**, 1173.

Peterson, N. C., Kurylo, M. J., Braun, W., Bass, A. M., and Keller, R. A. (1971). *J. Opt. Soc. Am.* **61**, 746.

Phillips, E., Lee, L. C., and Judge, D. L. (1976). *J. Chem. Phys.* **65**, 3118.

Phillips, E., Lee, L. C., and Judge, D. L. (1977). *J. Chem. Phys.* **66**, 3688.

Phillips, L. F. (1971). *Rev. Sci. Instrum.* **42**, 1078.

Phillips, L. F. (1975). *Prog. React. Kinet.* **7**, 83.

Pimentel, G. C., and Herr, K. C. (1964). *J. Chim. Phys.* **61**, 1509.

Poole, P. R., and Pimental, G. C. (1975). *J. Chem. Phys.* **63**, 1950.

Porter, G. (1950). *Proc. R. Soc. London, Ser. A* **200**, 284.

Porter, G., and Wright, F. J. (1953). *Discuss. Faraday Soc.* **14**, 23.

Porter, G., Sadkowski, P. J., and Tredwell, C. J. (1977). *Chem. Phys. Lett.* **49**, 416.

Porter, T. L., Mann, D. E., and Acquista, N. (1965). *J. Mol. Spectrosc.* **16**, 228.

Powell, H. T., and Kelley, J. D. (1974). *J. Chem. Phys.* **60**, 2191.

Powell, H. T., Murray, J. R., and Rhodes, C. K. (1974). *Appl. Phys. Lett.* **25**, 730.

Pressley, R. J., ed. (1971). "Handbook of Lasers." Chem. Rubber Publ. Co., Cleveland, Ohio.

Ramsay, D. A. (1952). *J. Chem. Phys.* **20**, 1920.

Ravishankara, A. R., Wagner, S., Fischer, S., Smith, G., Schiff, R., Watson, R. T., Tesi, G., and Davis, D. D. (1978). *Int. J. Chem. Kinet.* **10**, 783.

Rebbert, R. E., and Ausloss, P. (1972). *J. Photochem.* **1**, 171.

Reilly, J. P., Clark, J. H., Moore, C. B., and Pimentel, G. C. (1978). *J. Chem. Phys.* **69**, 4381.

Reinhardt, K., Wagner, H.Gg., and Wolfrum, J. (1969). *Ber. Bunsenges. Phys. Chem.* **73**, 638.

Rentzepis, P. M., and Bush, G. E. (1972). *Mol. Photochem.* **4**, 353.

Richardson, T. H., and Setser, D. W. (1977). *J. Am. Chem. Soc.* **81**, 2301.

Ridley, B. A., Davenport, J. A., Stief, L. J., and Welge, K. H. (1972). *J. Chem. Phys.* **57**, 520.

Ridley, B. A. *et al.* (1978).

Ritter, J. J., and Freund, S. M. (1976). *J. Chem. Soc., Chem. Commun.* p. 811.

Rosen, B. (1970). "International Tables of Selected Constants," vol. 17: "Spectroscopic Data Relative to Diatomic Molecules." Pergamon Press, Oxford.

Rosengren, L.-G. (1975). *Appl. Opt.* **14**, 1960.

Sabety-Dzvonik, M., and Cody, R. (1976). *J. Chem. Phys.* **64**, 4794.

Sabety-Dzvonik, M., Cody, R., and Jackson, W. M. (1976). *Chem. Phys. Lett.* **44**, 1311.

Sabety-Dzvonik, M., Cody, R., and Jackson, W. M. (1977). *J. Chem. Phys.* **66**, 125.

Schacke, H., Schamtjko, K. J., and Wolfrum, J. (1973). *Ber. Bunsenges. Physik. Chem.* **77**, 248.

Scheps, R., Florida, D., and Rice, S. A. (1974). *J. Chem. Phys.* **61**, 1730.

Schmatjko, K. J., and Wolfrum, J. (1975). *Ber. Bunsenges. Phys. Chem.* **79**, 696.

Schmatjko, K. J., and Wolfrum, J. (1977). *Ber. Bunsenges. Phys. Chem.* **82**, 419.

Schnell, W., and Fisher, G. (1978). *Opt. Lett.* **2**, 67.

Schröder, H., Neussen, H. J., and Schlag, E. W. (1975). *Opt. Commun.* **14**, 395.

Searles, S. K., and Djeu, N. (1973). *IEEE J. Quantum Electron.* **qe-9**, 116.

Selinger, B. K., and Ware, W. R. (1970). *J. Chem. Phys.* **53**, 3160.

Shapiro, S. L., Hyer, R. C., and Campillo, A. J. (1974). *Phys. Rev. Lett.* **33**, 513.

Shimoda, K. (1976). *Top. Appl. Phys.* **2**, 198.

Shortridge, R. G., and Lin, M. C. (1975). *Chem. Phys. Lett.* **35**, 146.

Shortridge, R. G., and Lin, M. C. (1976). *J. Chem. Phys.* **64**, 4076.

Singleton, D. L., Irwin, R. S., and Cvetanović, R. J. (1977). *Can. J. Chem.* **55**, 3321.

Slanger, T. G., and Black, G. (1970). *J. Chem. Phys.* **53**, 3717.

Slanger, T. G., and Black, G. (1973). *J. Chem. Phys.* **58**, 194.

Slanger, T. G., and Black, G. (1974). *J. Chem. Phys.* **60**, 468.

Slanger, T. G., and Black, G. (1976). *J. Chem. Phys.* **64**, 4442.

Slanger, T. G., Wood, B. J., and Black, G. (1971). *J. Geophys. Res.* **76**, 8430.

Slanger, T. G., Sharpless, R. L., Black, G., and Filseth, S. V. (1974). *J. Chem. Phys.* **61**, 5022.

Slobodskaya, P. V. (1948). *Izv. Akad. Nauk. SSSR, Fiz.* **12**, 656.

Smith, I. W. M. (1967). *Discuss. Faraday Soc.* **44**, 194.

Smith, I. W. M. (1973). *J. Chem. Soc., Faraday Trans 2* **69**, 1617.

Smith, I. W. M., and Zellner, R. (1973). *J. Chem. Soc., Faraday Trans.* **2 69**, 1617.

Smith, I. W. M., and Zellner, R. (1974). *J. Chem. Soc., Faraday Trans.* **2 70**, 1045.

Smith, I. W. M., and Zellner, R. (1975). *Int. J. Chem. Kinet.* **7**, Suppl., 341.

Smith, I. W. M., and Zellner, R. (1976). *J. Chem. Soc., Faraday Trans.* **2 72**, 1459.

Smith, K. K., Koo, J. Y., Schuster, G. B., and Kaufman, K. K. (1977). *Chem. Phys. Lett.* **48**, 267.

Smith, G. K., Butler, J. E., and Lin, M. C. (1979). *Chem. Phys. Lett.* **65**, 115.

Spears, K. G., and Hoffland, L. (1977). *J. Chem. Phys.* **66**, 1755.

Spears, K. G., and Rice, S. A. (1971). *J. Chem. Phys.* **55**, 5561.

Speiser, S., and Jortner, J. (1976). *Chem. Phys. Lett.* **44**, 399.

Stone, J., Goodman, M. F., and Dow, D. A. (1976). *Chem. Phys. Lett.* **44**, 411.

Streit, G. E., Howard, C. J., Schmeltekopf, A. L., Davidson, J. A., and Schiff, H. I. (1976). *J. Chem. Phys.* **65**, 4761.

Stuhl, F., and Niki, H. (1971). *J. Chem. Phys.* **55**, 3943.

Stuhl, F., and Niki, H. (1972). *J. Chem. Phys.* **57**, 3671.

Sudbø, Aa.S., Schulz, P. A., Grant, E. R., Shen, Y. R., and Lee, Y. T. (1978). *J. Chem. Phys.* **68**, 1306.

Tablas, F. M. G., and Pimentel, G. C. (1970). *IEEE J. Quantum Electron.* **qe-6**, 176.

Tan, Y. L., Winer, A. M., and Pimentel, G. C. (1972). *J. Chem. Phys.* **57**, 4028.

Tanaka, I., and McNesby, J. R. (1962). *J. Chem. Phys.* **36**, 3170.

Tanaka, I., Carrington, T., and Broida, H. P. (1961). *J. Chem. Phys.* **35**, 750.

Tatarczyk, T., Fink, E. F., and Becker, K. H. (1976). *Chem. Phys. Lett.* **40**, 126.

Terenin, A. (1926). *Z. Phys.* **37**, 98.

Terenin, A., and Prileshajewa, N. (1931). *Z. Phys. Chem., Abt. B* **13**, 72.

Thrush, B. A., and Zwolenik, J. J. (1963). *Trans. Faraday Soc.* **59**, 582.

Tohma, K. (1975). *Opt. Commun.* **15**, 17.

Tokue, I., Urisu, T., and Kuchitsu, K. (1974–1975). *J. Photochem.* **3**, 273.

Tomboulian, D. H., and Hartman, P. L. (1956). *Phys. Rev.* **102**, 1423.

Toth, R. A., Hunt, R. H., and Plyer, E. K. (1969). *J. Mol. Spectrosc.* **32**, 85.

Trainor, D. W., and von Rosenberg, C. W., Jr. (1974). *J. Chem. Phys.* **61**, 1010.

Tsang, W., Bauer, S. H., and Waelbroeck, F. (1962). *J. Phys. Chem.* **66**, 282.

Tyerman, W. J. R. (1969). *J. Chem. Soc. A* p. 2483.

Umstead, M. E., and Lin, M. C. (1977). *Chem. Phys.* **25**, 353.

van den Bergh, H. E., and Callear, A. B. (1971). *Trans. Faraday Soc.,* **67**, 2017.

Vear, C. J., Hendra, P. J., and Macfarlane, J. J. (1972). *J. Chem. Soc. Chem. Commun.* 381.

Venkateswarlu, P. (1950). *Phys. Rev.* **77**, 676.

Vervloet, M., and Merienne-Lafore, M. F. (1978). *J. Chem. Phys.* **69**, 1257.

Viengerov, M. L. (1940). *Izv. Akad. Nauk. SSSR, Fiz.* **4**, 94.

von Rosenberg, C. W., Jr., and Trainor, D. W. (1974). *J. Chem. Phys.* **61**, 2442.

Wang, C. C., Davis, L. I., Wu, C. H., Japar, S., Niki, A., and Weinstock, B. (1975). *Science* **189**, 797.

Ware, W. R. (1971). *In* "Creation and Detection of the Excited State" (A. A. Lamola, ed.), Vol. 1, p. 213, Dekker, New York.

Ware, W. R., and Garcia, A. M. (1974). *J. Chem. Phys.* **61**, 187.

Watson, R. T. (1977). *J. Chem. Phys., Ref. Data* **6**, 871.

Watson, R. T., Machado, G., Conaway, B., Wagner, S., and Davis, D. D. (1977). *J. Phys. Chem.* **81**, 256.

Weiz, E., and Flynn, G. (1974). *Annu. Rev. Phys. Chem.* **25**, 275.

Welge, K. H. (1966a). *J. Chem. Phys.* **45**, 1113.

Welge, K. H. (1966b). *J. Chem. Phys.* **45**, 4373.

Welge, K. H., and Atkinson, R. (1976). *J. Chem. Phys.* **64**, 531.

West, M. A. (1976). *In* "Creation and Detection of the Excited State" (W. R. Ware, ed.), Vol. 4, p. 218. Dekker, New York.

Willets, F. W. (1972). *Prog. React. Kinet.* **6**, 51.

Williams, D., Wenstrand, D. C., Brockman, R. J., and Curnutte, B. (1971). *Mol. Phys.* **20**, 769.

Wood, R. W. (1905). *Philos. Mag.* [5] **10**, 513.

Wurzberg, E., Grimley, A. J., and Houston, P. L. (1978). *Chem. Phys. Lett.* **57**, 373.

Yardley, J. T., and Moore, C. B. (1966). *J. Chem. Phys.* **45**, 1066.

Yardley, J. T., and Moore, C. B. (1968). *J. Chem. Phys.* **49**, 1111.

Yguerabide, J. (1965). *Rev. Sci. Instrum.* **36**, 1734.

Young, L. A., and Eachus, W. J. (1966). *J. Chem. Phys.* **44**, 4195.

Young, R. A., Black, G., and Slanger, T. G. (1968). *J. Chem. Phys.* **49**, 4769.

Yu, W., Pellegrino, F., Grant, M., and Alfano, R. R. (1977). *J. Chem. Phys.* **67**, 1766.

Zare, R. N., and Dagdigian, P. J. (1974). *Science* **185**, 739.

Zetzsch, C., and Stuhl, F. (1975). *Chem. Phys. Lett.* **33**, 375.

5

Production of Small Positive Ions in a Mass Spectrometer

LARRY I. BONE†

Department of Chemistry
Appalachian State University
Boone, North Carolina

I. INTRODUCTION

Ion–molecule reactions have been observed in mass spectrometers since the instrument was first developed (Dempster, 1916). In fact, early investigations found them to be the source of great difficulty. Since one of the most valuable pieces of information one gets from a mass spectrometer is the mass of the molecular ion, it was very difficult to identify a compound if reactions led to species that were heavier than the molecule under investigation or if apparent fragments were produced which were not very characteristic of the original molecule. Good analytical mass spectrometry had to wait on the

---

†Present address: Dow Chemical Company, Freeport, Texas.

development of better vacuum technology so that ion–molecule reactions could be eliminated.

Some of the first ion–molecule reaction studies, by modern standards, were carried out by Stevenson and Schissler (1955), Field, et al. (1957), and Tal'roze and Lyubimova (1952). For the most part they used instruments with magnetic analyzers which were designed for analytical purposes. Most of these studies were accomplished by increasing the pressure in a fairly conventional analytical ion source such that the ions produced by electron impact from a filament suffered reactive collisions as they were swept out of the source under the influence of a repeller voltage. One of the most difficult tasks was to relate a given product ion to a specific reactant ion and a major accomplishment of some of the early studies was unraveling a very complex scheme of reactions. In these early studies there were basically two parameters, pressure and repeller voltage, which could be manipulated to effect the extent of the reactions taking place. The pressure in the ion source could be increased which had the effect of increasing the extent of the reaction. Plots of the ion current (or relative ion currents) for various reactant and product ions as a function of the pressure in the source allowed one to deduce mechanisms and measure rate constants. The electric field strength in the source could be changed by varying the repeller voltage. This changed the residence time of the ions in the source and, in conjunction with the pressure, controlled the ion kinetic energy. The ratio of the field strength to the pressure (E/P) was a measure of this kinetic energy and thus by changing the repeller voltage one could vary a parameter closely related to the reaction temperature.

Later, various pulsing techniques were developed (Tal'roze and Frankevich, 1960; Tal'roze, 1962). If the repeller voltage in the source is turned off for brief periods of time, the residence time of the ions in the source is increased and consequently so is the extent of reaction. The relative ion abundances can then be studied as a function of the delay time and kinetic data can then be deduced. An additional advantage of this technique is that the ions react more nearly at thermal kinetic energies. If in addition to the repeller, the filament can be pulsed, even better kinetic data can be deduced because of the establishment of an actual time base.

Modern ion–molecule studies are frequently carried out on instruments designed, or modified, specifically for this purpose. They employ a variety of methods for producing the reactant ion. In addition to electron impact, photoionization, radioactive isotopes, sparks, discharges, and plasmas have been used. A variety of different instrument types have also been used. The quadrupole, time of flight, and ion cyclotron resonance instruments have all made significant contributions. A variety of detecting methods have also been used. The electron multiplier and various electrometers are most popular while pulse counting techniques, borrowed from the nuclear field, have allowed significant refinements to take place. Some of the more active areas of interest in the field of ion–molecule reactions in recent years involve the use of higher pressures to study clustering or condensation reactions. Of particular value in this area are the advances made in ion thermochemistry (Kebarle, 1977) by actually studying ionic equilibria in the gas phase. These equilibria involve proton, electron, and hydride transfer reactions as well as condensation. Another area receiving intense interest in the more recent literature involves the reactions of negative ions. The field of mass spectrometry has long been dominated by positive ions but that is changing rapidly, not only in the ion–molecule community, but the analytical community as well.

The purpose of this chapter is to provide a basic exposure to techniques used in the field, not to discuss ion–molecule reactions for anyone with experience in the field. Emphasis will be placed on techniques and instruments rather than the chemistry, physics, or theory involved. We also have tried to point out good review articles when they are available. Finally, we have concentrated most heavily on studies involving photoionization principally because of our own research interest and background.

II. BASIC INSTRUMENTATION

A variety of different mass analyzers have been used in ion–molecule mass spectrometers. Although the actual reactions take place in the source and the mass analyzer is simply used to weigh and count ions, there are a few characteristics of the different types of analyzers which have some bearing on the methods of ion production and reaction. Consequently each type of mass analyzer will be discussed briefly.

A. Magnetic Analyser

The magnetic analyzer has been most frequently used, at least until the last five years or so. Ions are separated by their mass-to-charge ratio according to the following equation:

$$m/e = H^2r^2/2V,$$

where m/e is the mass-to-charge ratio, H the magnetic field strength, V the accelerating voltage of the ion or actually its kinetic energy, and r the radius of curvature through the magnet. The most important aspect of this type of analyzer for ion–molecule studies is that in order to properly resolve ions of different m/e, the spread in their kinetic energy must be small compared to their total kinetic energy. This is unfortunate since collisons, especially reactive ones, cause spreads in kinetic energy and consequently, to reduce the relative spread, the total kinetic energy must be quite large. This means high accelerating voltages. The combination of higher source pressures (10^{-3} Torr to a few Torr is typical) and high voltages make things difficult and/or expensive. The transmission efficiency of ions through the instrument, however, is not a function of the mass of the ion as is the case with the quadrupole analyzer.

B. Quadrupole Analyzer

The use of quadrupole analyzers has shown a spectacular increase in recent years, both in ion–molecule as well as analytical applications. Ions are mass analyzed by their ability to follow a combined rf and dc voltage placed on four rods located in a square array. At a certain combination of rf frequency, rf amplitude, and dc voltage only one particular mass-to-charge ratio has a stable trajectory through the array. The spectra can be mass scanned by periodically varying the rf and dc voltages. The equations of motions in the three directions in space are

$$m\frac{d^2x}{dt^2} + \frac{2e(U + Vo\cos wt)x}{r_0^2} = 0,$$

$$m\frac{d^2y}{dt^2} + \frac{2e(U + Vo\cos wt)Y}{r_0^2} = 0,$$

$$m\frac{d^2z}{dt^2} = 0,$$

where m is the mass of the ion, e the unit charge, U the dc voltage, Vo the rf voltage, w the rf frequency, and r_0 is the radius of the rods. The z direction is that direction down the instrument parallel to the rods while x and y are perpendicular. One of the principle advantages of this type of analyzer is that the instrument performance is very insensitive to the velocity or kinetic energy in the z direction as can be seen from the equations. This means that low accelerating voltages (typically 20 V or lower) are used. The strong focusing power of the instrument also allows considerable energy spread in the x and y directions without loss of sensitivity.

The most significant disadvantage of the quadrupole instrument is its discrimination against ions of higher mass. Most quadrupole instruments simply transmits ions of lower mass with greater efficiency. The higher the resolution of the instrument (a factor which can be controlled by the rf-to-dc-voltage ratio), the greater the mass discrimination. Although some commercial analytical instruments have variable voltage ratios to correct for this; it must always be considered. Some researchers have corrected for it

(Sieck *et al.*, 1969) and others have operated at sufficiently low resolution (<30) to minimize the effect (Turner and Bone, 1974; Smets *et al.*, 1977).

C. Ion Cyclotron Resonance Spectrometer

A very popular instrument in recent years is the ion cyclotron resonance (ICR) spectrometer (Beauchamp, 1961). Ions are produced in a rectangular cell, usually by electron impact. The cell is exposed to crossed magnetic and rf fields such that an ion of a given mass-to-charge ratio (m/e) can absorb rf energy which corresponds to its cyclotron frequency ω according to the following equation:

$$\omega e = eH/mc,$$

where H is the strength of the magnetic field and c the velocity of light. Further refinements allow a reactant ion to be radiated with one oscillator while the spectrum is swept with another marginal oscillator. Since, in general, the cross section for an exothermic ion–molecule reaction decreases with energy, product ions which are reactively coupled decrease in intensity when the reactant from which they are produced is radiated, while products of endothermic reactions will increase when their precursor ion is irradiated.

A further refinement, particularly useful for the study of gas phase ionic equilibria, involves the adaptation of ion trapping techniques (McIver, 1970) to the ICR instrument. A trapped ion ICR with pulsed electron impact ionization has been used by McIver (1970) and McMahon and Beauchamp (1972). Ions are trapped by a magnetic field crossed by perpendicular static voltages applied to the sides of the cell. Ions are produced by a pulse from the filament, circle at their cyclotron frequency for a delay time which is experimentally measurable, and are subsequently detected by pulsing the magnetic field such that they are detected by a marginal oscillator at the appropriate frequency.

An advantage of the ICR spectrometer for ion–molecule studies is its ability to unambiguously determine reactant–product relationships and the added bonus of identifying exothermic or endothermic processes. Ion abundances can be determined without the use of a leak making pressure measurements easier. Long reaction times are possible at low pressures (10^{-6} Torr). One of the most common criticisms is that an ion must be accelerated to be measured and thus cannot be, at least at that instant, in thermal equilibrium. Recent experimental results indicate that thermal equilibrium is achieved (Aue *et al.*, 1976; Wolf *et al.*, 1977). Third-order processes have not been extensively studied by ICR because they are too slow at the low pressures used.

D. Tandem and Beam Instruments

A number of important contributions to ion–molecule chemistry have been made by tandem mass spectrometers (Futrell and Miller, 1966; Smith and Futrell, 1974) or crossed ion and molecular beams (Herman *et al.*, 1967; Mahan, 1968). A very prolific tandem instrument was the one used at Aerospace Research Laboratories (Bone and Futrell, 1967; Tiernan and Bhattacharya, 1970). The instrument allowed the investigator to select a given reactant ion, produced in the first mass spectrometer, and react it with a given neutral molecule in the source of the second instrument at translational energies variable down to near-thermal velocity. The EVA (Herman *et al.*, 1967) instrument at Harvard yielded some very interesting results by crossing molecular and ion beams and measuring the angular and energetic distribution of the reaction products. Information from this type of experiment allowed one to understand the role of activated complexes or stripping mechanisms in ion–molecule kinetics.

III. POSITIVE ION PRODUCTION

Positive ions have been produced for a study of ion–molecule reactions in a variety of ways. Some of the earliest ion–molecule kinetics were deduced from radiation chemistry studies (Ausloos and Lias, 1971) involving α, x, and γ rays as well as fast electrons from accelerators. Later vacuum ultraviolet photolysis (Ausloos and Lias, 1971) contributed a wealth of information to the field. Although some of these studies used a mass spectrometer to identify reaction products, they will not be discussed here. We will limit our discussion to studies where reactive ions, reactant and product, are subject to mass spectral analysis. Methods falling within this framework include ionization in the source of mass spectrometer by: electron impact, radioactive nuclide emissions, thermal ionization, and photoionization, as well as flowing afterglow studies.

IV. ELECTRON IMPACT

Most of the early work, as well as the majority of the work today, involves ions produced by electron impact. Electrons, emitted by a heated filament, are accelerated by a single, or a series of, lens elements to a voltage which is usually controllable. They are then focused into a source (a reaction chamber) containing a gas or mixture of gases to be studied. Initially a molecular ion (the molecule less an electron) is produced by the bombarding electron. Since it is found experimentally that the ionization cross section is a very strong function of electron energy up to an energy of 100 eV, many studies are carried out using electron energies far in excess of the minimum energy required to ionize the molecule under study, i.e., in excess of the ionization potential. Such high energies are often required to generate satisfactory intensities but lead to fragmentation and excited ions. The reactant ions produced in such a study then have variable amounts of internal energy of all types, as well as kinetic energies resulting from fragmentation processes. Additional kinetic energy results from electric fields such as a repeller or drawout potential.

A number of techniques are possible to control or define the energetics of an ion produced by electron impact. If sufficient ion detection sensitivity is available to overcome the loss in ion production, lower energy electrons can be used. Because of the Franck–Condon principle, lowering of the electron energy can reduce excitation of the initial molecular ion but usually does not eliminate excitation and frequently not even fragmentation. Further, the energy spread of electrons emitted from a hot filament is quite broad. Monoenergetic electrons, or at least beams with a narrow energy spread, can be produced by elaborate optics or by using electrostatic sectors (Marmet and Kerwin, 1960) to energy sort the ion beam. Such techniques, of course, are accompanied by a dramatic loss in intensity. For the most part, elaborate attempts to define or control ion energetics by controlling the energy of the ionizing electrons are not common. Some investigators (Foster and Beauchamp, 1975) however, do use lower energies to exercise some control.

Collisional deactivation (Kebarle, 1972) or long reaction times (Baldeschwieler and Woodgate, 1971) are the most successful ways to produce ground state ions by electron impact. Kebarle was an early pioneer in developing techniques to study the reactions of ground state ions. He used high pressures in his ion source, buffer gasses, and pulsing techniques to allow longer residence times of ions in the source and to assure that the reactions took place in a field-free region. Initially, these efforts were difficult with conventional filaments so radioactive isotopes (Kebarle et al., 1966), proton generators (Collins and Kebarle, 1967), and electron guns (Arshadi et al., 1970) were used.

Tandem mass spectrometers (Futrell and Miller, 1966; Smith and Futrell, 1974), which produce the reactant ion in one source and react it in another, have contributed a great deal to the understanding of the effects of ion energetics on ion–molecule reactions. In these instruments, since the kinetic energy of the reactant ion can be controlled, reactions can be studied as a function of this parameter. This type of study led to the first attempts to distinguish reactions which proceeded by either a "long lived" complex or a stripping mechanism (Ding et al., 1968), i.e. the reactant ion simply strips a particle from the neutral reactant without seriously disturbing the remainder of the molecule. If these instruments use electron impact, and most do because of intensity requirements, they are still subject to the internal energy considerations discussed above. Furthermore, thermal translational energies are, at best, barely accessible. As mentioned previously, instruments of this type, particularly those with crossed molecular beams, which are also capable of measuring the angular and energy dependence of the product ions (Mahan, 1970), have contributed a wealth of information to ion–molecule and basic chemical kinetics. These instruments are also valuable in that reactant–product relationships are unambiguous. This is not always the case with other instruments.

Cermak and Herman (1961) developed a particularly clever method of producing product ions without observing their precursor reactants. Electrons produced by an ordinary filament are passed through the source of a mass spectrometer at an energy insufficient to ionize the gas. After the electrons leave the source region they enter a region containing the ion trap. Here they experience an additional acceleration provided by the voltage difference between the ionization chamber and the trap which is sufficient to ionize the gas. Hence, all primary ions are formed in the region of the trap. These primary ions are accelerated by the chamber to trap voltage and move back into the ion chamber where they can react. Product ions are drawn out of the chamber by the extracting voltage while the movement of the primary ions is in a direction parallel to the slit and are prevented from being efficiently extracted from the source. The original authors used the technique to study the breakdown pattern of molecules resulting from charge transfer but it has been used for other ion–molecule studies as well.

A. Photoionization

Photoionization mass spectrometers were developed primarily to allow more precise determination and control of the ionizing energy. Photoionization cross sections are not subject to the dramatic energy dependence found for electron impact. The photoionization cross sections, as a function of energy, are usually step functions. That is, the cross section is zero at photon energies less than the ionization potential and jumps to some value, usually acceptable for mass spectrometric analysis, at the ionization potential. Further increases in photon energy increase the kinetic energy of the ejected electron without increasing the cross section. When new states become accessible there is usually a discontinuous increase in the cross section, i.e., a step function. If autoionization is possible a somewhat different behavior is observed. The ionization cross section as a function of photon energy resembles an ordinary absorption spectrum, i.e., bands or lines are observed and discrete ion states are produced (Chupka et al., 1970).

Photoionization thus allows the production of sufficient ion intensity at the ionization threshold which is not available by electron impact. Photoionization is compatible with higher source pressures where arcing and pyrolysis problems are often experienced with electron filaments. Photons also are not affected by electric or magnetic fields which can lead to pulsating electron currents from filaments resulting from surface buildup of a charge or leakage. Another important consideration is that the temperature of the source can be more easily measured and controlled, i.e., the ion source is not heated by a filament. The technique is capable of producing ions with a discrete internal energy such that the effects of internal energy on ion–molecule reactions can be studied. A distinct advantage of photoionization techniques is the capability of selectively ionizing one component of a mixture (McAdams and Bone, 1972). This simplifies reactant–product relationships.

The first paper actually using photoionization in a mass spectrometer was probably the work of Ditchburn and Arnot (1929), followed by a similar experiment by Terenin and Popov (1932). The latter experiment used a quartz prism in conjunction with an arc from cadmium, zinc, or aluminum to study the photo pair production in thallium halides. The study of metal halides was particularly accessible to photoionization techniques and a number of workers have concentrated on this area. Much of this work was reviewed by Berkowitz (1971).

Photoionization mass spectrometry, like the entire field of mass spectrometry a few decades earlier, had to await better vacuum technology. The first paper using energies capable of ionizing molecules other than high temperature species was written by Lossing and Tanaka (1956). They used a krypton discharge lamp with a 0.5 mm lithium fluoride window. Such a lamp produces two lines at 1236 Å (10.03 eV) and 1165 Å (10.63 eV). The authors recognized the potential of the technique which "permits a sharp separation of ion-forming processes without the loss of sensitivity inherent in low energy electron experiments." They also pointed out an important experimental difficulty observed in photoionization. One must prevent the production of photoelectrons in regions where they can be accelerated by an electric field or an electron impactlike spectra is observed.

The versatility of a photoionization mass spectrometer is greatly enhanced by the addition of a vacuum monochromator. This improvement was first demonstrated by Hurzeler et al. (1957). The first instrument used lithium fluoride windows but differential pumping allows a windowless system. This advance was first demonstrated in 1966 (Elder et al., 1966) and allows wavelengths of shorter than 1050 Å (11.81 eV) to be used. The expense and experimental difficulty of operating a windowless system is, of course, considerable. Thin films have been used to pass high energy radiation but they must be very thin and consequently very delicate to avoid serious loss of intensity. A thin aluminum foil (Sieck *et al.*, 1969) over a screen appears to be one of the best compromises between energy loss and strength.

A variety of light sources have been used in photoionization mass spectrometers. The choice of lamps is, of course, dictated by the experiment to be performed. If continuously variable photon energies are required for performing experiments such as cross section measurements or the variation in reaction rates with internal energy states, a continuum light source and a monochromator of high resolving power is required. If, on the other hand, variable energy but greater intensity is required, lamps with a line output and a monochromator with lower resolving power is useful. If one can settle for a constant energy (or a few energies), lamps with line outputs without a monochromator are preferred.

The simplest type of lamp is probably the rare gas resonance lamp. A practical design is described by Gordon *et al.* (1969). This type of lamp uses a microwave discharge to power a chamber filled with the rare gas at reduced pressure. The high purity of the gas which is required is obtained by using a titanium getter sealed into the body of the lamp. A very simple krypton lamp with LiF windows can be constructed by filling with gas at slightly greater than 1 atm pressure and trapping the gas in a cold finger at liquid nitrogen temperature when the lamp is in operation. The resulting vapor pressure is in an acceptable range for microwave excitation. The lamp is easy to build and has a long shelf life because it does not leak. There are a number of interesting kinetic experiments that can be done with these simple lamps and they are quite satisfactory for many analytical purposes. Warneck (1971) has compared the intensity with and without a monochromator (3 Å half-width) and finds a 10^3–10^4 increase in intensity if the monochromator can be avoided.

The most effective way to produce a continuum is with a pulsed discharge applied to a gas confined to a capillary. Chupka *et al.* (1975) have described techniques which resulted in 100-fold increased intensity. They have increased the pulse repetition rate in a helium Hopfield continuum lamp from 9–10 to 100 kHz and have increased the capillary length to 50 cm. It was found that the photon intensity increased almost linearly with a capillary length but improvement is limited by the voltage obtainable from the pulser. A He lamp can be used to cover the region between 580 and 1100 Å and an argon lamp between 1050 and 1550 Å. Because of the low intensity in the overlap region between these two sources, a hydrogen many-line spectrum generated by a dc discharge is used (Ajello *et al.*, 1976) to produce light wavelengths between 900 and 1100 Å.

Buttrill (1974) has described a time-resolved photoionization mass spectrometer which uses a pulsed light source. A hydrogen many-line spectrum lamp with 2 Torr of H_2 is pulsed with a high power pulse generator. The output pulse width is $10 \, \mu sec$ at a repetition rate of 359 Hz and a dc current of 0.75 A. This well-designed instrument allows kinetics studies on a time rather than a pressure basis, which is typical of many other photoionization instruments.

A few novel sources of vacuum uv energy have been used for photoionization. Andreyev et al. (1975) have used a vacuum uv laser with a wavelength of 1610 Å and a repetition rate of 10 Hz. This is sufficient to ionize 50% of the elements and many low ionization potential molecules. Taylor et al. (1974) have reported using synchrotron radiation from the Wisconsin 240 MeV storage ring.

Photoionization mass spectrometers have been used for a variety of experimental purposes, as indicated by Reid (1971) in an excellent review of the subject. These include appearance potential measurements and related energetics of fragmentation, analytical work, and chemical kinetics, especially studies designed to evaluate the effects of internal energy on reaction rates. An interesting use of a closely related instrument is the study of photodissociation of ion clusters. Burke and Wayne (1977) recently have studied the photodissociation of O_2^+, N_2^+, NO^+, and H^+ clusters with H_2O, N_2, and NO by continuous wave and pulsed laser radiation in the visible region. Related work has been carried out by Henderson and Schmeltekopf (1972) and Cosby et al. (1976).

The use of photoionization mass spectrometers for analytical purposes has probably been more extensive in the Soviet Union than in any other part of the world. Some of the early work was reviewed in the Soviet Union (Akopyan, 1969) and extensive work is continuing. Orlov and co-workers (1972, 1974) seem to be particularly active and some interesting work related to the subject of this chapter in the use of vacuum uv lasers was reported by a group at the Institute for Spectroscopy, Moscow. They report (Andreyev et al., 1975) that by measuring the photocurrent as a function of frequency one can distinguish molecular ions of the same mass but with different structure. They (Potapov et al., 1976) also report the use of two-photon processes.

The photoionization mass spectrometer is very well suited to the measurement of ionization efficiency curves and resulting related parameters such as ionization potentials, heats of formation of ions and radicals, internal energy states, Franck–Condon factors, and lifetimes of states. Such information is also used in the discussion of unimolecular decomposition theories and to verify ion structures.

Some of the groups most active in this type of research certainly include Chupka's group and Berkowitz's group both at Argonne National Laboratory, and Deibler at the National Bureau of Standards. A "bible" on the interpretation of ionization efficiency curves has been published by Guyon and Berkowitz (1971). A few other articles which are representative of the field and which are of interest to the author of this chapter are briefly summarized in the following paragraphs.

Traeger and McLaughlin (1977) report the threshold photoionization efficiency curves for the formation of $C_7H_7^+$ from toluene and cycloheptatriene. They conclude that the $C_7H_7^+$ ion formed from these two sources can not have the benzyl structure and must be tropylium ion. They further report the heat of formation of tropylium is 207 kcal/mole.

Buttrill (1974) has developed a technique to measure the time resolved fragmentation of excited ions by photoionization mass spectrometry. He shows that loss of hydrogen from the toluene molecular ion comes from only three states and that there is no evidence for a continuous distribution of lifetimes as had been earlier claimed.

Chupka et al. (1975) have studied pair production in H_2, HD, and D_2 which complements earlier studies on the same molecules (Chupka and Berkowitz, 1968, 1969; Berkowitz and Chupka, 1969). The peaks in the photoionization efficiency curves which are characteristic of pair production are measured at 78 K with a wavelength resolution of 0.035 Å. Together with the accurately known values of the dissociation energy of H_2 and the ionization potential of atomic hydrogen, they are able to calculate an electron affinity for the hydrogen atom of 0.754 eV. They also observe that HD dissociates into H^+ and D^- as compared to D^+ and H^- with H^-/D^- ratio of approximately two at threshold. The ratio decreases with increasing energy.

Smets et al. (1977) have investigated the ionization efficiency curves for P_4 decomposition into P_3^+ and P_2^+ in a photoionization mass spectrometer. They are able to evaluate the relative importance of direct and autoionization and calculate the ionization potential of P_3 (7.85 $\pm$ 0.20 eV) and the P_2-P_2 bond energy (2.37 $\pm$ 0.05 eV). Ajello et al. (1976), using a similar technique, have studied the photoionization efficiency curves for $CFCl_3$, CF_2Cl_2, and CF_3Cl. They are able to calculate heats of formation for some 12 different fragment ions.

The photoionization mass spectrometer has proven to be a very useful instrument for studying ion–molecule reactions. The most useful feature is the ability to know or control the internal energy states of the ion. The type of instrument used varies from the most simple, which use rare gas resonance lamps and thus are only capable of operating at a few select energies, to complex instruments capable of ionizing at any desired energy with a very narrow resolution.

An example of a simple instrument is the one constructed by Bone at East Texas State University. It mated a quadrupole mass spectrometer with a choice of rare gas resonance lamps. Ion–molecule reactions were allowed to proceed in an entirely field-free region and products were studied as a function of source pressure. Cluster reactions of NO^+ with H_2O (McAdams and Bone, 1972), NO^+ with methanol (Turner and Bone, 1974), and NO^+ with H_2S (Bone, 1974) were investigated by selectively ionizing NO with 10.03 eV photons from a Kr resonance lamp with a CaF_2 window to eliminate the 10.63 eV line. From this series of experiments we were able to suggest that clusters of NO^+ with molecules capable of hydrogen bonding had the following structure

$$NO^+ \; O\!\!\begin{array}{c} {}^{\displaystyle R} \\[-2pt] \diagup \\[-2pt] \diagdown \\[-2pt] {}_{\displaystyle H} \end{array} \; \cdots \; O\!\!\begin{array}{c} {}^{\displaystyle R} \\[-2pt] \diagup \\[-2pt] \diagdown \\[-2pt] {}_{\displaystyle H} \end{array}$$

That is, at least for the first few molecules, hydrogen bonding was the dominate bonding factor. Grimsued and Kebarle (1973) arrived at the same conclusion by studying clustering reactions involving the series H_2O, CH_3OH, and CH_3OCH_3.

The same instrument was used to measure relative proton affinities (Hopkins and Bone, 1973; Wei and Bone, 1974) and to investigate clustering reactions in formaldehyde (Bone and Garrett, 1976). The value of photoionization was exemplified in the proton affinity difference measured for the H_2O–H_2S pair (Hopkins and Bone, 1973). The value measured by photoionization differed from a value measured by electron impact ICR spectrometry and the difference presumably can be attributed to excited ions produced by electron impact.

A similar instrument built by Sieck at the National Bureau of Standards has been used to study ring opening in cyclobutane (Sieck et al., 1969), H^- transfer from hydrocarbons to NO^+ (Searles and Sieck, 1970), reactions of the acetone cation with acetone (Sieck and Ausloos, 1972), clustering reactions in NO_2 (Sieck et al., 1973), and reactions in C_2F_6 (Sieck et al., 1972) to cite only a few. Another similar instrument has been used (Kronberg and Stone, 1977) to measure ion–molecule reactions in $(CH_2S)_2$, CH_3SH, and C_2H_5SH. The authors present evidence for dimer ions involving S—S bonding for the first two molecules in the series and proton transfer followed by weak hydrogen bound clusters for the mercaptans.

Chupka's group has contributed extensively to the understanding of ion–molecule reactions by studying a number of reactions between ground state neutral molecules and ions in various internal (particularly vibrational) energy states (Chupka et al., 1970). They have studied

$$H_2^+ + H_2 \longrightarrow H_3^+ + H \qquad \text{(Chupka et al., 1968)},$$
$$NH_3^+ + NH_3 \longrightarrow NH_4^+ + NH_2 \qquad \text{(Chupka and Russell, 1968a), and}$$
$$H_2^+ + He \longrightarrow HeH^+ + H \qquad \text{(Chupka and Russell, 1968a)}$$

as a function of vibrational energy. They found that for exothermic reactions at low translational energies, the effect of vibrational energy was small and the cross section decreased as the vibrational energy increased. This is predicted by theory (Light, 1965; Light and Lin, 1968). Endothermic reactions, however, show a larger variation of cross section with vibrational energy and the effect is in the opposite direction. Vibrational energy can be effectively used to make up the deficit in the reaction energetics. In fact, these studies show that vibrational energy is more effective than translational energy in promoting reactions.

Le Breton *et al.* (1975) have used both a photoionization and an ICR mass spectrometer to study the following reaction as a function of internal energy

$$C_2H_4^+ + C_2H_4 \longrightarrow C_3H_5^+ + CH_3.$$

They find that the cross section for this exothermic reaction decreases with increasing vibrational energy and that the rate of decrease is greater above the third vibrational level. They also find that the cross section for an ion in the $^2B_3$ state is much lower but that this electronic energy can be easily relaxed in collisions. The same group has measured ion–molecule reactions in ketene as a functional of vibrational energy and find that the cross section for the endothermic reaction

$$CH_2CO^+ + CH_2CO \longrightarrow C_2H_4^+ + 2CO$$

becomes finite when the internal energy equals the endothermicity and increases with vibrational energy thereafter. In contrast, the cross section for the exothermic reaction

$$CH_2CO^+ + CH_2CO \longrightarrow C_3H_4O^+ + CO$$

decreases slightly with increasing vibrational energy.

B. Thermal Ionic Emission

Thermal ionic emission has been used successfully to study gas phase ion–molecule reactions involving metal ions such as Pb, Bi, Sr, and the alkali and alkaline earth elements. Metal ions are produced by painting a suitable salt containing the desired ion onto a filament. The filament is heated by passing a current through it and the ions are drawn off by a negative potential between the filament and the reaction chamber. Sometimes adequate beams ($\sim 10^{-7}$ A) can be obtained by using either a melt or an aqueous paste of a simple salt although a more elaborate mixture (Jenkins and Trodden, 1965; Blewett and Jones, 1936) can be used to enhance ion production.

The first study of ion–molecule reactions of metal ions produced by thermal ionic emission was reported by Searles and Kebarle (1969) and involved hydration of Li^+, Na^+, Rb^+, and Cs^+. They used salts painted on a platinum gauze filament, an enclosed filament region such that the ions could be collisionally thermalized prior to reaction, and a field-free reaction chamber. They were able to measure equilibrium constants, enthalpies, and entropies for the successive clustering reactions

$$M^+(H_2O)n - 1 + H_2O \longrightarrow M^+(H_2O)n,$$

by observing ion abundances as a function of water pressure and temperature. Further studies of this type have been reported by the same research group (Dzidic and Kebarle, 1970; Davidson and Kebarle, 1976).

Tang and Castleman (1972, 1974, 1975) have also been active in this area. They have studied hydration reactions with Pb^+, Bi^+, and Sr^+ and report clustering of up to eleven water molecules. They are particularly interested in the effect of valence electrons of ions with open shell structures on the binding of molecules of hydration. They use an apparatus very similar to the one reported by Searles and Kebarle (1969). Tang et al. (1976) report that an ion source with a painted filament can produce a steady Sr^+ current for months with continued daily operations.

The author of this chapter has used a source of a similar design to study clustering reactions of Na^+, K^+, Cs^+, Sc^+, and $(CH_3)_4N^+$. A practical observation is that we found it necessary to use a buffer gas in the region of the filament to produce a thermalized ion beam suitable for reaction studies. Further, higher ion intensities and longer source lifetimes can be obtained by using a screen or mesh filament rather than a wire (we used a tungsten screen). Although a new source usually emits low ionization potential ions (Na^+ and K^+) immediately after it has been prepared, after a few hours of operation the trace impurities will be used up and the peaks will go away.

C. Flowing Afterglow

One of the most highly developed techniques for the study of ion–molecule reactions is the flowing afterglow method (Ferguson et al., 1969) or the related flow–drift techniques (McFarland et al., 1973). Ions of either polarity are produced by a filament or electron gun in the presence of a buffer or carrier gas (usually He) at one end of the apparatus. These ions are then swept down the tube by the carrier gas (velocity $\cong 10^4$ cm/sec.). Other gases may be added through inlet ports downstream in order to produce the desired reactant ion and supply the desired neutral reactant. The tube is terminated by a sampling orifice and a quadrupole mass spectrometer.

Some examples of the chemical versatility of instruments of this type are contained in a review by Ferguson (1975). The technique can be used to measure ion drift velocity or mobilities by pulsing the ions entering a drift region. Rate constants can be measured as a function of translational energy by varying the field in the tube or as a function of temperature (Fehsenfeld, 1975). Equilibrium constants can be measured by observing the forward and reverse reaction separately or by allowing equilibrium conditions to be attained (Bohme, 1975).

REFERENCES

Ajello, J. M., Huntress, W. T., Jr., and Rayerman, P. (1976). *J. Chem. Phys.* **64**, 4746.

Akopyan, M. E. (1969). *Usp. Fotoniki* **1**, 46–77; *Chem. Abstr.* **72**, 37670f (1970).

Andreyev, S. V., Antonov, U. S., Kryazev, I. N., Letokhov, U. S., and Moushev, V. G. (1975). *Phys. Lett. A* **54**, 91.

Arshadi, M., Yamdagni, R., and Kebarle, P. (1970). *J. Phys. Chem.* **74**, 1475.

Aue, D. H., Webb, H. M., and Bowers, M. T., *J. Am. Chem. Soc.* **98**, 311.

Ausloos, P. J., and Lias, S. G. (1967). "Actions chimiques et biologiques des radiations," Chapter I, p. 1. Masson, Paris.

Ausloos, P. J., and Lias, S. G. (1971). *Annu. Rev. Phys. Chem.* **22**, 85.

Baldeschwieler, J. D., and Woodgate, S. S. (1971). *Acc. Chem. Res.* **4**, 114.

Beauchamp, J. L. (1961). *Annu. Rev. Phys. Chem.* **22**, 527.

Berkowitz, J. (1971). *Adv. High Temp. Chem.* **3**, 123.

Berkowitz, J., and Chupka, W. A. (1969). *J. Chem. Phys.* **51**, 2341.

Blewett, J. P., and Jones, E. J. (1936). *Phys. Rev.* **50**, 464.

Bohme, D. K. (1975). *In* "Interactions of Ions with Molecules" (P. Ausloos, ed.), p. 489. Plenum, New York.

Bone, L. I. (1974). *Adv. Mass Spectrom.* **6**, 753.

Bone, L. I., and Futrell, J. H. (1967). *J. Chem. Phys.* **46**, 4084.

Bone, L. I., and Garrett, M. A. (1976). *J. Chem. Phys.* **64**, 3892.

Burke, R. R., and Wayne, R. P. (1977). *Int. J. Mass Spectrom. Ion Phys.* **25**, 199.

Buttrill, S. E., Jr. (1974). *J. Chem. Phys.* **61**, 609.

Cermak, V., and Herman, Z. (1961). *Nucleonics* **19**, 106.

Chupka, W. A., and Berkowitz, J. (1968). *J. Chem. Phys.* **48**, 5726.

Chupka, W. A., and Berkowitz, J. (1969). *J. Chem. Phys.* **51**, 4244.

Chupka, W. A., Berkowitz, J., and Russell, M. E. (1970). *Polym. Prep. Am. Chem. Soc., Div. Polym. Chem.* **15**, D58.

Chupka, W. A., and Russell, M. E. (1968a). *J. Chem. Phys.* **48**, 1527.

Chupka, W. A., and Russell, M. E. (1968b). *J. Chem. Phys.* **49**, 5426.

Chupka, W. A., Russell, M. E., and Refaey, K. (1968). *J. Chem. Phys.* **48**, 1518.

Chupka, W. A., Dehmer, P. M., and Jivery, W. T. (1975). *J. Chem. Phys.* **63**, 3929.

Collins, J. G., and Kebarle, P. (1967). *J. Chem. Phys.* **46**, 1082.

Cosby, P. C., Ling, J. H., Peterson, J. R., and Moseley, J. T. (1976). *J. Chem. Phys.* **65**, 5267.

Davidson, W. R., and Kebarle, P. (1976). *J. Am. Chem. Soc.* **98**, 6125.

Dempster, A. J. (1916). *Philos.* [6] *Mag.* **31**, 438.

Ding, A., Henglein, A., Hyatt, D., and Laemann, K. (1968). *Z. Naturforsch., Teil A* **23**, 2084.

Ditchburn, R. W., and Arnot, F. L. (1929). *Proc. R. Soc. London, Ser. A* **123**, 516.

Dzidic, I., and Kebarle, P. (1970). *J. Phys. Chem.* **74**, 1466.

Elder, F. A., Villarejo, D., and Inghram, M. G. (1966). *J. Chem. Phys.* **43**, 758.

Fehsenfeld, F. C. (1975). *Int. J. Mass Spectrom. Ion Phys.* **16**, 151.

Ferguson, E. E. (1975). *Annu. Rev. Phys. Chem.* **26**, 17.

Ferguson, E. E., Fehsenfeld, F. C., and Schmeltekopf. A. L. (1969). *Adv. At. Mol. Phys.* **5**, 1.

Field, F. H., Franklin, J. L., and Lampe, F. W. (1957). *J. Am. Chem. Soc.* **79**, 2419.

Foster, M. S., and Beauchamp, J. L. (1975). *J. Am. Chem. Soc.* **97**, 17.

Futrell, J. H., and Miller, C. D. (1966). *Rev. Sci. Instrum.* **37**, 1521.

Gordon, R., Jr., Rebbert, R. E., and Ausloos, P. (1969). *Natl. Bur. Stand. (U.S.), Tech. Note* **496**.

Grimsued, E. P., and Kebarle, P. (1973). *J. Am. Chem. Soc.* **95**, 7939.

Guyon, P. M., and Berkowitz, J. (1971). *J. Chem. Phys.* **54**, 1814.

Henderson, W. R., and Schmeltekopf. A. J. (1972). *J. Chem. Phys.* **57**, 4502.

Herman, Z., Kerstetter, J. D., Rose, T. L., and Wolfgang, R. (1967). *J. Chem. Phys.* **47**, 1856.

Hopkins, J. M., and Bone, L. I. (1973). *J. Chem. Phys.* **58**, 1473.

Hurzeler, H., Inghram, M. G., and Morrison, J. D. (1957). *J. Chem. Phys.* **27**, 313.

Jenkins, R. O., and Trodden, W. G. (1965). "Electron and Ion Emission from Solids," Chapter 7. Routledge, Kegan Paul, London.

Kebarle, P. (1972). *In* "Ion-Molecule Reactions" (J. L. Franklin, ed.), Chapter 7. Plenum, New York.

Kebarle, P. (1977). *Annu. Rev. Phys. Chem.* **28**, 445.

Kebarle, P., Haynes, R. M., and Searles, S. K. (1966). *Adv. Chem. Ser.* **58**, 210.

Kronberg, J. E., and Stone, J. A. (1977). *Int. J. Mass Spectrom. Ion Phys.* **24**, 373.

Le Breton, P. R., Williamson, A. D., Beauchamp, J. L. and Huntress, W. T. (1975). *J. Chem. Phys.* **62**, 1623.

Light, J. C. (1965). *J. Chem. Phys.* **40**, 3209.

Light, J. C., and Lin, J. (1968). *J. Chem. Phys.* **49**, 5426.

Lossing, F. P., and Tanaka, I. (1956). *J. Chem. Phys.* **25**, 1031.

McAdams, M. J., and Bone, L. I. (1972). *J. Chem. Phys.* **57**, 2173.

McFarland, M., Albritton, D. L., Fehsenfeld, F. C., Ferguson, E. E., and Schmeltekopf. A. L. (1973). *J. Chem. Phys.* **59**, 6610.

McIver, R. T. (1970). *Rev. Sci. Instrum.* **41**, 555.

McMahon, T. B., and Beauchamp, J. L. (1972). *Rev. Sci. Instrum.* **43**, 509.

Mahan, B. H. (1968). *Acc. Chem. Res.* **1**, 217.

Mahan, B. H. (1970). *Acc. Chem. Res.* **3**, 393.

Marmet and Kerwin (1960).

Orlov, V. M., Varshavsky, Y. M., and Kiryushkin, A. A. (1972). *Org. Mass Spectrom.* **6**, 9.

Orlov, V. M., Varshavsky, Y. M., and Miroshnikov, A. I. (1974). *Org. Mass Spectrom.* **9**, 801.

Potapov, V. K., Movshev, V. G., Letokhov, U. S., Knyazev, I. N., and Evlasheva, T. I. (1976). *Kvantovaya Elektron. (Moscow)* **3**, 2610; *Chem. Abstr.* **86**, 164225d (1976).

Reid, N. W. (1971). *Int. J. Mass Spectrom. Ion Phys.* **6**, 1.

Searles, S. K., and Kebarle, P. (1969). *Can. J. Chem.* **47**, 2620.

Searles, S. K., and Sieck, L. W. (1970). *J. Chem. Phys.* **53**, 795.

Sieck, L. W., and Ausloos, P. (1972). *Radiat. Res.* **52**, 47.

Sieck, L. W., Searles, S. H., and Ausloos, P. (1969). *J. Am. Chem. Soc.* **91**, 7627.

Sieck, L. W., Gordon, R., Jr., and Ausloos, P. (1972). *J. Res. Stand.* **52**, 47.

Sieck, L. W., Gordon, R., Jr., Ausloos, P., Lias, S. G., and Field, F. (1973). *Radiat. Res.* **56**, 441.

Smets, J., Coppens, P., and Drowart, J., (1977). *Chem. Phys.* **20**, 243.

Smith, D. L., and Futrell, J. H. (1974). *Int. J. Mass Spectrom. Ion Phys.* **14**, 171.

Stevenson, D. P., and Schissler, D. O. (1955). *J. Chem. Phys.* **23**, 1353.

Tal'roze, V. L. (1962). *Pure Appl. Chem.* **5**, 455.

Tal'roze, V. L., and Frankevich, E. L. (1960). *Zh. Fiz. Khim.* **34**, 2709.

Tal'roze, V. L., and Lyubimova, A. K. (1952). *Dokl. Akad. Nauk. SSSR* **86**, 909.

Tang, I. N., and Castleman, A. W. (1972). *J. Chem. Phys.* **57**, 3638.

Tang, I. N., and Castleman, A. W. (1974). *J. Chem. Phys.* **60**, 3981.

Tang, I. N., and Castleman, A. W. (1975). *J. Chem. Phys.* **62**, 4576.

Tang, I. N., Lian, M. S., and Castleman, A. W. (1976). *J. Chem. Phys.* **65**, 4022.

Taylor, J. W., Parr, G. R., and Jones, G. G. (1974). *Proc. Int. Conf. Vac. UV Radiat. Phys., 4th, 1974* p. 197.

Terenin, A., and Popov, B. (1932). *Phys. Z. Sowjetunion* **2**, 299.

Tiernan, T. O., and Bhattacharya, A. K. (1970). "Recent Developments in Mass Spectrometry." University Park Press, Baltimore, Maryland.

Traeger, J. C., and McLoughlin, R. G. (1977). *J. Am. Chem. Soc.* **99**, 7352.

Turner, D. L., and Bone, L. I. (1974). *J. Phys. Chem.* **78**, 501.

Warneck, P. (1971). *SPIE Annu. Tech. Symp., Proc., 15th, 1970* Vol. 9, p. 149.

Wei, L. Y., and Bone, L. I. (1974). *J. Phys. Chem.* **78**, 2527.

Wolf, J. F., Stanley, R. H., Koppel, I., Taagepera, M., and McIver, R. T. (1977). *J. Am. Chem. Soc.* **99**, 5417.

6

Discharge-Excited Rare Gas Halide Lasers

N. DJEU

Laser Physics Branch
Naval Research Laboratory
Washington, D.C.

I. INTRODUCTION

For many years now lasers have been finding their way into the laboratories of spectroscopists, kineticists, and other atomic scientists working in related areas. In view of the many special properties of the laser this "invasion" has hardly been unexpected. The spectral purity and tunability of the device permit the monitoring of specific atomic state populations as well as the resolution of spectral details, while its high output flux makes

323

possible the creation of large concentrations of excited species. In the visible–ultraviolet region a strong impetus to these applications was provided by the invention of the dye laser. The high optical gain and broad tunability of the dye laser have made it an extremely versatile instrument in the laboratory. However, due to both the lack of efficient short wavelength pumps and the instability of dyes at short wavelengths, the tuning range of dye lasers has been mostly limited to above 350 nm. To go below that wavelength, one must resort to second- or higher-order harmonic generation in nonlinear media which, besides introducing additional system complexities, often reduces the output to intolerably small fractions. The recent emergence of the rare gas halide lasers, with output wavelengths reaching below 200 nm and single pulse energy easily in the neighborhood of 100 mJ, therefore represents an important step in the evolution of coherent ultraviolet sources. The rare gas halide lasers take on an extra significance as some of the potentially most interesting laser applications, in fact, require wavelengths shorter than one can obtain with the dye lasers. While the ultimate rare gas halide laser may have yet to be engineered, a system of its present sophistication clearly is already useful in numerous situations. It is to familiarize the interested researcher with the state-of-the-art construction and operation of this new class of lasers that the present chapter is being written.

The exploitation of the rare gas halide molecules for laser applications was first suggested by Velazco and Setser (1975). After their initial observation of high chemiluminescence yields in reactions between $Xe(^3P_2)$ and halogen compounds, it was only a matter of months before the first lasers based on these and similar reactions were demonstrated by Searles and Hart (1975) and Ewing and Brau (1975). It was clear even in these earliest laser experiments that the rare gas halide systems had tremendous potentials in terms of both efficiency and power output. However, the excitation source in all these initial experiments had been the electron beam machine which, bulky in size and ejecting large quantities of high energy electrons, is not ideally suited as a laboratory tool. The first step in turning the rare gas halide lasers into practical laboratory devices was taken by Burnham et al. (1976a) when they showed that these lasers can be excited in the avalanche mode in a simple discharge device having a fast rising current pulse. Subsequent studies led to two variations of this basic discharge excitation technique which proved to be superior. They are the Blumlein (Wang et al., 1976; Burnham et al., 1976b) and uv-preionized (Burnham and Djeu, 1976) discharges, and their description will occupy most of the space in this chapter.

The rare gas halide (RgX) states responsible for the laser transitions are most likely the $^2\Sigma_{1/2}^+$ excited state correlating to the atomic ions and the $^2\Sigma_{1/2}^+$ ground state (Brau and Ewing, 1975; Hay and Dunning, 1977; Tellinghuisen et al., 1976a). While the excited state is bound by several thousand wavenumbers for most of the RgX's, the ground state bond energy varies from nearly zero to about 1000 cm^{-1} (Tellinghuisen *et al.*, 1976b). A strongly bound ground state is undesirable for two reasons. First of all, it impedes the rapid removal of molecules entering the lower laser level if the latter should lie substantially below the dissociation limit. The resulting bottle-necking effect would then lead to both a smaller inversion and a lower saturation intensity of the laser. A second consequence of a deep ground state potential well is that it gives rise to a finely structured emission spectrum which may in turn seriously limit the tunability of the laser. The reverse arguments which would apply to molecules with weakly bound ground states are, of course, precisely the ones which aroused great interest in exciplex laser systems in the first place.

Other factors which determine the performance of any specific rare gas halide laser are the transparency of the excited medium at the laser wavelength and the effectiveness of the discharge in the production of excited RgX molecules. Although the neutral channel reaction

$$Rg^* + XM \longrightarrow RgX^* + M$$

has been found to be extremely efficient in producing RgX* in numerous instances (Velazco and Setser, 1975; Velazco *et al.*, 1976) it is quite possible that in an avalanche discharge recombination reactions such as

$$Rg^+ + X^- + M \longrightarrow RgX^* + M$$

play an equally important role. The detailed spectroscopy and kinetics of the rare gas halide systems are still being actively investigated, and their discussion will not be pursued any further here.

So far, the avalanche discharge technique has been shown to be efficacious in the excitation of six distinct rare gas halide laser systems. In order of decreasing wavelength they are XeF (351 nm, 353 nm), XeCl (308 nm), KrF (249 nm), KrCl (222 nm), ArF (193 nm), and ArCl (175 nm). In terms of sheer power output, the KrF laser is without question the most interesting of the group. Understandably then, most of the development to date has evolved around that particular system. At this writing, output energy as much as 800 mJ per pulse (Sarjeant *et al.*, 1978) and average power as much as 40 W have been reported for the KrF discharge laser. It is almost certain that these records will be shortly surpassed, especially if attempts to prolong the pulse duration should prove to be fruitful. Because of this practical importance of the KrF laser, it will serve as a prototype system for discussion toward the end of the chapter. It is expected, however, that the beam controlling and gas recycling techniques used for the KrF laser should be equally effective when applied to the other rare gas halide systems.

We shall start with a description of the more successful discharge excitation techniques realized to date. A special effort will be made to point out the tolerances of the various aspects of the designs, so that the reader would not have to unnecessarily spend time to rediscover them. Next, the individual systems will be discussed with regard to gas mixture composition and output characteristics. This will probably be the section in which we rely most heavily on results obtained outside of our own laboratory. We will then turn our attention to some recent developments on the specific system of KrF. Results on output frequency control and recycling of the rare gases will be described in turn. Finally, the application of the rare gas halide lasers to the generation of coherent radiation at other wavelengths will be briefly mentioned.

II. CONSTRUCTION AND OPERATION OF DISCHARGE RARE GAS HALIDE LASERS

In this section, we shall describe two kinds of discharge devices that are capable of effectively exciting the rare gas halide lasers. Although vastly dissimilar in appearance, they, in fact, share the same basic circuitry and differ only in the speed with which the current pulse builds up in the laser medium. In the Blumlein design, the rapid rise of the current pulse is the all-important factor for the successful operation of the laser. Consequently, the circuit inductances must be kept extremely low, which, in turn, implies that the technique is not readily scalable to large lateral dimensions. Even so, output energy on the order of 100 mJ can be obtained in both XeF and KrF from a 1 m Blumlein device. Its major advantages are ease of construction and affordability. In the uv-preionized technique, the risetime of the current pulse does not appear to be as crucial. Hence, the cross sectional area enclosed by the electrodes and the capacitors can be made quite large. A further distinction between the two techniques is that in the Blumlein method discharge uniformity begins to suffer severely at pressures above about 1 atm, while the uv-preionized discharge can be run up to at least 6 atm without deleterious effects (Sarjeant *et al.*, 1978). The ability to utilize larger active volumes and higher gas pressures has permitted the attainment of output energy from the uv-preionized laser almost an order of magnitude greater than that from the Blumlein laser.

Both the Blumlein method and the uv-preionized method produce what is known as an avalanche discharge. In that mode of operation the discharge current is assumed to develop in an "uncontrolled" manner. After initiation, it continues to build up until the energy stored in the capacitors is appreciably depleted, at which point it drops precipitously back to zero. The precise shape of the current pulse is determined entirely by the composition of the gas mixture once the charging voltage is specified. The duration of the current pulse in these devices is typically 10–20 nsec. Although reaction kinetics stretches the fluorescence pulse out to roughly twice that duration, the actual laser pulse width under optimal cavity coupling conditions is generally no more than 20 nsec due to stimulated emission buildup time effects. Laser pulses as long as 50 nsec may be obtained at the expense of reduced output energy when low-loss cavities are employed.

Attempts to lengthen the laser pulse are underway in several laboratories. Two approaches are being most actively pursued. The first is to apply an L–C pulse forming network power supply to the uv-preionized discharge. If the discharge in existing devices is not terminated by the growth of some local instability, a pulse forming network with the proper characteristic impedance could in principle stabilize and lengthen it. In the other approach, stability of the discharge is sought by placing the electrodes between two dielectric surfaces separated by a small gap. This method has been found to be effective for a number of other laser systems.

Before going on to a description of the actual devices, some remarks on the choice of material for fabrication may be appropriate here. Although one would think that the presence of fluorine in the laser gas mixtures might place severe restrictions on the materials, experience has in fact shown otherwise. Lucite, teflon, PVC, and laminated epoxy have all been found to be acceptable as materials for the fabrication of the laser body. The number of discharges one can get on a static fill of gas before the output drops appreciably does not seem to vary much from one wall material to another. Since monitoring the visual appearance of the discharge can be exceedingly helpful in the optimization of these lasers, the transparency of lucite makes it the preferred material. The reader should be cautioned, however, to anneal the lucite both before and after any machining work is done on it. Failure to do so could cause cracks to develop around the worked parts. If small cracks should appear in spite of this precautionary measure, some MDC solvent adhesive may be used to seal them. Since only a moderately good vacuum is necessary for the rare gas halide lasers, minute leaks in the apparatus are readily tolerated.

A. The Blumlein Laser

The Blumlein circuit has been used for years as an effective and inexpensive way of building the nitrogen laser. The rare gas halide Blumlein laser differs from the nitrogen Blumlein laser only in that it requires much greater care in the fabrication of the switch and the electrodes. It should be noted, however, that whereas the Blumlein design with its associated short current pulse is selected for the nitrogen laser mainly for kinetic reasons, its adoption here is solely for the purpose of arc suppression.

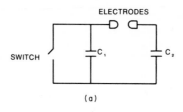

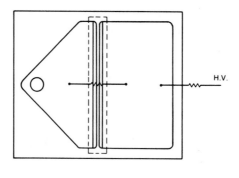

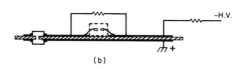

Fig. 1. The circuit diagram for the Blumlein laser is shown in (a), while its top and side views are shown in (b).

The basic Blumlein circuit is given in Fig. 1a. The two capacitors are initially charged to the same voltage V. As the switch is closed, the voltage across C_1 begins to oscillate. At the maximum of its excursion, in the absence of discharge breakdown, the voltage across the electrodes would be 2 V. If the breakdown voltage for the laser medium lies below 2 V, at some point during the polarity reversal of C_1 a discharge will be initiated between the electrodes. In a true Blumlein circuit parallel plate capacitors and solid dielectric switches are used to minimize the risetime of the current pulse (Shipman, 1967).

Blumlein devices of various sizes and refinement have been used for the excitation of the rare gas halides lasers (Wang *et al.*, 1976; Burnham *et al.*, 1976b; Godard and Vannier, 1976; McKee *et al.*, 1977; Christensen, 1977). The design to be given here was developed at NRL and has been found to be very effective by ourselves as well as others. It readily yields output energies of 80 and 100 mJ for KrF and XeF, respectively. Figure 1b shows the top and end views of the device. The storage capacitors here are made from a 0.16-cm-thick 100×120 cm^2 copper-clad circuit board with a total capacitance of 20 nF. The outermost 5 cm of the conducting surface is etched away on both sides to prevent arcing. In addition, on the top side a 5 cm strip at the middle as well as two corners of the conducting surface are removed as shown. Near the apex of the tapered end a 1.3-cm-diameter hole is made and an annular region of 0.8-cm width etched on both sides in preparation for the installation of the spark gap. In practice, the etching is accomplished by taping all but the areas to be stripped with overlapping acid resistant tape (e.g., 3M No. 471) and dipping the masked board in sulfuric dichromate solution.

The limiting factor in the overall risetime of the Blumlein circuit is generally the risetime of the switch. Therefore, every effort must be made to minimize the self-inductance of the switch and avoid the creation of additional loops in its coupling to the capacitor to be switched. The detailed construction of a low inductance triggered spark gap switch is given in the scaled drawing of Fig. 2. Both the drill rod electrode and the tool steel conical outer conductor should be press fitted into the brass housings for good electrical contact. The tungsten center pin is welded onto the threaded stainless steel stem and should be adjusted to be flush with the edge of the outer conductor. The flanges on the housings should be no thicker than 1 mm to facilitate their soldering onto the circuit board. The latter is accomplished by first wetting both surfaces to be joined with solder and then with the housing in place heating around the flange with a soldering iron.

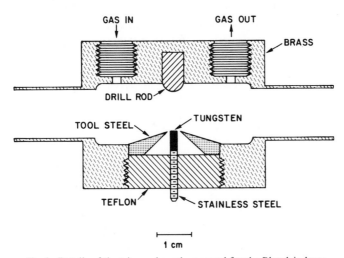

Fig. 2. Details of the triggered spark gap used for the Blumlein laser.

Each capacitor is connected to its electrode by a smoothly bent copper sheet. The electrodes, 90-cm long and 1-cm in height, are slotted in the back for a tight fit onto the copper sheet. It has been found that the best profiles for the electrodes are a flat face with rounded corners for the anode and a tapered edge with a 1-mm-radius tip for the cathode. Polished nickel plated electrodes have worked satisfactorily from both lifetime and discharge uniformity standpoints. The addition of a simple preionizing wire as shown in Fig. 1a of the paper by Burnham *et al.* (1976b) further improves the appearance of the discharge at elevated pressures. Finally, the laser body itself, represented by the dashed box, may be built out of any of the dielectric materials mentioned earlier. Ultraviolet grade quartz should be used for the windows, which may be set either at Brewster's angle or just slightly tilted. Both RTV and epoxy are adequate for making seals where needed. An ultimate vacuum of about 0.1 Torr is necessary for the satisfactory performance of the rare gas halide lasers.

In operation, the two capacitors are simultaneously charged to the same voltage through a charging resistor and a connecting resistor. Provided the time constant presented by the connecting resistor is much greater than that of the laser discharge gap, a negligible amount of power will be dissipated in it during the discharging process. The spark gap switch is slowly flushed with 5% SF_6 in N_2 at 1 atm, and is triggered by a pulse generator (e.g., EG&G TM-11). Up to the circuit board breakdown voltage of slightly greater than 30 kV, a monotonic increase of laser output is observed as a function of stored energy for a 2 cm electrode spacing in both XeF and KrF.

Pulse repetition rates of up to 200 Hz have been achieved on smaller Blumlein rare gas halide lasers than the one described when a rapid gas flow and pulse charging are employed (Christensen, 1977). There is no reason why the same techniques should not be applicable to the device specified here. Pulse charging should have the additional advantage of prolonging the lifetime of the circuit board, which under CW charging conditions can be expected to last for about 10^5 discharges at 30 kV.

B. The uv-Preionized Laser

When an ultraviolet radiation source is present to generate a low (but orders of magnitude greater than background) density of electrons through photoionization of the laser gas mixture just prior to the application of the discharge high voltage, the uniformity of the discharge and, hence, the laser performance can be greatly improved. As pointed out earlier, the incorporation of uv preionization provides one with the capability for scaling in both the beam area and the gas mixture pressure of the laser. At the same time, the latitude afforded by preionization on the composition of gas mixtures in which a glow discharge can be achieved has enabled the realization of several other rare gas halide laser systems. Some of these have proved to be almost as powerful and efficient as the XeF and KrF systems.

While it has been found quite universally that a delay of about 1 μsec between the preionization pulse and the main discharge pulse results in the best performance of the lasers, the exact mechanism by which uv preionization promotes the uniformity of discharges in the rare gas halide systems is still poorly understood. A possible explanation involves the formation of X^- ions which are rapidly produced by the dissociative attachment of the photoelectrons by the halogen donors and the subsequent release of electrons by these negative ions when the main discharge field is applied (Hsia, 1977). The delay is postulated to be necessary for the smoothing of the initially nonuniform X^- density via the nonlinearity of the three body ion recombination reaction. Although the numbers used in the proposed model do not appear to be realistic, it is nevertheless the most plausible argument advanced thus far. For the purpose of building and operating a uv-preionized rare gas halide laser, it is sufficient to note that the optimal delay time appears to be quite insensitive to the intensity of the uv-preionization source as well as the rate of change of the rising discharge current.

A variety of discharge circuits have been tried in conjunction with uv preionization for the excitation of the rare gas halide lasers. They range from a simple capacitive transfer circuit with a single triggered spark gap switch (Andrews et al., 1977) to very elaborate pulse forming network circuits operating at extremely high voltages (Sarjeant et al., 1977, 1978). As would be expected, the obtainable output from the laser varies roughly in proportion to the complexity of the design. The version to be described below represents a compromise between high laser output and ease of construction. It does not require any but the most ordinary familiarity with high voltage engineering on the part of the builder. Furthermore, since the only commercially available rare gas halide laser at present (from Tachisto, Inc.) is based on a nearly identical design, an option is provided whereby as many of the components as economically practicable may be purchased directly.

The main discharge circuitry of the uv-preionized device to be described is the same as that of the Blumlein device discussed earlier, as a comparison of the left-hand half of Fig. 3 with Fig. 1a at once shows. Here the area between the dashed lines represents the laser enclosure, and we note that in addition to the laser electrodes there is now a sparkboard (S) to provide the desired uv-preionization. Lasers of this design have been constructed in our laboratory with three different active lengths: 30, 60, and 120 cm. For the same charging voltage, capacitance per unit length, and electrode profile, the laser output scales essentially linearly with the active length.

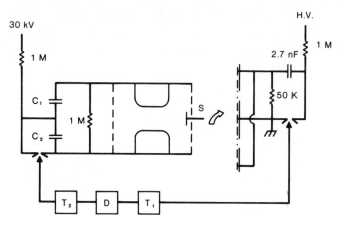

Fig. 3. Schematic for the uv-preionized laser.

Since results to be discussed in Sections IV and V were obtained on the shortest of the three lasers, only its specifications will be given below. Scaling to any desired active length involves only changes in the longitudinal dimensions of the electrodes and the sparkboard and proportional changes in C_1 and C_2.

A photograph of the 30 cm active length uv-preionized rare gas halide laser is shown in Fig. 4. The top of the laser chamber has been removed to show the interior configuration of the electrodes and the sparkboard. In its normal state, the top plate is connected to the top row of capacitors by means of a conducting sheet along its entire length. The laser body is made from lucite and has 8.7 × 13.2 × 39.5 cm outside dimensions and 2.5-cm-thick walls. Whereas the one shown in the photograph was machined out of a solid block of lucite, we have also been able to fabricate it by carefully bonding four pieces of lucite plates together with MDC solvent adhesive. The vacuum integrity of the system did not appear to be compromised by the latter approach. The aluminum electrodes measure 2.8 cm in width and 3.4 cm in height, leaving a gap of 1.9 cm between them when in place. They are flat over a 6 mm strip in the middle and have uniform field Rogowski profiles on the sides.† The electrodes, centered in the middle of the lucite frame, are secured onto 6-mm-thick aluminum plates. When 6-mm-thick quartz flats are used for windows over 3-cm-diameter areas together with sufficiently sturdy retainers, this enclosure has successfully withstood internal pressures of up to 4 atm.

The aluminum plate at the bottom extends beyond the lucite frame on

† Obtained from Tachisto, Inc.

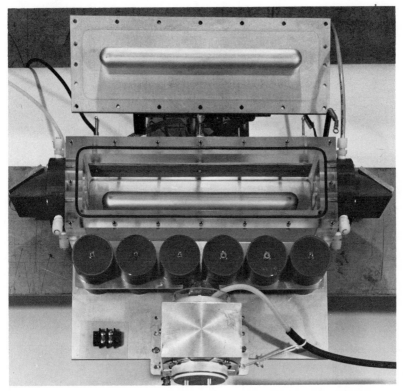

Fig. 4. Photograph of the 30-cm-active-length uv-preionized laser with its top removed to show the interior of the laser.

one side to provide a platform for the main discharge capacitors and switch. Six capacitors of 2.7 nF each are connected in parallel for both C_1 and C_2. They are stacked in two rows with an aluminum plate in between. Good contact between the capacitors and the common plate is insured by the judicious use of springy washers. Both Sprague 715C-Z and Murata 272 Z doorknob capacitors have performed satisfactorily. The latter are preferred on the ground that they have a tendency to split apart when damaged, relieving one from the tedious task of individually testing all the capacitors when malfunction occurs. The sparkboard, positioned in midplane between the electrodes, is fabricated from a 4 × 33 cm piece of circuit board of 1.6 mm thickness. The conducting surface is totally removed from one side, and is etched on the other side to yield an array of 0.6 × 1.5 cm rectangles spaced by 2 mm gaps. Holes of 3-mm diameter are drilled in the middle as well as the end units, and through them brass rods are inserted from the back and soldered onto the conducting surfaces. After being fed through Swagelok fittings, the center lead is connected to ground and the outside ones to the capacitor. For longer sparkboards it may be necessary to use separate capacitors for the two halves.

For the switching of the capacitors, either triggered spark gaps or thyratrons may be used. The latter are recommended if low jitter and high repetition rate operation ($\gtrsim 30$ pps) are desired. If triggering by thyratons is chosen, EG&G HY-3202 and HY-1102 (less costly) may be used for the main discharge and the sparkboard circuits respectively. A separate power source is then required to keep the high voltage on the sparkboard circuit at about 15 kV. Both thyratrons may be triggered by EG&G TM-27 thyratron drivers. If spark gap switches are selected, they are obtainable from Tachisto and may be triggered by EG&G TM-11 trigger modules. With either mode of operation, it is advisable to start with a variable pulse delay unit (D) between $T1$ and $T2$. As noted earlier, a typical delay between the sparkboard current pulse and the main discharge current pulse of about 1 μsec has been found to be ideal for the rare gas halide lasers. Once the correct interval has been determined, the variable pulse delay unit may be replaced by a fixed time delay element.

The width of the discharge typically does not extend over the entire flat portion of the electrodes, but instead only fills the region on the near side of the sparkboard. This pulling effect is probably caused by strong attenuation of the ionizing radiation which then results in an uneven distribution of the photoelectrons needed for the buildup of the avalanche. The width of the laser beam on the output mirror is generally between 2 and 3 mm. As the delay generated by D is increased from zero, a sharp transition point is observed around which the uniformity of the discharge improves dramatically. It should be noted that some fine filaments almost always appear together with the glow in a rare gas halide avalanche discharge. Therefore, one's goal ought to be the reduction in the severity of the filamentation rather than its total elimination. When the delay is further increased, the visible appearance of the discharge is only very slightly altered. However, by monitoring the energy of the laser pulse, one sees that the optimal delay time is quite sharply peaked and is located near the point of transition of the discharge from numerous arcs to a glow. The observation of a reasonably uniform glow discharge with zero delay provided by D would indicate an excessive intrinsic delay associated with the main discharge triggering circuit. One may then want to reverse the triggering order or simply leave out the delay element. Threshold voltage for most rare gas halide systems falls between 20 and 25 kV, and laser output is generally saturated before 30 kV is reached under normal operating conditions.

III. SPECIFIC RARE GAS HALIDE LASER SYSTEMS

Before proceeding to a discussion of the specific rare gas halide systems, a few general remarks on gas handling and optics may be in order. Because of the corrosive nature of most halogen bearing compounds, care must be taken not only to guarantee their survival during the preparation of the laser gas mixtures but also to prevent their escape in any appreciable quantity into the laboratory. Both of these tasks are made easier by the use of a diluted supply of the reactive gas in He, which is the major constituent in all rare gas halide laser gas mixtures. As an example, in the case of fluorine, we have found it expedient to use a 2 % mixture in He premixed by the gas supplier. For transferring the halogen donor gas from its source to the mixing manifold, a teflon vacuum line is suggested. Otherwise, the mixing manifold itself is constructed from just stainless steel pipes, Swageloks, and vacuum valves. Measured quantities of the halogen donor gas, the desired rare gas, and He are admitted, in that order, into a 1 liter stainless steel high pressure container to give a final pressure of about 10 atm. The mixture is then immediately ready for filling the previously evacuated laser chamber. Flushing the latter with a small amount of the laser gas mixture prior to filling may be helpful if a sufficiently low background pressure cannot be achieved. Ordinary forepumps, with vent lines attached to their exhausts, are quite adequate for evacuating both the laser chamber and the gas mixing system.

For the laser cavity, dielectric coated mirrors with large radii of curvature (e.g., 5 m) are generally used. At the energy levels normally encountered in the lasers of interest here (~ 100 mJ cm^{-2}), damage of the dielectric coatings does not appear to be a problem. While a number of manufacturers are capable of supplying coatings for wavelengths above 200 nm, at present only Acton Research Corp. has the facilities for producing coatings at shorter wavelengths.

Although the two discharge techniques just outlined in the last section have been widely used for their simplicity and effectiveness, they are by no means the only ones capable of exciting the rare gas halide systems. The fact is that a variety of other designs, each with its own special features, have now been investigated. For the sake of presenting as complete a picture as possible on the capabilities of each rare gas halide system, noteworthy results produced by some of these other lasers will be discussed as well. This is done for two reasons. First, comparable results from either of the two designs given here may not be available for some systems simply because no attempt has been made to reproduce them yet. Second, in cases where higher output was obtained on more elaborate devices, it is felt that the reader should be made aware of them. A qualitative summary of the capabilities of

Table I

	Wavelength (nm)	Output	Tunability (nm)
XeF	351.1, 353.2	Good	discrete lines
XeCl	308.0, 308.2	Excellent	0.1
KrF	248.5	Excellent	2.2
KrCl	222.0	Fair	Good[a]
ArF	193.2	Good	2.2
ArCl	175.0	Poor[b]	Good[a]

[a] Although no tuning of the KrCl or ArCl system has yet been attempted, one anticipates good tunability from their smooth emission profiles.

[b] This opinion is likely to be revised by further work on the ArCl system.

each of the rare gas halide systems with regards to laser output and tunability is given in Table I. It is hoped that, as far as the overall picture is concerned, the information given in this section will not have to be appreciably modified as further advances in this field are made.

A. XeF

The XeF system has the highest gain among all the rare gas halides and therefore serves as a good test case for any newly instrumented device. When the 90-cm-long Blumlein device is used for its excitation, super-fluorescent emission is readily observed. The best fluorine donor for the XeF laser appears to be NF_3. For a mixture of $NF_3 : Xe : He(1.2 : 1.2 : 97.6)$ at a total pressure of 1 atm and a cavity consisting of a curved total reflector and a quartz flat, an XeF laser output of 100 mJ may be obtained from the Blumlein device for a 30 kV charging voltage. Since the laser pulse duration under those conditions is only 4 nsec, this corresponds to a very respectable peak power of 25 MW.

The optimal mixture of the uv-preionized device is much leaner in NF_3. On a gas fill of $NF_3 : Xe : He(0.3 : 1 : 98.7)$ at 2 atm, a 60 cm active length uv-preionized device equipped with a total reflector and a 70% output coupler yields a 20 nsec pulse of 65 mJ at saturation current. The replacement of NF_3 by F_2 results in a slightly lower output. Here, as well as in the discussion of some of the other systems below, the output figures quoted are appropriate for a uv-preionized laser equipped with electrodes supplied by Tachisto. It is possible that improved electrode profiles might lead to broader discharges and, hence, higher output for the same active length.

The highest output energy from the XeF laser to date was reported by Sarjeant *et al.* (1977). Using uv preionization and a pulse forming network for the main discharge, they were able to obtain a uniform discharge over a width of greater than 1 cm between electrodes separated by 2.4 cm. An output energy of 290 mJ was measured from a 60 cm active length apparatus containing $NF_3 : Xe : He$ (0.04 : 0.1 : 99.86) at a total pressure of 4.7 atm and charged by a 76 kV high voltage supply.

With regard to laser gas mixtures, in general, the reader will notice some discrepancies in the specification of optimal compositions from one work to another. This is partly due to the fact that different incremental matrices have been used by the various workers in the optimization process. In cases where F_2 is the halogen donor, there is the further possibility that the percentage of dilution in He may be quite different from its nominal value because of its reactivity. For these reasons, the gas ratios given here as well as below should be regarded only as guidelines. It is suggested that the reader test a few other mixtures which are slight variations of the specified one to pinpoint the one most suitable for his laser.

The XeF laser emission spectrum consists of two bands at 351.1 and 353.2 nm, as shown in Fig. 5a. The considerable amount of structure within each of these bands is caused by the unusually deep well in the ground state potential curve of the molecule. While the presence of sharp peaks in the XeF gain profile obviously precludes the continuous tuning over any appreciable range, efficient narrow band extraction at a number of fixed frequencies appears to be possible (Goldhar *et al.*, 1977).

B. XeCl

This rare gas halide system had been largely ignored since its initial investigation in electron-beam pumped devices which indicated a lack of promise. Interest in XeCl was revived by the work of Ishchenko *et al.* (1977), which showed that the system's efficiency can be much greater when excited by a discharge. The difference observed for the behavior of the XeCl laser between the two excitation techniques apparently arises from the choice of Ar as the diluent in the electronbeam approach. It is believed that Ar_2^+, which is produced in substantial quantities by the electron beam, may have a strong absorption band near the XeCl laser wavelength and thereby severely reduce its output.

Maximum output from a XeCl laser pumped by the 60-cm-long uv-preionized device is 110 mJ (Burnham, 1978). This was obtained by using an HCl : Xe : He (0.2 : 1 : 98.8) mixture at 2.7 atm, a charging voltage of 30 kV, and a cavity formed by a total reflector and a 70% output coupler. Although no Blumlein excited XeCl laser has been reported yet, it is expected that not

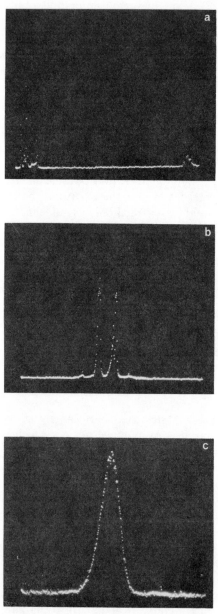

Fig. 5. Untuned output spectra of (a) the XeF laser, (b) the XeCl laser, and (c) the KrF laser. The scale is approximately 0.4 nm per division, with increasing wavelength to the right.

only can such a laser be made but that it would probably yield a fairly high output. This projection is based on the small percentages of HCl and Xe required and the low minimum pressure (~ 1 atm) at which appreciable output has been obtained with uv-preionization.

Nearly half of the XeCl laser's output appears in each of the "lines" at 308.0 and 308.2 nm, as shown in Fig. 5b. A slight amount of tuning (~ 0.1 nm) is possible around each of these peaks (Burnham, 1978). The most interesting aspect of this system is the stability of the laser gas mixture. Whereas the output of all the other rare gas halide systems, when operated on a static fill, would decrease noticeably after a few hundred discharges, the XeCl laser output has been observed to remain constant for as many as 7×10^3 pulses (Sze and Scott, 1978). This extraordinary durability of the XeCl laser gas mixture could very well eliminate the need for gas recycling for a variety of experiments.

C. KrF

By far the greatest effort in the exploration of the rare gas halide lasers has been expended on the KrF system. The tremendous interest in the KrF laser has been generated primarily by its demonstrated high electrical efficiency and high specific energy content. However, for some researchers its relatively broad tuning range may represent a source of even greater attraction.

The Blumlein circuit described earlier provides a moderately effective means of exciting the KrF laser. From a 50% partially reflecting mirror, an output of 80 mJ can be realized on a $F_2 : Kr : He$ (0.2 : 6 : 93.8) mixture at 1 atm. This is to be compared with 150 mJ from the 60 cm uv-preionized device operating on essentially the same gas mixture at 2.5 atm in an identical optical resonator. Substitution of F_2 by NF_3 generally reduces the output by more than a factor of two in either approach. The record for output energy from an avalanche discharge device in the case of KrF belongs again to Sargeant *et al.* (1978). From a uv-preionized volume of 180 cm^3, they were able to get laser pulses measuring up to 800 mJ with a 6 atm gas mixture and charging to 92 kV.

Unlike XeF and XeCl, the spectrum of the KrF laser output consists of only one broad peak centered at 248.5 nm and having a FWHM of 0.4 nm (see Fig. 5c). Continuous tuning of the KrF laser output has been achieved, with a narrow dip near the peak, over the entire emission bandwidth of 2 nm. The cause of the dip has been attributed to absorption by CF_2 at 248.8 nm (Goldhar and Murray, 1977). If this conjecture proves to be correct, one should then, in principle, be able to tune through that region smoothly with the use of properly purified laser gases. Quantitative data concerning the tuning behavior of the KrF laser will be given in the following section.

D. KrCl

A distinguishing feature of the KrCl laser is that its threshold occurs at a relatively high filling pressure of 2 atm. The record reported output of 40 mJ was obtained for an $HCl : Kr : He$ (0.15 : 10 : 89.85) mixture at 4 atm (Sze and Scott, 1978). Although the uv-preionized laser which yielded this result follows a slightly different design from ours, it is expected that after an appropriate reduction in the spacing between electrodes to maintain the necessary E/N, the uv-preionized device described in the previous section should perform very similarly.

The output spectrum of the KrCl laser resembles that of the KrF laser in shape and is centered at 222 nm. If medium absorption at that wavelength proves to be as serious as suspected (Sze and Scott, 1978), the output may be tunable over only a limited portion of the spontaneous emission bandwidth of 2 nm.

E. ArF

The ArF system holds the distinction of being the only powerful rare gas halide laser which works in the vacuum ultraviolet region. From the 60-cm-long uv-preionized device ArF laser pulses of 60 mJ can be extracted with a gas mixture of $F_2 : Ar : He$ (0.3 : 30 : 69.7) at 2 atm and a 50% output mirror. Because of the large absorption ordinary quartz presents at the ArF wavelength, laser windows made from the highest quality Suprasil are mandatory.

A densitometer trace of the ArF laser output spectrum can be seen in Fig. 3 of the paper by Burnham and Djeu (1976). Absorption by the Schumann–Runge system of O_2 near the output wavelength of 193.2 nm is evident even with the use of internal mirrors. Tuning of the ArF laser has recently been achieved by Loree et al. (1978), and will be discussed in the following section.

F. ArCl

The ArCl system is being mentioned here for its potential rather than proven capability. Thus far, only an output of 0.2 mJ has been realized from a 160-cm-long traveling-wave Blumlein apparatus (Waynant, 1977). Since Cl_2 was used in that work as the chlorine donor and since HCl has been found to be vastly superior for the other rare gas chloride systems, further work on the ArCl laser would be necessary before a final judgement can be made.

The emission of the ArCl laser is centered at 175.0 nm. The output should be tunable over most of the fluorescence bandwidth of 5 nm (Golde and Thrust, 1974) if sufficiently low-loss optics can be found.

IV. FREQUENCY CONTROL OF THE KrF LASER

By frequency control is meant both the coarse tuning of a laser's output and the reduction in bandwidth about a tuned frequency. These capabilities can be achieved through the introduction of optical filters such as diffraction gratings and etalons inside the laser cavity. The coarse tuning element is used to shift the center of the output spectrum to a wavelength either only weakly or not at all present in the untuned emission. If all the excited molecules have essentially the same emission profile or if the coupling between molecules contributing to different portions of the gain spectrum is sufficiently strong, substantial extraction in the wings of the gain profile is possible. While the coarse tuning element may at the same time sharpen the output spectrum to some extent, additional intracavity etalons are often necessary to further frequency narrow the output. Provided that the gain of the system is high enough and that the pulse duration is long enough, one can through successive stages of filtering force the laser to eventually oscillate on a single mode of the resonator formed by the end mirrors. A useful rule of thumb to remember in the selection of etalons is that the free spectral range of every additional one should be greater than the bandwidth of the laser output before its insertion. In this manner, single frequency operation will always be assured.

For the short pulse lasers under consideration, the gain lasts only long enough for the photons to make a few round trips inside the cavity. Since the discrimination offered by any intracavity filtering element becomes more pronounced with each passage, the use of a short cavity is clearly indicated. For the sake of efficient extraction, however, the active medium must not be so short that the net gain is changed drastically by the introduction of the filtering elements. For the results given below, a 30 cm active length uv-preionized KrF device was used. The laser was equipped with Brewster's angle windows and produced an output of 40 mJ when a 5 m radius of curvature 30 % output coupler was used together with a 5 m total reflector separated by 75 cm. The spectrum of emission from this untuned cavity, as can be seen in Fig. 5c, has a full-width at half-maximum of about 0.4 nm. When the total reflector was replaced by a 98.76 lines/mm echelle grating blazed at 63°26′ and oriented with its grooves parallel to the long dimension of the discharge cross section, the emission bandwidth was narrowed to 2×10^{-2} nm with a corresponding energy of 15 mJ. The output was continuously tunable from 247.5 to 249.5 nm, with a dip in power near 248.8 nm which degrades to a hole of increasing width as the laser gas mixture becomes more used. Whether the origin of this gap is in the absorption by CF_2 (Goldhar and Murray, 1977) or some other species, it would appear to be removable only in a system which is able to continuously purify the gas mixture. It is worth noting that when the same grating is used for a 90 cm active length device with a resonator spacing of 120 cm, the output bandwidth is increased by approximately a factor of two. This result shows that the extra angular dispersion provided by the longer cavity does not quite compensate for the reduction in the number of reflections the photons are allowed at the grating.

When a solid quartz etalon of $\frac{1}{2}$ mm thickness coated for 70% reflectance on each surface was inserted between the grating and the laser, the output bandwidth was further reduced to less than 2×10^{-3} nm, with a corresponding output energy of 6 mJ. The bandwidth given here is an upper limit since the measured value appeared to be equal to the theoretical minimum resolvable bandwidth of the spectrum analyzer used. While the incorporation of the grating and the etalon cuts the output energy by a factor of seven, it reduces the output bandwidth by a much larger factor of at least 200. In terms of the spectral brightness of the laser emission, therefore, one has made a vast improvement. An extrapolation of these results suggests that single mode operation may well obtain with the introduction of a second etalon and appropriate aperturing to establish oscillation on the fundamental transverse mode only.

The drop in power output accompanying the frequency narrowing procedure is primarily due to a longer buildup time necessitated by the increase in cavity loss. It appears that the only way to obtain narrow band KrF laser output without any significant reduction in energy compared with broad band operation is through the use of a second laser. With a small signal gain of several percent per centimeter, it is clear that just a straightforward amplifier would easily provide factors of 10 to 100 gain. But the second laser can be put to much better use through the technique of injection locking (Goldhar and Murray, 1977). In this mode of operation the output from the narrow band oscillator is injected into an unstable resonator containing the second laser. The injected signal causes a perferential buildup of the second laser's flux in the frequency range identical to its own. In the experiment of Goldhar and Murray, they showed that the output from the second laser with or without injection by a signal of 2×10^{-2} nm bandwidth are exactly the same, proving the existence of strong coupling between all the excited KrF molecules and the narrow band of radiation injected. The injection locking method is superior to direct amplification in that the energy in the second laser is more efficiently extracted through the multiple pass effect created by the resonator.

The tuning and extraction behavior of the other rare gas halide systems with "true" continuum emission profiles should be similar to that of KrF. The tuning of the ArF laser has been investigated by Loree et al. (1978). Using a pair of quartz prisms as filtering elements, they were able to tune across a region of 2.2 nm centered at 193.3 nm. With a bandwidth of less than 5×10^{-2} nm, the output at the peak of the gain curve contained as much as 80% of the energy from the untuned cavity. The same optical arrangement was also applied to the KrF system, but due to the higher gain of the medium the results were not nearly as good.

V. RECYCLING OF THE RARE GASES

As laboratory tools, the rare gas halide lasers suffer from the drawback that the halogen donor gases are either extremely reactive or not subject to complete reconstitution after discharge excitation. In practice, this means that on a static fill of the laser chamber only a limited number of discharges can be made before the output drops to a fraction of its initial value. The experimenter is thus forced to either frequently refill the laser or operate on a flowing basis. With either approach, for those rare gas halide lasers based on Kr or Xe few can afford their high cost for any extended period of time. (In some countries even the cost of He may be prohibitive.) It therefore becomes economically compelling to develop some simple, reliable way of reclaiming as much of the rare gases as possible once the gas mixture itself is no longer functional.

There are basically two approaches one can take in the extraction of the rare gases from the used laser mixtures. The first is to remove on a continuous basis all the contaminants from the flowing stream of gas mixture and then add just upstream of the laser chamber the fresh halogen donor gas. The second is to recirculate a large volume of the laser mixture until the output drops below some acceptable level and then perform the gas cleanup when the laser is not being operated. While in the first approach one has the clear advantage of being able to run the laser uninterrupted for virtually indefinite periods of time, its implementation is complicated by the requirement of a highly efficient one pass cleanup system. In addition, for systems using F_2 as the halogen donor gas it would entail the injection of pure fluorine, a hazard one would like to avoid. As of this writing, there has been no report of a continuously running rare gas halide laser with recycling of the rare gases.

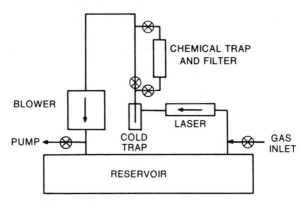

Fig. 6. Block Diagram for the gas recycling system.

The removal of the contaminants can be effected either chemically or cryogenically. The salvaging of Xe from a discharged He–Xe–NF$_3$ mixture by cryogenic means has been demonstrated by Christensen (1977). Using two cold traps at different temperatures, he was able to separate the Xe and NF$_3$ from a multitude of reaction products. In general, however, a complete cryogenic separation of the rare gases from the unwanted gases may not always be a straightforward matter. Presented below is a gas cleanup system based on a chemical procedure which is both simple and versatile.

A schematic of the chemical gas recycling system is shown in Fig. 6. The recirculation of the gas mixture in the direction indicated by the arrows is provided by a blower. In the particular unit assembled in our laboratory a modified 100 liter sec^{-1} blower is used for that purpose. The modification consists of a change from its original direct motor drive to a gear drive to reduce the speed of rotation to about four revolutions per second. The choice of the blower was made solely on the basis of its availability; any other vacuum tight recirculating fan would probably work just as well. The blower is mounted directly on a large aluminum tank measuring approximately 100 liters in volume. After passing through the laser chamber, the gas mixture is directed through an arc welded stainless steel cold trap which is to be used for the condensation of the Kr. The cold trap has a diameter of 5 cm, and its center tube is made to terminate 2.5 cm above the bottom of the trap in order to accommodate the substantial volume of the condensed Kr. From there the gas can be flowed through either a straight empty tube or the chemical trap. The valve which leads to the straight section is an on–off type and has a through diameter of 2.5 cm. Stainless steel pipes of 3.5-cm diameter are used throughout the system, except for the two sidearms connecting to the chemical trap and the teflon leads connecting to the laser chamber. The latter are 90 cm in length and have an inner diameter of 6 mm.

The chemical trap itself is nothing more than a stainless steel pipe containing a 30 cm column of calcium pellets and heated on the outside by a pair of 15-cm-long semicylindrical heaters. The calcium pellets are about 5 mm in diameter and are supported on a stainless steel wire screen positioned a few centimeters above the gas inlet sidearm. Since a sizable amount of heat is imparted to the gas flow during its passage through the heated calcium bed, water cooling is provided a small distance above the heating elements to keep the rest of the apparatus cool. Finally, just below the outlet sidearm a section of the pipe is stuffed with copper turnings to filter out any particulate matters.

In a typical run, the entire system is first evacuated. Then the chemical trap is isolated from the remainder of the system by the valves on its sidearms, and the laser gases are admitted with the blower running. If the gases are introduced in the order of Kr, F_2, and then He, complete mixing is achieved in less than 1 min. The output energy from the laser then becomes identical to that obtained under static fill conditions. However, whereas on a static fill the output would drop by 50% in roughly 5 min at a pulse repetition rate of 1 sec^{-1}, the halflife is now extended to about 2 hr with the aide of the large reservoir.

For the cleanup of the exhausted gas mixture, the through valve in the main line is closed and the valves leading to the chemical trap are opened. The heaters are set at 650°C and the gas is allowed to circulate through the calcium column overnight, although shorter cleanup times may well suffice. Once all the impurities have been removed, the calcium column is again isolated and the cold trap is immersed in liquid nitrogen. From the drop in total pressure, it is determined that more than 90% of the Kr can be condensed out in the first two hours. Then, with the liquid nitrogen dewar still in place, approximately 10% of the He is pumped out to allow replacement by the He–F_2 mixture. This brings the system to the start of another cycle. In the longest series of tests performed to date, this recycling procedure was carried out five times with no discernable decrease in either laser energy output or laser running time.

There is no reason to believe that the recycling system just described would not work equally well for the other rare gas halide laser mixtures. At a temperature of 800°C, the rare gases are the only gases which would go through the calcium bed unreacted. Taken a step further, one may also consider a continuous cleanup system with the chemical trap if a sufficiently long residence time in the column can be arranged.

VI. GENERATION OF SECONDARY WAVELENGTHS

The high power output of the rare gas halide lasers has made possible their use as optical pumps in the generation of coherent radiation at new wavelengths, thereby greatly extending the usefulness of this class of lasers. Although numerous schemes for wavelength conversion are known, only optical pumping of dyes and Raman conversion will be mentioned here, since these are the only areas where some work has been done in connection with the rare gas halide lasers.

Of all the short wavelength dyes investigated to date, paraterphenyl has given by far the most spectacular performance. When pumped by a KrF laser, the conversion efficiency in this dye has been measured to be as high as 28% (Godard and de Witte, 1976). The untuned output from para-terphenyl is centered at 340 nm and has a bandwidth of 2.5 nm. With the incorporation of a grating as one end mirror, tuning between 323 and 364 nm has been achieved. A list of additional KrF laser pumped uv dye lasers with untuned output wavelengths ranging from 347 to 359 nm can be found in the paper by Rulliere et al. (1977).

Stimulated Raman scattering offers another means of converting the frequencies of the rare gas halide lasers. Since gain at the Raman transition is confined to the bandwidth of the pump laser, the Raman output is tunable only to the extent that the pump laser output is tunable. Raman shifting experiments with rare gas halide lasers as pumps have been carried out in both molecular gases and metal vapors. Loree et al. (1977) have been able to shift the ArF laser output in H_2 and KrF output in H_2, D_2, and CH_4, generating a large set of new uv lines at the various Stokes and anti-Stokes frequencies. Electronic Raman scattering in metal vapors could yield very high conversion efficiencies if a near-resonant intermediate state with strong transition probabilities to both the initial and final states can be identified. An example is the Raman conversion of the XeF laser output in Ba vapor (Djeu and Burnham, 1977).

The rare gas halide lasers are still only in their infancy of development. As their spatial and temporal coherence are improved with time, the conversion of their wavelengths through other schemes such as harmonic generation and sum frequency mixing would become possible. One would then have at his disposal tunable coherent radiation sources far into the vacuum ultraviolet.

REFERENCES

Andrews, A. J., Kearsley, A. J., Webb, C. E., and Haydon, S. C. (1977). *Opt. Commun*, **20**, 265.
Brau, C. A., and Ewing, J. J. (1975). *J. Chem. Phys.* **63**, 4640.

Burnham, R. (1978). *Opt. Commun.* **24**, 161.

Burnham, R., and Djeu, N. (1976). *Appl. Phys. Lett.* **29**, 707.

Burnham, R., Harris, N. W., and Djeu, N. (1976a). *Appl. Phys. Lett.* **28**, 86.

Burnham, R., Powell, F. X., and Djeu, N. (1976b). *Appl. Phys. Lett.* **29**, 30.

Christensen, C. P. (1977). *Appl. Phys. Lett.* **30**, 483.

Djeu, N., and Burnham, R. (1977). *Appl. Phys. Lett.* **30**, 473.

Ewing, J. J., and Brau, C. A. (1975). *Appl. Phys. Lett.* **27**, 350.

Godard, B., and de Witte, O. (1976). *Opt. Commun.* **19**, 325.

Godard, B., and Vannier, M. (1976). *Opt. Commun.* **18**, 206.

Golde, M. F., and Thrush, B. A. (1974). *Chem. Phys. Lett.* **29**, 486.

Goldhar, J., and Murray, J. R. (1977). *Opt. Lett.* **1**, 199.

Goldhar, J., Dickie, J., Bradley, L. P., and Pleasance, L. D. (1977). *Appl. Phys. Lett.* **31**, 677.

Hay, P. J., and Dunning, J. H., Jr. (1977). *J. Chem. Phys.* **66**, 1306.

Hsia, J. (1977). *Appl. Phys. Lett.* **30**, 101.

Ishchenko, V. N., Lisitsyn, V. N., and Razhev, A. M. (1977). *Opt. Commun.* **21**, 30.

Loree, T. R., Sze, R. C., and Barker, D. L. (1977). *Appl. Phys. Lett.* **31**, 37.

Loree, T. R., Butterfield, K. B., and Barker, D. L. (1978). *Appl. Phys. Lett.* **32**, 171.

McKee, T. J., Stoicheff, B. P., and Wallace, S. C. (1977). *Appl. Phys. Lett.* **30**, 278.

Rulliere, C., Morand, J. P., and de Witte, O. (1977). *Opt. Commun.* **20**, 339.

Sarjeant, W. J., Alcock, A. J., and Leopold, K. E. (1977). *Appl. Phys. Lett.* **30**, 635.

Sarjeant, W. J., Alcock, A. J., and Leopold, K. E. (1978). *IEEE J. Quantum Electron.* **QE-14**, 177.

Searles, S. K., and Hart, G. A. (1975). *Appl. Phys. Lett.* **27**, 243.

Shipman, J. D., Jr. (1967). *Appl. Phys. Lett.* **10**, 3.

Sze, R. C., and Scott, P. B. (1978). *Appl. Phys. Lett.* **33**, 419.

Tellinghuisen, J., Hoffman, J. M., Tisone, G. C., and Hays, A. K. (1976a). *J. Chem. Phys.* **64**, 2484.

Tellinghuisen, J., Tisone, G. C., Hoffman, J. M., and Hays, A. K. (1976b). *J. Chem. Phys.* **64**, 4796.

Velazco, J. E., and Setser, D. W. (1975). *J. Chem. Phys.* **62**, 1990.

Velazco, J. E., Kolts, J. H., and Setser, D. W. (1976). *J. Chem. Phys.* **65**, 3468.

Wang, C. P., Mirels, H., Sutton, D. G., and Suchard, S. N. (1976). *Appl. Phys. Lett.* **28**, 326.

Waynant, R. W. (1977). *Appl. Phys. Lett.* **30**, 234.

Index